D1105638

SELF-SIMILAR PROCESSES IN TELECOMMUNICATIONS

Oleg I. Sheluhin
Moscow State Technical University of Service (MSTUS), Russia

Sergey M. Smolskiy
Moscow Power Engineering Institute (MPEI), Russia

Andrey V. Osin
Moscow State Technical University of Service (MSTUS), Russia

John Wiley & Sons, Ltd

Copyright © 2007 John Wiley & Sons Ltd, The Atrium, Southern Gate, Chichester, West Sussex PO19 8SQ, England

Telephone (+44) 1243 779777

Email (for orders and customer service enquiries): cs-books@wiley.co.uk
Visit our Home Page on www.wiley.com

All Rights Reserved. No part of this publication may be reproduced, stored in a retrieval system or transmitted in any form or by any means, electronic, mechanical, photocopying, recording, scanning or otherwise, except under the terms of the Copyright, Designs and Patents Act 1988 or under the terms of a licence issued by the Copyright Licensing Agency Ltd, 90 Tottenham Court Road, London W1T 4LP, UK, without the permission in writing of the Publisher. Requests to the Publisher should be addressed to the Permissions Department, John Wiley & Sons Ltd, The Atrium, Southern Gate, Chichester, West Sussex PO19 8SQ, England, or emailed to permreq@wiley.co.uk, or faxed to (+44) 1243 770620.

Designations used by companies to distinguish their products are often claimed as trademarks. All brand names and product names used in this book are trade names, service marks, trademarks or registered trademarks of their respective owners. The Publisher is not associated with any product or vendor mentioned in this book.

This publication is designed to provide accurate and authoritative information in regard to the subject matter covered. It is sold on the understanding that the Publisher is not engaged in rendering professional services. If professional advice or other expert assistance is required, the services of a competent professional should be sought.

Other Wiley Editorial Offices

John Wiley & Sons Inc., 111 River Street, Hoboken, NJ 07030, USA

Jossey-Bass, 989 Market Street, San Francisco, CA 94103-1741, USA

Wiley-VCH Verlag GmbH, Boschstr. 12, D-69469 Weinheim, Germany

John Wiley & Sons Australia Ltd, 42 McDougall Street, Milton, Queensland 4064, Australia

John Wiley & Sons (Asia) Pte Ltd, 2 Clementi Loop #02-01, Jin Xing Distripark, Singapore 129809

John Wiley & Sons Canada Ltd, 6045 Freemont Blvd, Mississauga, ONT, L5R 4J3, Canada

Wiley also publishes its books in a variety of electronic formats. Some content that appears in print may not be available in electronic books.

Anniversary Logo Design: Richard J. Pacifico

British Library Cataloguing in Publication Data

A catalogue record for this book is available from the British Library

ISBN 978-0-470-01486-8 (HB)

Typeset in 10/12 pt. Times by Thomson Press (India) Limited, New Delhi
Printed and bound in Great Britain by Antony Rowe Ltd, Chippenham, England.
This book is printed on acid-free paper responsibly manufactured from sustainable forestry in which at least two trees are planted for each one used for paper production.

Contents

Foreword

At the very beginning of development of the telephony technique twin-wire telephone lines started to entangle our world. At first the signals were transferred by the human voice and the data on the called number, and the spectrum width did not exceed 3.4 kHz. The popularity and the necessity to improve the telephone communication lines immediately attracted the attention of communications engineers and experts dreaming of increasing the activity factor of the already laid communication lines and aspiring to (as they say today) 'multiplex' the messages and to use the channel simultaneously and frequently.

As radio engineering and radio communication developed, the problem to increase the information capacity for radio channels became as acute as that for wire communication. The initial studies in the 1930s were oriented towards analysis of the discretization (the first stage when numeralizing the analogous signals) of the transferred message: the analogous message transformation (e.g. the slow voice) into the digital signal and further digital transfer in the multiplexed mode, with the transformation into the initial analogous form at the receiving side. As a result, the so-called 'sampling theorem' acquired special significance for the process of numeralization and further reconstruction of the initial message, the 70th anniversary of which was celebrated in Russia in 2003. Various experts have connected this theorem with the names of V.A. Kotelnikov, Cl.E. Shannon, H. Nyquist, H. Raabe, W.R. Bennett, I. Someya, and E.T. Whittaker. In 1999 the German Professor Hans Dieter Luke (from Aachen University)[1] recognized the importance of the Russian expert in radio engineering and radio physics, Vladimir A. Kotelnikov, and published an excellent research on the history of this problem. The sampling theorem (Kotelnikov's theorem) allowed researchers and engineers to approach extreme possibilities in the communication network and to initiate the development of various branches of science and technology, such as, for example, the theory of the potential noise immunity of the radio and wire communication channels.

The twentieth century can easily be characterized by the growth of the need for informational exchange in accordance with the geometric series, which naturally required the channels to transfer this information. When finally high-capacity and branching communication networks were formed, the researchers and engineers of the telecommunication networks faced absolutely new problems. Under conditions of rapid technological development, leading to growth of the processing speed of both the computer systems and the communication channels and systems as a whole, the number of users steadily increased. Since the users download the communication networks in their professional activity (remote job, distant education, IP-telephony, etc.) as well as in their spare time (web, music, games, chats, etc.), the list of claimed services using the telecommunication networks and their information capacity grew very rapidly.

[1] H.L. Luke, 'The origins of the sampling theorem', *IEEE Communication Magazine*, **37**(4), April 1999, 106–108.

Unfortunately, technological improvement does not keep pace with users' needs and the situations of communication channel overload occur more and more frequently, which leads to information transfer delays and in the worst cases to its loss. The users cannot and should not know the reasons for the caused discomfort: they have concluded the agreement, paid for the service and they have the right to demand a high-quality service.

In order to find a compromise between the growing needs in communication network resources and their limited possibilities it is necessary to apply well-engineered algorithms of control and regulation of the informational flows. Therefore, the problems of the optimal use of telecommunication channels acquire a different aspect, and the priority allocation and the queue in inquiries and answers become first and foremost. The message volume among users grows considerably or, as the specialists say, the traffic dynamics in the channels becomes multiplexed and complicated. Therefore, the problems of optimal traffic control and the investigation of new traffic features caused by the huge users and services volume in the networks are becoming especially important.

One of these features is connected with the nature of the traffic as a time process, which more and more acquires the features of so-called 'fractals'. Many fractals have self-similar characteristics and, generally speaking, these concepts are closely related to each other. In mathematical language the self-similar feature results in an exact or probabilistic replication of the object characteristics when considered on different scales. The self-similar feature leads to definite regularities in the traffic statistical behaviour and to the necessity to consider the probability of complicated stochastic processes. Then the traffic itself heads to be described as a peculiar dynamic system by so-called 'fractal' or chaotic models. In worldly language this can mean that the traffic possesses the features to save the basic distinctive patterns irrespective of the periods when it is analysed. The process becomes 'similar to itself', just as a fern leaf looks so much like the other leaves that they seem to us to be absolutely similar. The fern leaf pattern can be found in almost every book as being related to the fractal phenomena because this example has already become canonical for an explanation of fractal features, such as the UK coastline or Koch curves.

Chaotic consideration of the most plentiful processes in our lives has probably become one of the most attractive and 'fashionable' scientific tendencies in the past decades. These are the processes in biology, in medicine, in mathematics, in economics, in forecasting and in telecommunications. It is most likely that in future it will be impossible to analyse any complicated systems without using the chaotic approach.

The aim of this book is to try to investigate the self-similar processes in the telecommunication network application, to present some more or less generalized understanding of many publications of the past 10 to 20 years connected with it, to acquaint readers having an active interest in the main approaches in this interesting and complicated direction, to give a review of previously obtained and these new results and, 'to open the door' to versatile specialists in this new and fascinating field of research activity.

The authors are very well aware of the fact that the desire to explain visually the new and complicated ideas in this area, where even the terminology is hardly settled and we ourselves are yet very far from a full understanding, may not be rewarding. Even more so, the authors are oriented towards a wide audience: the students, the engineers, the researchers, the communications experts and the communication network equipment designers. Naturally, this book can arouse a sharply critical assessment by many experts on traffic and the evident displeasure of those whose scientific research lines are either not reflected in the book or described taciturnly. Others may be dissatisfied who had decided that they could understand the issues within a couple of evenings and for them the considered problems may have turned out to be hard to perceive and mathematically complicated. Nevertheless, we shall consider our aim fulfilled if the reader becomes interested in self-similar processes and the number of experts in this perspective and steadily developing field increases. It often happens that various experienced specialists working alongside in a certain direction for some reason enrich each other, causing the most unexpected 'singular' processes to occur, which stepwise could lead to definite revolutions in standard scientific approaches. There is every expectation that this will happen in the promising field of self-similar processes.

The term 'fractal' was first introduced by Benua Mandelbrot. As we have already mentioned, self-similar processes closely follow the fractals. They describe the phenomenon in which some object feature (e.g. some image, voice, digital telecommunication message, time series) is preserved with varying space or time scales. If the analysed object is self-similar (or fractal), its parts (fractions) are similar when increased (in a certain sense) to their full image. In contrast to the deterministic (sharply and unambiguously defined) fractals, the stochastic fractal processes have no evident similarity to the component parts in the finest detail, but in spite of this, stochastic self-similarity is the feature that can be illustrated visually and can be estimated mathematically definitely enough (H.E. Hurst).

In most cases it is enough to use the statistical characteristics of the second order, well known to telecommunication networks experts, for a quantitative estimation and description of the bursty (pulsating) structure (or changeability) of the stochastic fractal processes. As a result, the usual correlation function of the process plays rather an important part, being essentially the main criterion, with the help of which the scaling invariance of similar processes, i.e. self-similarity, is successfully determined. The existence of the correlation 'within range' can usually be characterized by the term 'long-range dependence'. The distinctive difference of the self-similar process correlation function compared to the usual process correlation function is that for the former the correlation, as the time delay function, assumes the polynomial delay rather than the exponential delay.

In the telecommunication applications the measured traffic traces (routes) correspond to the stochastic self-similarity (fractality) features. It is assumed here that the traffic form with the corresponding amplitude normalization is the conformity measure. It is difficult to observe the clear structure of the measured traffic traces, but self-similarity allows consideration to be taken of the stochastic nature of many network devices and events that together influence the network traffic. If the viewpoint that the traffic series is a sample of the stochastic process realization is accepted and the conformity degree is weakened, i.e. some statistical characteristic of the re-scaled time series is chosen, it would then possible to obtain the exact similarity of the mathematical objects and the asymptotic similarity of its specific samples regarding this weakened similarity criteria.

The telecommunication traffic self-similarity as an independent scientific direction was formed very recently. The essential contributions in this direction were made by J. Beran, M. Crovella, K. Park, W. Willinger, P. Abry, M.S. Taqqu, V. Teverovsky, W.E. Leland, J.R. Wallis, P.M. Robinson, C.F. Chung, V. Paxson, S. Floyd, S.I. Resnick, R. Riedi, J.B. Levy, J.W. Roberts, S.B Lowen, I. Norros, B.K. Ryu, G. Samorodnitsky and many others. The investigations fulfilled by these authors are quite extensive and the results are significant.

The book is divided into two parts.

In the first part (Chapters 1 and 2) the theoretical aspects of the self-similar (fractal and multifractal) random processes are considered. The main definitions necessary to understand the rest of the book are given. The current state as well as the problems related to the self-similar process description are analysed, and it is explained why the traffic in modern telecommunication systems should be considered fractal.

The second part (Chapters 3 to 5) is devoted to the theoretical aspects of the best-known models demonstrating self-similar features. These models are considered also from the point of view of their software realization (the algorithms and modelling results, etc., are given). The main theoretical results relating to each model are presented and discussed. Various approaches used to estimate the traffic fractal and multifractal features are considered.

In Chapters 3 and 4 the traffic of the real telecommunication and computer network is analysed in detail. Chapter 3 is devoted to the self-similarity research of the real time traffic to which the traffic, created by the voice and video services, is referred. On the basis of an analysis of the traffic experimental results the characteristics of the description and self-similar properties are studied, including mono- and multifractal characteristics. The traffic self-similarity in LAN (Ethernet) and WAN (Internet) is analysed in Chapter 4 with an account of the transport (TCP/IP) and application (HTTP, UDP, SMTP, etc.) protocol levels.

In Chapter 5 the features of the self-similarity influence on the quality of service estimates on the examples of voice services are analysed. The traffic control aspects under conditions of its self-similarity and the long-range dependence are discussed. To do this, the information extracted over large time scales is used, which can be applied to correct the control mechanisms of the network resources.

In particular, it is shown that the queue length distribution in the infinite system buffer in the long-range dependent input process decays slower than exponentially (or subexponentially). Conversely, for short-range dependence at the input the decay has an exponential character. The queue length distribution illustrates that the buffering (as the strategy providing the resources) is ineffective from the point of view of the occurrence of disproportionate delays, when the input traffic is self-similar.

From the position of traffic control, self-similarity implies the existence of the correlation structure over the time interval, which can be used for traffic control. To do this the information extracted from the large time scales can be used to correct the overload control mechanisms.

In spite of the obviously mathematical consideration of many aspects of the self-similarity and stochastic phenomena used by many authors, this book, in the authors' opinion, is not overloaded with mathematical expressions, and in a number of cases it will act as a reference book for specialists. That is why it can be recommended to readers at large who are interested in telecommunication and computer technologies. The interest of potential students in this book can be related to the specific lecture courses (standard or short) or parts of other courses devoted to self-similar processes. Acronyms used in the book are explained at the end in appendix B.

The authors would appreciate any comments concerning this book.

About the Authors

OLEG I. SHELUHIN

Oleg I. Sheluhin was born in 1952 in Moscow, Russia. In 1974 he graduated from the Moscow Institute of Transport Engineers (MITE) with a Master of Science Degree in Radio Engineering. After that he entered the Lomonosov State University (Moscow) and graduated in 1979 with a Second Diploma of Mathematics. He received a PhD (Techn.) at MITE in 1979 in Radio Engineering and Dr Sci (Techn.) at Kharkov Aviation Institute in 1990. The title of his PhD thesis was 'Investigation of interfering factors influence on the structure and activity of noise short-range radar' and the Dr Sci thesis, 'Synthesis, analysis and realisation of short-range radio detectors and measuring systems'.

Oleg I. Sheluhin is a member of the International Academy of Sciences of Higher Educational Institutions. He has published 15 scientific books and textbooks for universities and more than 250 scientific papers. Since 1990 he has been the Head of the Radio Engineering and Radio Systems Department of Moscow State Technical University of Service (MSTUS).

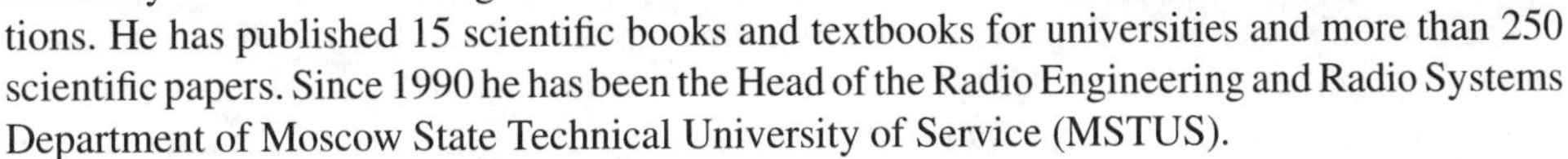

Oleg I. Sheluhin is the Chief Editor of the scientific journal *Electrical and Informational Complexes and Systems* and a member of Editorial Boards of various scientific journals. In 2004 he was awarded the honorary title 'Honoured scientific worker of the Russian Federation' by the Russian President.

His scientific interests are radio and telecommunication systems and devices.

SERGEY M. SMOLSKIY

Sergey M. Smolskiy was born in 1946 in Moscow. In 1970 he graduated from the Radio Engineering Faculty of the Moscow Power Engineering Institute (MPEI). In the same year he began work at the Department of Radio Transmitting Devices of MPEI. After concluding his postgraduate study and his PhD thesis in 1974 ('Quasi-harmonic oscillations stability in autonomous and synchronized high-frequency transistor oscillators') he continued research at the Department of Radio Transmitting Devices, where he was engaged in theoretical and practical questions concerning the transmitting stages of short-range radar development and in various questions concerning transistor oscillators theory and microwave oscillations stability. He was nominated as the scientific supervisor of many scientific projects, which were carried out under the decrees of the USSR Government. In 1993 he presented his thesis for the Dr Sci (Techn.) degree ('Short-range radar systems on the basis of controlled oscillators') and became a Full Professor.

He has been the Chairman of the Radio Receivers Department of MPEI since 1995. His pedagogical experience extends over twenty years. He is a lecturer in the following courses: 'Radio Transmitting Devices', 'Systems of Generation and Control of Oscillations', 'Non-linear Oscillations Theory in Radio Engineering', 'Analysis Methods for Non-linear Radio-Electronic systems'and 'Autodyne Short-Range Radar'.

The list of his scientific publications and inventions contains 170 scientific papers, ten books, three copyright certificates of USSR inventions and more than 90 scientific and technical reports at various conferences, including international ones. He is a member of the International Academy of Informatization, the International Academy of Electrical Engineering Sciences, the International Academy of Sciences of Higher Educational Institutions, a member of the IEEE and an Honorary Doctor of several foreign universities. He was awarded the State Order of Poland for merits in preparation of the scientific staff, the title of 'Honoured Radio Engineer' and the title of 'Honoured worker of universities'.

His scientific work during the last ten years has been connected with conversion directions of short-range radar system engineering, radio-measuring systems for the fuel and energy industry, systems of information acquisition and transfer for industrial purposes with the use of wireless channels and systems of medical electronics.

ANDREY V. OSIN

Andrey V. Osin was born in 1980, received a Bachelor Degree in 2001 and an Engineer Degree in 2002 in Radio Engineering at the Moscow State Technical University of Service. He entered a three-year PhD course and successfully presented his PhD thesis ('The influence of voice traffic self-similarity on quality of service in telecommunication networks') in 2005 in the speciality 'Telecommunications systems, networks and devices' at Moscow Power Engineering Institute (Technical University).

At present Dr. A. Osin works as the senior lecturer at Moscow State Technical University of Service (Radio Engineering and Radio Systems Department) and delivers the lecture courses 'Radio Engineering System Modelling', 'The Bases of Computer Modelling and Computer Design of the Radio Engineering Sets' and 'CAD Systems in Service Activity'. He works with postgraduate and PhD students on mutual research. Three Bachelor final projects and five Engineering projects were fulfilled under his supervision.

He has published 11 scientific papers and is the co-author of two books in the field of telecommunication systems modelling. He prepared 12 scientific reports at various conferences held in Russia.

Acknowledgements

The authors would like to express their thanks to colleagues from the Department of Radio Engineering and Radio Systems at the Moscow State Technical University of Service and from the Department of Radio Receivers at the Moscow Power Engineering Institute (MPEI) for useful discussions of manuscript material. They appreciate the help given by Lydia Grishaeva in correcting the English text of the manuscript.

1

Principal Concepts of Fractal Theory and Self-Similar Processes

1.1 Fractals and Multifractals

B. Mandelbrot introduced the term 'fractal' for geometrical objects: lines, surfaces and spatial bodies having a strongly irregular form. These objects can possess the property of self-similarity. The term 'fractal' comes from the Latin word *fractus* and can be translated as fractional or broken. The fractional object has an infinite length, which essentially singles it out on the traditional Euclidean geometry background. As the fractal has the self-similar property it is more or less uniformly arranged in a wide scale range; i.e. there is a characteristic similarity of the fractal when considered for different resolutions. In the ideal case self-similarity leads to the fractional object being invariant when the scale is changed. The fractional object may not be self-similar, but self-similar properties of the fractals considered in this book are observed everywhere. Therefore, when self-similar traffic is mentioned, it will be assumed that its time realizations are fractals.

There is some minimal length $l_{\min}$ for the naturally originated fractal such that at the $l \approx l_{\min}$ scale its fractional structure is not ensured. Moreover, at rather a large scale $l > l_{\max}$, where $l_{\max}$ is the typical geometrical size for the object in a considered environment, the fractional structure is also violated. That is why the natural fractal properties are analysed for l scales only, which satisfies the relation $l_{\min} \ll l \ll l_{\max}$.

These restrictions become understandable when the broken (nonsmooth) trajectory of a Brownian particle is used as an example of the fractal. On a small scale the Brownian particle mass and size finiteness affects this trajectory as well as the collision time finiteness. Taking these circumstances into consideration the Brownian particle trajectory becomes a smooth curve and loses its fractal properties. This means that the scale ($l_{\min}$) at which the Brownian motion can be examined in the fractal theory context is limited by the mentioned factors. When speaking about the scale restrictions from above ($l_{\max}$) it is obvious that the Brownian particle motion is limited by some space in which this particle is located, e.g. the tank with the liquid into which the paint particles are injected during the classical experiment of Brownian motion identification.

Self-Similar Processes in Telecommunications O. I. Sheluhin, S. M. Smolskiy and A. V. Osin
© 2007 John Wiley & Sons, Ltd

It is noteworthy that the exact self-similarity property is typical for regular fractals only. If some element of chance is included in its creation algorithm instead of the deterministic approach, so-called random (stochastic) fractals occur. Their main difference from the regular ones consists in the fact that self-similarity properties are correct only after corresponding averaging has taken place over all statistically independent object realizations. At the same time the enlarged fractal part is not fully identical to the initial fragment, but their statistical characteristics are the same. Network (telecommunications) traffic is often referred to as a class of self-similar stochastic fractals. That is why in the scientific literature the concepts of fractal and self-similar traffic are used synonymously when this does not lead to confusion.

1.1.1 Fractal Dimension of a Set

It was mentioned earlier that the fractional dimension presence is a distinctive fractal property. The fractional dimension concept is now formalized and its calculation approach is evaluated.

In accordance with the algorithm from Reference [1] for determination of the Hausdorff dimension D_f of the set occupying the area with volume L^{D_f} in D-dimensional space, this set is now covered by cubes having the volume ε^{D_f}. The minimum number of nonempty cubes covering the set is $M(\varepsilon) = L^{D_f}(1/\varepsilon)^{D_f}$. From this expression an approximate estimation of D_f can be obtained:

$$D_f = \lim_{\varepsilon \to 0} \left[\frac{\ln M(\varepsilon)}{\ln(1/\varepsilon)} \right] \tag{1.1}$$

In practice, to estimate this dimension it is more convenient to use the mathematical structure well-known as the Renji dimension D_q related to the probability p^i of the test point presence in the ith cell to power q:

$$D_q = \lim_{l \to 0} \left(\frac{1}{q-1} \right) \frac{\ln \left[\sum_{i=0}^{M(t)} p(\varepsilon)_i^q \right]}{\ln \varepsilon}, \quad q = 0, 1, 2, \ldots \tag{1.2}$$

As $q \to 0$, using Equation (1.2) gives

$$D_0 = \frac{\lim_{\varepsilon \to 0} \left(\ln \sum_{i=1}^{M(t)} 1 \right)}{\ln \varepsilon} = -\lim_{\varepsilon \to 0} \frac{\ln M(\varepsilon)}{\ln \varepsilon} = D_f \tag{1.3}$$

i.e. the Renji dimension D_0 coincides with the Hausdorff dimension (1.1). Due to D_q monotony as a function of q, the Renji dimension decreases as a power function and therefore the following inequality is fulfilled: $D_2 \leq D_0 = D$. Thus the largest low border of the Hausdorff dimension can be presented as

$$D_2 = \lim_{l \to 0} \frac{\ln \left[\sum_{i=1}^{M(t)} p(\varepsilon)_i^2 \right]}{\ln \varepsilon} \tag{1.4}$$

Taking this into consideration, the probability p_i of the test point in the presence of the ith set can be estimated as

$$p_i(\varepsilon) = \lim_{N \to \infty} \frac{N_i(\varepsilon)}{N} \tag{1.5}$$

where N is the total number of test points over $1/L$ intervals and N_i is the number of points in the ith set.

The expression (1.4) can be calculated on the basis of the experimentally measured segment duration. In practice, the largest low border of the D_2 dimension can be calculated as the tangent of the linear regression inclination angle for the next points $[\ln 1/N^2(\sum_{i=1}^{M(t)} N_i^2); \ln(\varepsilon)]$ defined for different ε values.

1.1.2 Multifractals

Multifractals are heterogeneous fractional objects where, compared to regular fractals, it is not sufficient to introduce only one magnitude, its fractal dimension D_f, in a detailed description, the full spectrum of the dimensions whose number is infinite in the general case must be given. The reason for this lies in the fact that, together with purely geometrical features defined by the D_f magnitude, these fractals have some statistical properties.

Multifractal objects can be described from the formal point of view. The fractal object is considered to occupy the boundary area £ defined by the size L in Euclidean space with the D dimension. The fractal is represented by a set of $N \gg 1$ points distributed in a certain manner in this area at some formation stage. It is supposed that $N \to \infty$. The full £ area is divided into cells with $l \ll L$ sides that cover ε^D units of the examined space. Only the occupied cells will be considered, in which at least one point from K points belonging to this fractal is contained. Let the occupied cell index i be changed within $i = 1, 2, \ldots, N(\varepsilon)$, where $N(\varepsilon)$ is the total quantity of occupied cells depending on the size of the cell side ε. Assuming that $N_i(\varepsilon)$ is the number of points in the cell with index i, the magnitude can be found where

$$p_i(\varepsilon) = \lim_{K \to \infty} \frac{N_i(\varepsilon)}{K} \tag{1.6}$$

represents the probability that the point chosen at random from the set is located in cell i. Due to probability normalization the conclusion can be drawn that $\sum_{i=1}^{N(\varepsilon)} p_i(\varepsilon) = 1$. The generalized statistical sum $Z(q, \varepsilon)$, characterized by exponent q, can possess any values in the interval $-\infty < q < +\infty$ as follows:

$$Z(q, \varepsilon) = \sum_{i=1}^{N(\varepsilon)} p_i^q(\varepsilon) \tag{1.7}$$

Definition 1.1
The set of values

$$D_q = \frac{\tau(q)}{q - 1} \tag{1.8a}$$

where

$$\tau(q) = \lim_{\varepsilon \to 0} \frac{\ln Z(q, \varepsilon)}{\ln \varepsilon} \tag{1.8b}$$

can be called the spectrum of generalized fractal Renji dimensions D_q describing the distribution of points in the £ area.

If $D_q = D_f =$ constant, i.e. does not depend on q, the given set of points represents the normal regular fractal, which can be defined by only one magnitude: the fractal dimension D_f. Conversely, if function D_q changes with q, the set of points under consideration is the multifractal. Thus, a multifractal can be characterized in a general case by some nonlinear function $\tau(q)$ defining the behaviour of the statistical sum $Z(q, \varepsilon)$ as $\varepsilon \to 0$:

$$Z(q, \varepsilon) = \sum_{i=1}^{N(\varepsilon)} p_i^q(\varepsilon) \approx \varepsilon^{\tau(q)} \tag{1.9}$$

The behaviour of the generalized statistical sum will be examined in the case of a normal regular fractal with the D_f fractional dimension. In this case there is an equal number of points $n_i(\varepsilon) = N/N(\varepsilon)$ in all occupied cells, i.e. the fractal is homogeneous. It is obvious that relative occupancies $p(\varepsilon) = 1/N(\varepsilon)$ are also the same and the generalized statistical sum becomes

$$Z(q, \varepsilon) = N^{1-q}(\varepsilon) \tag{1.10}$$

It is clear that according to the fractal dimension definition D_f, the number of occupied cells at small enough ε behaves as follows:

$$N(\varepsilon) \approx \varepsilon^{-D_f} \tag{1.11}$$

Substituting (1.11) into Equation (1.10) and comparing to (1.9) it is found that in the case of the regular fractal the function

$$\tau(q) = (q - 1)D_f \tag{1.12}$$

i.e. it is linear. In this case all $D_q = q$ and does not really depend on q. In the case of the fractal for which all generalized fractal dimensions D_q coincide the term *monofractal* is often used.

If distribution of the points among the cells is unequal the fractal is referred to as heterogeneous, i.e. it represents a multifractal. For its characteristic it is necessary to contain the whole spectrum of generalized fractional dimensions D_q, the number of which in the general case is infinite. To determine the point distribution it is necessary not only to define the $\tau(q)$ function but also its derivative, which is calculated directly from expressions (1.8b) and (1.7):

$$\frac{d\tau(q)}{dq} = \lim_{\varepsilon \to 0} \frac{\sum_{i=1}^{N(\varepsilon)} p_i^q \ln p_i}{\left(\sum_{i=1}^{N(\varepsilon)} p_i^q\right) \ln \varepsilon} \tag{1.13}$$

This derivative has an important physical characteristic. If it is not constant and changes with q, it is a multifractal.

1.1.3 Fractal Dimension D_0 and Informational Dimension D_1

The physical sense of generalized fractal dimensions D_q will be analysed for several q values. At $q = 0$ it follows from Equations (1.7) that $Z(0, \varepsilon) = N(\varepsilon)$. On the other hand, according to Equations (1.9) and (1.8a),

$$Z(0, \varepsilon) \approx \varepsilon^{\tau(0)} = \varepsilon^{-D_0} \tag{1.14}$$

Comparing these two equalities gives the relation $N(\varepsilon) \approx \varepsilon^{-D_0}$. This shows that the D_0 magnitude represents the usual Hausdorff dimension of the £ set, which is the roughest fractal characteristic and provides no information about its statistical properties. For the physical sense of the D_1 magnitude it can be shown that

$$D_1 = \lim_{\varepsilon \to 0} \frac{\sum_{i=1}^{N(\varepsilon)} p_i \ln p_i}{\ln \varepsilon} \tag{1.15}$$

The numerator of this expression represents (with the accuracy of a sign) the entropy $S(\varepsilon)$ of the fractal set: $S(\varepsilon) = -\sum_{i=1}^{N(\varepsilon)} p_i \ln p_i$. As a result, the generalized fractional dimension D_1 relates to the entropy $S(\varepsilon)$ by

$$D_1 = -\lim_{\varepsilon \to 0} \frac{S(\varepsilon)}{\ln \varepsilon} \tag{1.16}$$

On the basis of a similar understanding Claude Shannon generalized the entropy concept S for abstract problems of information transmission and processing theory. For these problems the entropy became a measure of the information quantity necessary for system determination at some state i. In other words, it is a measure of the lack of knowledge about a system. Returning to the initial problem about the distribution of points in the fractal set £, it is possible to conclude that since

$$S(\varepsilon) \approx \varepsilon^{-D_1} \tag{1.17}$$

the D_1 quantity characterizes the information required for determination of the point location in some cell. In this connection the generalized fractal dimension D_1 is often called an *informational dimension*. It shows how the information, required to determine the point location, increases as the cell size ε tends to zero.

1.1.3.1 Properties of the D_q Function

As stated above, a multifractal is characterized by the heterogeneous point distribution among the cells. At the same time, if the points that constitute the multifractal are distributed uniformly among all $N(\varepsilon)$ cells with the probability $p_i = 1/N(\varepsilon)$, the entropy of such a distribution would be maximal and equal to

$$S_{\max}(\varepsilon) = -\sum_{i=1}^{N(\varepsilon)} p_i \ln p_i = \ln N(\varepsilon) \approx -D_0 \ln \varepsilon \tag{1.18}$$

In other words, it would be larger than the actual value of fractal energy calculated for the true heterogeneous point distribution $S(\varepsilon) = -D_1 \ln \varepsilon$. Therefore an important conclusion can be made that the informational dimension D_1 of the multifractal is always smaller or equal to its Hausdorff dimension D_0. This inequality can be generalized to an arbitrary exponent q and it can be proved that the generalized fractional dimension D_q always monotonically decreases (or at least remains constant) as q rises: $D_q \geq D_{q'}$ at $q' > q$. The sign of equality occurs, for example, for the uniform fractal. D_q achieves the maximal value $D_{\max} = D_{-\infty}$ as $q \to -\infty$ and the minimal value $D_{\min} = D_{\infty}$ as $q \to \infty$.

1.1.3.2 Spectrum of Fractal Dimensions

The multifractal concept formulated above represents the heterogeneous fractal. In order to describe it, the set of generalized fractional dimensions D_q, where q possesses any value in the interval $-\infty < q < \infty$, was introduced. However, D_q values are not, strictly speaking, the fractional dimensions in the conventional sense of the word. For this reason they are called *generalized* dimensions.

Together with these terms used to characterize the multifractal set the function of the multifractional spectrum $f(\alpha)$ (the spectrum of multifractal singularities) is often used, with the term *fractional dimension* describing it best. It can be shown that the $f(\alpha)$ value is actually equal to the Hausdorff dimension of some homogeneous fractional subset from the initial £ set, which ensures that it is the dominating contribution in the statistical sum at a given q value.

The set of probabilities p_i, showing the relative occupation ε of the cells with which the set under examination can be covered, may be regarded as one of the main multifractal features. The smaller the size of the cells the smaller is the occupation value. For self-similar sets the dependence of p_i on the cell size ε has the power character

$$p_i(\varepsilon) \approx \varepsilon^{\alpha_i} \tag{1.19}$$

where α_i represents some exponent (which is different for different cells i). It is known that for the regular fractal all exponents α_i are the same and equal to the fractional dimension D_f:

$$p_i = \frac{1}{N(\varepsilon)} \approx \varepsilon^{D_f} \tag{1.20}$$

In this case the statistical sum (1.7) is given by

$$Z(q, \varepsilon) = \sum_{i=1}^{N(\varepsilon)} p_i^q(\varepsilon) = N(\varepsilon)\varepsilon^{D_f(q-1)} \tag{1.21}$$

Therefore, $\tau(q) = D_f(q-1)$ and all generalized fractional dimensions $D_q = D_f$ in this case coincide and do not depend on q.

However, for such a complex object as the multifractal, due to its heterogeneity, the probabilities of the cell occupation p_i in the common case are different and the α_i exponent for different cells can possess different values. In the case of the monofractal, where all α_i are the same (and equal to the fractional dimension D_f), the number $N(\varepsilon)$ obviously depends on cell size ε in accordance with the power function. Therefore, $N(\varepsilon) \approx \varepsilon^{-D_f}$. The exponent in this case is defined by the set dimension D_f.

For the multifractal this situation is not typical and different values of α_i can be met with the probability characterized not by the same D_f value but by different (depending on α) values of the $f(\alpha)$ exponent:

$$n(\alpha) \approx \varepsilon^{-f(\alpha)}$$

Thus the physical sense of the $f(\alpha)$ function consists in the fact that it represents the Hausdorff dimension of some homogeneous fractional subset £$_\alpha$ from the initial set £ characterized by the same probabilities of cell occupancy $p_i \approx \varepsilon^{\alpha}$. Since the fractional dimension of the subset is obviously always smaller or equal to the fractional dimension of the initial set D_0, the important inequality for the $f(\alpha)$ function takes place: $f(\alpha) \leq D_0$.

As a result it can be concluded that the set of different values of the $f(\alpha)$ function (at different α) represents the spectrum of the fractional dimensions for homogeneous subsets £$_\alpha$ of the initial set £, each of which has its own values of the fractional dimension $f(\alpha)$. Since any subset has only part of the total number of cells $N(\varepsilon)$, into which the initial set £ is divided, the probability normalization condition $\sum_{i=1}^{N(\varepsilon)} p_i(\varepsilon) = 1$ is not fulfilled when summing on this subset only. The sum of these probabilities is less than 1. Therefore the probabilities p_i having the same α_i value are obviously less (or at least have the same order) than the $\varepsilon^{f(\alpha_i)}$ value, which is inversely proportional to the number of used cells covering this subset (it should be noted that in the case of the monofractal $p_i \approx 1/N(\varepsilon)$). This results in the important inequality for the $f(\alpha)$ function and it follows that at all values of α then $f(\alpha) \leq \alpha$.

The sign of equality takes place, for example, for the fully homogeneous fractal, where $f(\alpha) = \alpha = D_f$.

1.1.4 Legendre Transform

If the interconnection between the $f(\alpha)$ function and the $\tau(q)$ function introduced earlier is established, it can be shown that the expression for the statistical sum takes the form

$$Z(q,\varepsilon) \approx \varepsilon^{q\alpha(q)-f(\alpha(q))} \tag{1.22}$$

Comparing (1.22) with (1.8) gives the conclusion that

$$\tau(q) = q\alpha(q) - f(\alpha(q)) \tag{1.23}$$

According to Definition 1.1,

$$D_q = \frac{1}{q-1}[q\alpha(q) - f(\alpha(q))] \tag{1.24}$$

Therefore, if the multifractional spectrum function $f(\alpha)$ is known, it is possible to obtain the function D_q, and vice versa, knowing D_q, it is possible to obtain function $\alpha(q)$ with the help of the equation

$$\alpha(q) = \frac{\mathrm{d}}{\mathrm{d}q}[(q-1)D_q] \tag{1.25}$$

and after that to find the function $f(\alpha(q))$ from Equation (1.24). These two equations define (in parametrical form) function $f(\alpha)$.

Equations (1.23) and (1.25) define the Legendre transform from $\{q, \tau(q)\}$ variables to $\{\alpha, f(\alpha)\}$ variables:

$$\begin{aligned} \alpha &= \frac{\mathrm{d}\tau(q)}{\mathrm{d}q} \\ f(\alpha) &= q\frac{\mathrm{d}\tau}{\mathrm{d}q} - \tau \end{aligned} \tag{1.26}$$

1.2 Self-Similar Processes

1.2.1 Definitions and Properties of Self-Similar Processes

A random process in discrete time or the time series $X(t)$, $t \in Z$, is considered where $X(t)$ is interpreted as the traffic volume (measured in packets, bytes or bits) up to moment t.

Definition 1.2
Consider that a real-valued process $\{X(t), t \in R\}$ has stationary increments if

$$\{X(t+\Delta t) - X(\Delta t), t \in \mathbb{R}\} \stackrel{d}{=} \{X(t) - X(0), t \in \mathbb{R}\}$$

for all $\Delta t \in \mathbb{R}$.

The sequence of increments for $\{X(t),\ t \in \mathbb{R}\}$ at discrete time is defined as $Y_k = X(k+1) - X(k)$, $k \in \mathbb{Z}$. For the purpose of traffic simulation the process $X(t)$ is considered as *stationary* in the wide sense, applying the restriction that the covariance function $R(t_1, t_2) = M[(X(t_1) - m)(X(t_2) - m)]$ is invariant with regard to the shift, i.e. $R(t_1, t_2) = R(t_1 + k, t_2 + k)$ for any t_1, t_2, $k \in Z$. It is assumed that two first moments $m = M[X(t)]$, $\sigma^2 = M[X(t) - m^2]$ exist and are finite for any $t \in \mathbb{Z}$. Here $M[\cdot]$ is the averaging operation, m is the initial moment (mathematical expectation) and σ^2 is the variance (dispersion) of the $X(t)$ process. Suppose for convenience that $m = 0$. Since at the stationary condition $R(t_1, t_2) = R(t_1 - t_2, 0)$, the covariance function will be designated as $R(k)$ and the correlation factor as $r(k) = R(k)/R(0) = R(k)/\sigma^2$.

Definition 1.3 [2]
The real-valued process $\{X(t), t \in \mathbb{R}\}$ is self-similar with the exponent $H > 0$ (H-ss) if, for any $a > 0$ finite-dimensional distributions for $\{X(at),\ t \in \mathbb{R}\}$ are identical to finite-dimensional

distributions $\{a^H X(t), t \in \mathbb{R}\}$, i.e. if, for any $k \geq 1$, $t_1, t_2, \ldots, t_k \in \mathbb{R}$ and any $a > 0$,

$$(X(at_1), X(at_2), \ldots, X(at_k)) \overset{d}{=} (a^H X(t_1), a^H X(t_2), \ldots, a^H X(t_k)) \tag{1.27}$$

Equation (1.27) can be rewritten more briefly as

$$\{X(at), t \in \mathbb{R}\} \overset{d}{=} \{a^H X(t), t \in \mathbb{R}\} \tag{1.28}$$

Equation (1.27) shows that the time scale variation is equivalent to changing the spatial scale of conditions. Therefore, the typical self-similar process realizations look alike irrespective of the time scale in which they are considered. This does not mean that the process is precisely repeated but that there is likely to be a similarity of statistical properties because of the fact that the statistical characteristics do not change at scaling [2]. The exponent H, known as the *Hurst exponent*, is of vital importance for the self-similar process theory because it serves as an indicator of random process self-similarity and defines the property of so-called long-range dependence (LRD).

Self-similar processes with the self-similar parameter H are marked as H-ss (H-self-similar) in the literature. The nondegenerate self-similar H-ss process cannot be stationary. Nevertheless, there is an important connection between self-similar and stationary processes.

Theorem 1.1 [2]
If $\{X(t), 0 < t < \infty\}$ is an H-ss process, the process

$$Y(t) = \mathrm{e}^{-tH} X(\mathrm{e}^t), \quad -\infty < t < \infty \tag{1.29}$$

is stationary. Conversely, if the process $\{Y(t), -\infty < t < \infty\}$ is stationary, the process

$$X(t) = t^H Y(\ln t), \quad 0 < t < \infty \tag{1.30}$$

is an H-ss process.

Theorem 1.1 shows that there is a set of different self-similar processes. From the point of view of practical implementation those with stationary increments are of special interest, as they lead to stationary sequences with a special behaviour.

An H-ss process having stationary increments is specifically marked as H-sssi (self-similar process with the self-similarity parameter H and stationary increments).

Definition 1.4 [2]
Process $\{X(t), t \in \mathbb{R}\}$ is called H-sssi, if it is self-similar with parameter H and possesses stationary increments.

Lemma 1.1 [2]
Assume that $\{X(t), t \in \mathbb{R}\}$ is the (nondegenerate) H-sssi process with an infinite variance. Then $0 < H \leq 1$, $X(0) = 0$ are nearly everywhere and the covariance is defined from the expression

$$R(t_1, t_2) = \frac{1}{2}\{[|t_1|^{2H} + |t_2|^{2H} - |t_1 - t_2|^{2H}]\sigma_x^2\} \tag{1.31}$$

If $X(t)$ is a (nondegenerate) H-sssi process with finite dispersion, then $0 < H < 1$. During the traffic simulation the range $0.5 < H < 1$ is of special interest since the H-sssi process $X(t)$ with $H < 0$ cannot be measured and corresponds to the pathologic case, whereas the case $H > 1$ is prohibited due to the stationary condition on the incremental process. The range $0 < H < 0.5$ can be excluded in practice because in this case the increment process is so-called short-range dependence (SRD). For practical goals only the range $0.5 < H < 1$ is important. In this range the correlation factor for the increment process $Y(t)$

$$Y_k = X(k) - X(k-1), \quad k \in \mathbb{Z} \tag{1.32}$$

has the following form:

$$r(k) = \frac{1}{2}[(k+1)^{2H} - 2k^{2H} + (k-1)^{2H}] \tag{1.33}$$

1.2.1.1 Aggregated Process

Let $Y = \{Y_i, i \in \mathbb{Z}\}$ be a stationary process with the covariance function $R(k)$. The m-aggregated time series $Y^{(m)}$ is defined by averaging the initial series over nonoverlapping intervals with m size and substituting its mean value for each interval, i.e.

$$Y_i^{(m)} = \frac{1}{m}(Y_{im-m+1} + Y_{im-m+2} + \cdots + Y_{im}), \quad m = 1, 2, \ldots$$

or in a more convenient form

$$Y_k^{(m)} = \frac{1}{m^H} \sum_{i=(k-1)m+1}^{km} Y_i, \quad k \in \mathbb{Z}, \quad 0 < H < 1 \tag{1.34}$$

and designating the covariance function corresponding to it as $R^{(m)}(k)$.

Definition 1.5
The discrete random process $\{Y_k,\ k \in \mathbb{Z}\}$ is strictly self-similar in the wide sense (exactly second-order self-similar) with the self-similarity parameter $H(\frac{1}{2} < H < 1)$ if

$$R(k) = \frac{\sigma^2}{2}[(k+1)^{2H} - 2k^{2H} + (k-1)^{2H}] \tag{1.35}$$

for any $k \geq 1$. $X(t)$ is asymptotically self-similar in the wide sense (second-order asymptotical self-similarity H-ssa) if

$$\lim_{m \to \infty} R^{(m)}(k) = \frac{\sigma^2}{2}[(k+1)^{2H} - 2k^{2H} + (k-1)^{2H}] \tag{1.36}$$

It can easily be verified that Equation (1.36) presupposes $R(k) = R^{(m)}(k)$ for any $m \geq 1$. Therefore, self-similarity in the wide sense means that a covariational structure – the exact condition (1.35) or the approximate (less strict) condition (1.36) – is preserved when the time series is aggregated.

The form $R(k) = [(k+1)^{2H} - 2k^{2H} + (k-1)^{2H}]\,\sigma^2/2$ is not fortuitous and presupposes an additional structure (a long-range dependence), which will be discussed later. The self-similarity of the second order (in an exact or approximate sense) is the main structure to simulate the network traffic.

There is a connection between the process that is strictly self-similar in the wide sense and the process that is self-similar in the narrow sense.

Definition 1.6

A process X is called self-similar in the narrow sense (strictly self-similarity), with parameter $H = 1 - (\beta/2)$, $0 < \beta < 1$, if $m^{1-H}X^{(m)} = X$, $m \in \mathbb{N}$, where the sign '=' designates the equality of finite-dimensional distributions. $X^{(m)} = (X_1^{(m)}, X_2^{(m)}, \ldots)$ is the process X averaged over intervals with length m, the components $X^{(m)}$ of which are defined by the expression

$$X_k^{(m)} \triangleq \frac{1}{m}(X_{km-m+1} + \cdots + X_{km}), \quad m, k \in \mathbb{N}$$

The connection between the self-similar process in the wide sense and the self-similar process in the narrow sense is analogous to the connection between processes being stationary in the wide and narrow senses.

In addition to the statistical similarity when scaling, the self-similar processes can be detected due to several equivalent features:

1. They have a hyperbolically decaying covariance function of the following form:

$$R(k) \cong k^{(2H-2)}L(t), \quad \text{as } k \to \infty \tag{1.37}$$

   where $L(t)$ is the function slowly variable at infinity (i.e. $\lim_{t\to\infty} L(tx)/L(t) = 1$ for all $x > 0$). Therefore, the covariance function is nonsummable and the series formed by sequential values of the covariance function diverges as

$$\sum_k R(k) = \infty \tag{1.38}$$

   This infinite sum is another definition of the long-range dependence, which is why almost all self-similar processes are long-range dependent. Consequences of this fact are very considerable because the cumulative effect of a wide range of delays can essentially differ from those observed in a short-range dependent process (e.g. Poisson, Markovian or autoregressive (AR) processes).

   Although in the past the teletraffic analysis was mainly based on SRD models, the consequences of LRD can be very serious. Since LRD is the reason for the appearance of bursty traffic which exceeds average levels of traffic, this property leads to buffer overflow and generates losses and/or delays.
2. The sample variance of aggregated processes decreases more slowly than the magnitude and is inversely proportional to the sample size. If the new time sequence $\{X_i^{(m)};\ i = 1, 2, \ldots\}$, obtained by averaging the initial sequence $\{X_i;\ i = 1, 2, \ldots\}$ over the nonoverlapping

consecutive intervals of size m, is introduced for analysis, for self-similar processes slower sample variance decay according to the law

$$\sigma^2(X^{(m)}) \sim m^{(2H-2)}, \text{ as } m \to \infty \tag{1.39}$$

will be typical, while for the traditional (not self-similar) stationary random processes $\sigma^2(\{X_i^{(m)};\ i = 1, 2, \ldots\}) = \sigma^2 m^{-1}$, i.e. the decay is inversely proportional to the sample length. This indicates that the statistical characteristics of samples, such as the sample mean value and the sample variance, will converge very slowly, especially as $H \to 1$. This property is reflected in all the measures of self-similar processes and will be considered in more detail when evaluating statistical characteristics.

3. If self-similar processes are examined in the frequency domain the long-range dependence phenomenon leads to the power character of the spectral density near zero:

$$S(\omega) \sim \omega^{-\gamma} L_2(\omega), \quad \text{as } \omega \to 0 \tag{1.40}$$

where $0 < \gamma < 1, L_2$ is a slowly varying function at 0 and $S(\omega) = \Sigma_k R(k) e^{ik\omega}$ is the spectral density. Therefore, from a spectral analysis position the long-range dependence means that $S(0) = \Sigma_k R(k) = \infty$, i.e. the spectral density tends to $+\infty$ when the frequency ω tends to 0 (a similar event is further called $1/f$ noise). Conversely, the processes with short-range dependence can be characterized by the spectral density, having a positive and finite value at $\omega = 0$.

Equations (1.37), (1.39) and (1.40) are related to parameter H, which is called the Hurst exponent. The Hurst exponent for self-similar processes is situated between 0.5 and 1. As H approaches 1 the series becomes more and more self-similar, revealing itself in a slower covariance decay, as can be seen from Equation (1.37).

1.2.2 Multifractal Processes

In contrast to self-similar processes, multiscaling or multifractal processes ensure a more flexible law of scaling behaviour. The class of multifractal processes embraces all the processes with the scaling property including self-similar monoscaled and multiscaled processes.

Definition 1.7 [3]
The stochastic process $X(t)$ is called multifractional if it has stationary increments and satisfies the equality

$$M[|X(t)|^q] = c(q) t^{\tau(q)+1} \tag{1.41}$$

for some positive $q \in Q,\ [0, 1] \subset Q$, where $\tau(q)$ is called the mass parameter (scale function) and the moment factor $c(q)$ does not depend on t.

The obvious consequence is that $\tau(q)$ is a convex function. If $\tau(q)$ depends on q linearly, the process is referred to as one-scaled or monofractal; otherwise it is multifractal. It can be shown that in the specific case of the self-similar process with parameter H, then $\tau(q) = qH - 1$ and $c(q) = M[|X(1)|^q]$. The class of multifractal processes also includes monofractal and self-similar cases.

1.2.3 Long-Range and Short-Range Dependence

So far the role of self-similarity in the stationarity of the second order has been discussed but not much has been mentioned about the role of H and its limiting values. The definition of long-range dependence and its interconnection with the correlation factor $r(k)$ will now be discussed again.

Among the large number of SRD processes in probability theory and in simulation the self-similar processes are of special interest due to their relation to the limiting theorems and their rather simple structure.

Definition 1.8 [2]
The process $\{Y_i,\ i \in \mathbb{Z}\}$ is called the stationary process with long-range dependence if there is a constant $\tilde{c}_r > 0$ and a real number $\alpha \in (0;1)$, $\alpha = 2 - 2H$, such that

$$\lim_{k\to\infty} \frac{r(k)}{c_r k^{-\alpha}} = 1 \tag{1.42}$$

The process $\{Y_i,\ i \in \mathbb{Z}\}$ is called the stationary process with short-range dependence if there is a constant $0 < c_0 < 1$ such that

$$\lim_{k\to\infty} \frac{r(k)}{c_0^k} = 1$$

The given definition of long-range dependence is an asymptotic one and shows only some limiting behaviour of correlation factors and the tendency to delay to infinity. It defines the convergence degree but not an absolute value, which is defined by constant $\tilde{c}_r$ which makes it difficult to show the long-range dependence [2].

The asymptotic behaviour of factor $r(k)$ results, with the help of expansion in the Taylor series, in

$$r(k) = H(2H-1)k^{2H-2} + O(k^{2H-2}) \quad \text{as } k \to \infty \tag{1.43}$$

In accordance with Definition 1.8, the process $\{Y_i,\ i \in \mathbb{Z}\}$ for $0.5 < H < 1$ is long-range dependent with exponent $\alpha = 2 - 2H$ in Equation (1.42). This also means that the correlations are nonsummable:

$$\sum_{k=-\infty}^{\infty} r(k) = \infty \tag{1.44}$$

Therefore, when $r(k)$ is hyperbolically decaying the appropriate stationary process $\{Y_i, i \in \mathbb{Z}\}$, as can be seen from Equation (1.42), is long-range dependent.

Accordingly, the $\{Y_i, i \in \mathbb{Z}\}$ process is short-range dependent if the normalized correlation function is summable ($\sum_{k=-\infty}^{\infty} r(k) = \text{finite} < \infty$). It should also be possible to offer an equivalent definition of long-range dependence in the frequency domain, where it is necessary that the process spectral density

$$S(\omega) = \frac{1}{2\pi}\sum_{k=-\infty}^{\infty} r(k)e^{ik\omega}, \quad \omega \in [-\pi, \pi], \quad i = \sqrt{-1}$$

satisfies this definition.

Definition 1.9 [2]
The process $\{Y_i, i \in \mathbb{Z}\}$ is called a stationary process with long-range dependence if there are real numbers $\beta \in (0;1)$ and constant $c_f > 0$ such that

$$\lim_{\lambda \to 0} \frac{S(\omega)}{|\omega|^{-\beta}} = 1 \tag{1.45}$$

In accordance with this definition the process $\{Y_i, i \in Z\}$ with $0.5 < H < 1$ is long-range dependent with exponent $\beta = 2H - 1$. The behaviour of $S(\omega)$ near the origin of coordinates is well described by the function behaviour at zero:

$$S(\omega) = c_f|\omega|^{1-2H} + O(|\omega|^{\min(3-2H,2)}) \tag{1.46}$$

Here

$$c_f = \frac{1}{2\pi}\sin(\pi H)\Gamma(2H+1)\sigma^2 \qquad \text{and} \qquad \Gamma(z) = \int_0^{+\infty} x^{z-1}e^{-x}dx, \quad z > 0$$

The approximation $S(\omega) \sim c_f|\omega|^{-\beta}$, $\omega \to 0$, $0 < \beta = 2H - 1 < 1$ is very good, even for relatively large frequencies. As a result, Equation (1.46) can be used in the evaluation of H in the frequency domain.

1.2.4 Slowly Decaying Variance

As shown above, for a self-similar process the variance of sample mean decays more slowly as compared with the magnitude inverse to the sample size:

$$\sigma^2(X_t^{(m)}) \sim m^{-\beta}, \quad 0 < \beta < 2H - 1 < 1 \tag{1.47}$$

for rather large m. Conversely, for SRD processes the parameter $\beta = 1$ and

$$\sigma^2(X_t^{(m)}) \sim m^{-1} \tag{1.48}$$

The property of *slowly decaying variance* can easily be discovered by plotting the m function $\sigma^2(X_t^{(m)})$ on a log–log diagram (variance–time plot). A straight line with a negative (less than 1) slope in the wide range of m demonstrates slowly decaying variance. This property can also be defined with the help of the index of dispersion for counts (IDC).

For a given time interval with length t the IDC is defined as a variance of entrance quantity A_t during length interval t divided by the expectation value of the same magnitude:

$$\text{IDC} = \frac{\sigma^2(X_t^{(m)})}{M(X_t^{(m)})} \tag{1.49}$$

For a finite data set the variance $\sigma^2(A_t)$ can be calculated by dividing all series by the nonoverlapping intervals with length t considered as different copies of A_t.

1.3 'Heavy Tails'

1.3.1 Distribution with 'Heavy Tails' (DHT)

There is a close connection between long-range dependence and DHT, which will be repeatedly mentioned in this book. To begin with some definitions will be introduced and the most typical cases will be examined. A random variable Z has a *distribution with a 'heavy tail'* if the probability

$$P[Z > x] \sim cx^{-\alpha} \quad \text{as} \quad x \to \infty \tag{1.50}$$

where $0 < \alpha < 2$ is called the *'tail' index* or *form parameter* and c is a positive constant. In other words, the distribution 'tail' decays according to the hyperbolic law. Conversely, distributions with 'light tails' (DLTs), e.g. exponential and Gaussian, have an exponentially decaying 'tail'. The specific property of DHT consists in the fact that its dispersion for $0 < \alpha < 2$ is infinite, but for $0 < \alpha \leq 1$ it also has an infinite mean value. When considering network traffic, the case $1 < \alpha < 2$ is of most interest. Pareto distribution is a frequently used DHT. There are some distributions (e.g. Weibull and lognormal) that have subexponentially decaying tails but a finite variance. More generally, it can be said that X has a distribution with a 'heavy tail' F if

$$P[X > x] = 1 - F(x) = x^{-\alpha}L(x) \tag{1.51}$$

where L is a function that is slowly varying at infinity (see Equation (1.37)), i.e. for $x > 0$,

$$\lim_{t\to\infty} \frac{L(tx)}{L(t)} = 1 \tag{1.52}$$

The following are examples of slowly varying functions:

$$\begin{aligned} L(x) &= c + O(1), \quad x > 0 \\ &= \log x, \quad x < 0 \\ &= \log(\log x), \quad x \gg 1 \\ &= 1/\log x, \quad x > 1 \end{aligned}$$

In the first example for L, where $L(x) = c + O(1)$, the term $O(1)$ may look completely harmless but can lead to significant distinctions, e.g. at Pareto tails determination and, say, stable distributions tails.

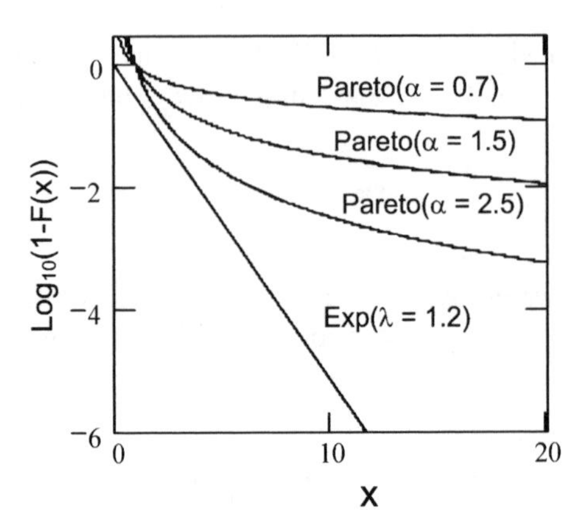

Figure 1.1 Tails on exponential and Pareto distributions with parameters λ and α respectively

The simplest case of a distribution with a 'heavy tail' is the so-called Pareto distribution. In this case $L(x) \equiv 1$; therefore the distribution function for the Pareto distribution can be written as $F(x) = 1 - x^{-\alpha}$. The difference between exponential tails and heavy tails can be seen in Figure 1.1.

The main distinctive feature of the random variable subordinate to DHT consists in the fact that it shows extraordinary changeability. In other words, DHT leads to very large values with finite, in the common case, nonnegligible probability. Therefore samples from such distribution have for the most part 'small' values, but a small number of 'very large' values occurs as well. It is not surprising that as $\alpha \to 1$ the influence of the 'heavy tail' has an effect on the sample by lowering the convergence speed of the sample mean value to the expectation value. For example, when the sample size is m the sample mean value for the Pareto distributed random variable Z may deviate widely from the expectation value, often underestimating it. The error magnitude for estimating $|\bar{Z}_m - M(Z)|$ actually behaves approximately as $m^{(1/\alpha)-1}$ [4]. Therefore, when the sample yields to DHT for α values close to 1 there should be concern that the conclusions about the network behaviour and performance related to sample error could be wrong. It can be shown that the DHT parameters directly linked to a network (e.g. file sizes and duration (lifetime) of the connection) are the reason for long-range dependence and self-similarity in network traffic.

A simple case is examined of the predictability related to the 'heavy tails' of random variables [5]. Z is considered as the variable having DHT and can be interpreted as the *duration* or *lifetime* of a connection, e.g. the TCP (transmission control protocol) connection or the IP (Internet protocol) stream or session. Since the connection duration is a physically measurable event it can be supposed that the connection is active during $\tau > 0$. To simplify the discussion, the time can be changed to the discrete form $(t \in Z_+)$ and $A : Z_+ \to \{0; 1\}$ is a point at $A(t) = 1$ only when $Z \geq t$. The conditional probability P that the connection continues to exist in the future, supporting its activity τ, can be estimated as

$$\Lambda(\tau) = 1 - \frac{P\{Z = \tau\}}{P\{Z \geq \tau\}} \tag{1.53}$$

First $\Lambda(\tau)$ is estimated for 'light tails', in particular for the distribution of approximately exponential tails $P\{Z > x\} \sim c_1 \mathrm{e}^{-c_2 x}$, where $c_1, c_2 > 0$ are constants. The second term in

Equation (1.53) can be calculated as

$$\frac{P\{Z=\tau\}}{P\{Z\geq\tau\}}\sim\frac{c_1\mathrm{e}^{-c_2\tau}-c_1\mathrm{e}^{-c_2(\tau+1)}}{c_1\mathrm{e}^{-c_2\tau}}=1-\mathrm{e}^{-c_2}$$

where for large τ, $\Lambda(\tau)\sim\mathrm{e}^{-c_2}$ is obtained. For heavy tails the same calculations will lead to

$$\frac{P\{Z=\tau\}}{P\{Z\geq\tau\}}\sim\frac{c\tau^{-\alpha}-c(\tau+1)^{-\alpha}}{c\tau^{-\alpha}}=1-\left(\frac{\tau}{\tau+1}\right)^{\alpha} \tag{1.54}$$

which gives $\Lambda(\tau)\rightarrow 1$ as $\tau\rightarrow\infty$.

If Z is considered as DHT variable and interpreted as the *connection duration* or *lifetime*, e.g. the TCP (transmission control protocol) connection or the IP (Internet protocol) stream or session, as shown in Reference [5], the larger the period of observed activity, the higher the probability that connection/session will continue to occur in the future. As a result, the forecast error can be reduced to an arbitrarily small value. Therefore 'heavy tails' lead to predictability and as a consequence they are the reason for long-range dependence in network traffic.

1.3.2 'Heavy Tails' Estimation

DHT processes can be characterized by the fact that the tail of the distribution decays much more slowly than in the case of an exponential distribution. It is the main starting point for methods using heavy tail detection. Moreover, various research graphic approaches are available, to estimate parameter α.

1.3.2.1 Hill's Estimation

$X_1, X_2, \ldots, X_n$ are designated as independent and identically distributed (i.i.d.) random numbers having the distribution function F and $X_{1,n}\geq X_{2,n}\geq\cdots\geq X_{n,n}$ are the order statistics. If F is DHT, Hill's estimation for index α takes the following form [6]:

$$\hat{\alpha}=\hat{\alpha}_{k,n}=\left(\frac{1}{k}\sum_{j=1}^{n}\log X_{j,n}-\log X_{k,n}\right)^{-1} \tag{1.55}$$

Hill's estimation for a set of real network data is shown in Figure 1.2. The graph rapidly turns into a stable condition equal to 1.65. This value is really the α index estimation for the distribution tail.

1.3.2.2 Modified QQ graph

The main principle of the QQ (quantile–quantile) graph is based on the following assumption: if $X_1\geq X_2\geq\cdots\geq X_k$ are the samples from the process with the distribution function F, and k is large enough, F at $x=X_j$ can be estimated in the form

$$P[x<X_j]=F(X_j)\approx 1-\frac{j}{k+1}$$

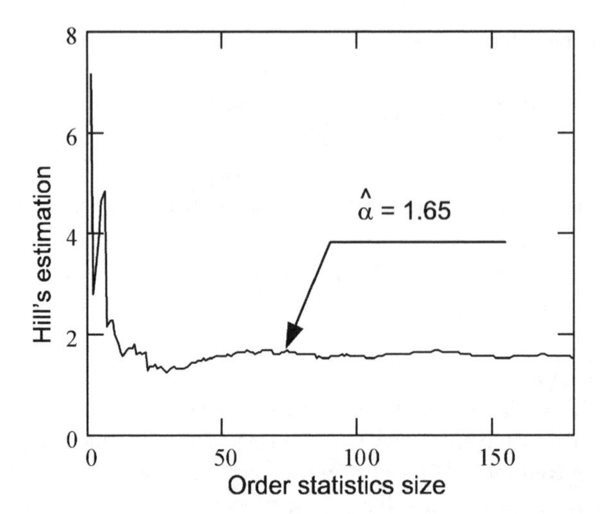

Figure 1.2 Example of Hill's estimation for real network data

This assumption allows the modified QQ graph to be defined in the following manner [7]. Let $X_1 \geq X_2 \geq \cdots \geq X_k = u$ be the order statistics of the distribution, which is approximately a Pareto distribution. Then the graph of $\{[\log X_j - \log u; -\log(j/k+1)], 1 \leq j \leq k\}$ has the form of a straight line with slope α.

In Figure 1.3 the perfected QQ graph for real network data is shown. It can be seen that this graph is not exactly a straight line, but for the point with a small deviation it is possible to select a regression line. The slope of the straight line gives an estimation for α equal to 1.472.

1.4 Hurst Exponent Estimation

In practice, testing for self-similarity and estimating the Hurst exponent produces a complicated problem. This problem in essence consists of the fact that under real conditions it is usual

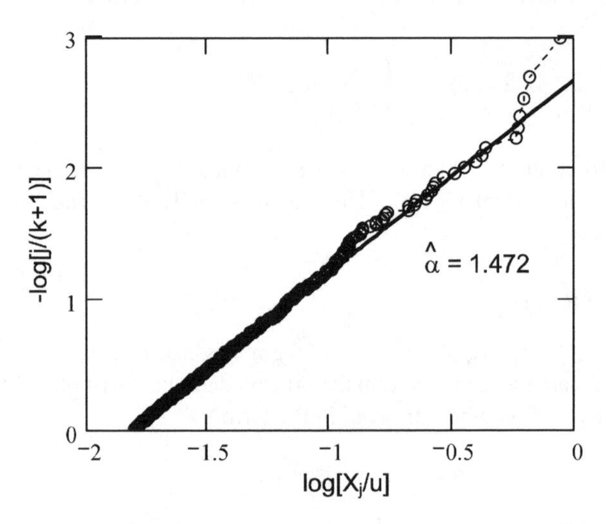

Figure 1.3 Example of modified QQ graph for real network data

to operate with the finite data set and therefore it is impossible to check whether a traffic trace is self-similar or not. This shows that it is necessary to examine various properties of self-similarity in real traffic.

The first problem usually arises because, even if some of the above-mentioned self-similarity properties are confirmed, it is impossible to make the conclusion that the analysed data have a self-similar structure because there are other influences that can lead to the same properties (e.g. a nonstationarity presence). Since the analysis was based only on tests that can lead to the mistake, it is quite reasonable to speak about a self-similar structure in a given scale range for a given data set.

The second problem concerns the fact that estimation of the Hurst exponent depends on many factors (e.g. the estimation procedure, the sample size, the correlation structure, etc.). This causes essential troubles when finding the most appropriate 'H estimation' for the set problem.

The third problem arises when using the Hurst exponent for practical purposes (e.g. definition of buffer sizes) and comes from the fact that Hurst exponent interpretation (which is evident for theoretical self-similar processes) is not obvious for real traffic and therefore can never be considered as a theoretically self-similar process.

At present several methods used to estimate self-similarity from the time series are known [2,8–10]. The most popular approaches are the following: analysis of R/S (rescaled adjusted range) statistics, analysis of the variance–time plot, analysis based on specific properties of $S(\omega)$, the Whittle estimation and analysis based on wavelet functions. Reviews of the statistical testing approaches of self-similar models and random processes with long-range dependence can be found in References [2] and [11]. Publications [12] to [14] describe the additional methods. R/S analysis is discussed in References [2] and [15] to [32], analysis of the variance–time plot in References [2], [17], [22] and [31] to [35] and the frequency domain methods in References [2], [17] and [36] to [61].

Examples of the new statistical approaches in this field are discussed in References [2] and [62] to [86]. Practical estimations of different methods are analysed in References [12] to [14]. The publication [87] gives a general review of the statistical analysis of the time series and References [88] to [90] explain some shortcomings of the time series traditional analysis in the presence of large sets of traffic measurements. The problem of the linear or polynomial regression estimation when errors have long-range dependence is described in References [91] to [96]. Prediction problems in the context of long-range dependence are considered in References [2] and [97] to [99].

The theoretical foundation for many of these statistical toolkits is based on the central and noncentral limit theorems for random sequences with long-range dependence [100–118]. The proof requires an understanding of moment structure for nonlinear functions of Gaussian random variables and linear processes [100, 119–124]. In publications [125] to [133] some of the results were extended to multidimensional processes.

1.4.1 Time Domain Methods of Hurst Exponent Estimation

1.4.1.1 Analysis of the Rescaled Adjusted Range

On the basis of research into various phenomena (e.g. the water level changing in a river) Hurst developed a normalized nondimensional measure capable of describing changeability. He called this measure a *rescaled adjusted range* (R/S). For a given observation set

$X = \{X_n, n \in \mathbb{Z}^+\}$ with a sampling mean value $\bar{X} = 1/n \sum_{j=1}^{n} X_j$, the concept of range $R(n)$ is introduced as

$$R(n) = \max_{1 \leq j \leq n} \Delta_j - \min_{1 \leq j \leq n} \Delta_j \tag{1.56}$$

where

$$\Delta_k = \sum_{i=1}^{k} X_i - k\bar{X}, \quad \forall k = \overline{1, n} \tag{1.57}$$

i.e. the difference between the maximal and minimal deviations.

This characteristic differs from the range of the time sequence of the random variable X, which is equal to

$$\max_{1 \leq j \leq N} X_j - \min_{1 \leq j \leq N} X_j \tag{1.58}$$

The mean value was chosen instead of the characteristic including the accumulation Δ_j and describing the changeability of the variable X. To describe the changeability the following rescaled nondimensional characteristic was found to be more suitable:

$$\frac{R(n)}{S(n)} = \frac{\max\limits_{1 \leq j \leq n} \Delta_j - \min\limits_{1 \leq j \leq n} \Delta_j}{[1/(n-1)] \sum\limits_{j=1}^{n} (X_j - \overline{X})^2} = \frac{\max(0, \Delta_1, \Delta_2, \ldots, \Delta_n) - \min(0, \Delta_1, \Delta_2, \ldots, \Delta_n)}{S(n)} \tag{1.59}$$

Hurst referred to this ratio as the *rescaled adjusted range* and showed that for many natural phenomena the following empiric relation is satisfied:

$$M\left[\frac{R(n)}{S(n)}\right] \sim cn^H \quad \text{as} \quad n \to \infty \tag{1.60}$$

where c is a positive finite constant not depending on n. Taking the logarithm of the two parts gives

$$\log\left\{M\left[\frac{R(n)}{S(n)}\right]\right\} \sim H \log(n) + \log(c) \quad \text{as} \quad n \to \infty \tag{1.61}$$

Thus the H parameter can be estimated by placing the graph of $\log\{M[R(n)/S(n)]\}$ on $\log(n)$ and using the obtained points to select a straight line with slope H based on the least-squares method.

The R/S approach is not very precise since it gives an estimation of the self-similarity degree in the time series. Therefore, this approach can be applied only to test whether the time series is self-similar and, when the answer is positive, to obtain a rough estimation of H (Figure 1.4).

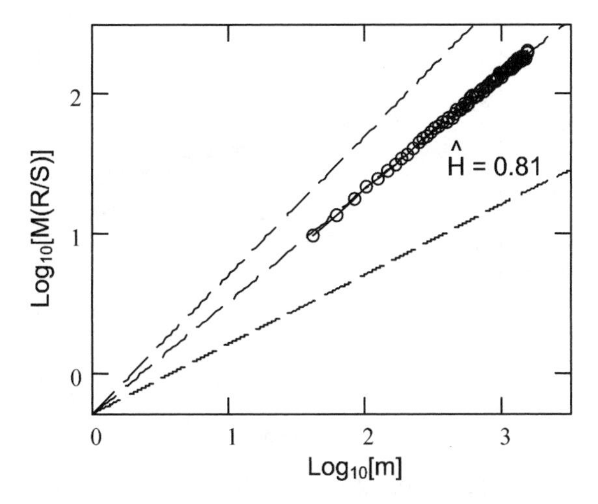

Figure 1.4 Graph of R/S statistics for Ethernet traffic

This result can be used to estimate the Hurst exponent of the given observation set. However, if the observations are taken from a short-range dependent process, as shown in Reference [134], then

$$M\left[\frac{R(n)}{S(n)}\right] \sim dn^{0.5}, \quad \text{as } n \to \infty \tag{1.62}$$

where d is a positive finite constant not depending on n. This case can be considered as the characteristic of the process not possessing the self-similarity property.

1.4.1.2 Variance–Time Plot

As shown above for the self-similar process, the relationship between the variance of the combined process $X^{(m)}$ and the interval size m can be formulated as (1.47)

$$\sigma^2(X_t^{(m)}) \sim am^{-\beta}, \quad \text{as } m \to \infty \tag{1.63}$$

where a is some finite positive constant. Taking the logarithm from both parts of (1.63) gives

$$\log[\sigma^2(X_t^{(m)})] \sim -\beta \log(m) + \log(a), \quad \text{as} \quad m \to \infty \tag{1.64}$$

Therefore, it is possible to obtain an estimation for β, calculating $\log[\sigma^2(X_t^{(m)})]$ for different m values and presenting the result graphically on $\log(m)$, and then to draw a straight line through these points on the basis of the least-squares technique. The estimation for β can be obtained as a negative slope of the straight line chosen according to the least-squares method. Since it is known that H is related to β by the relation $H = 1 - \beta/2$, this gives an estimation for H, which is equal to $1 - \hat{\beta}/2$, where $\hat{\beta}$ is an estimation of the β slope. The result of this approach applied to the measured trace is shown in Figure 1.5, where the logarithmic scale of variable—was chosen and $\log m$ was drawn on the background of $\log[\sigma^2(X_t^{(m)})]$.

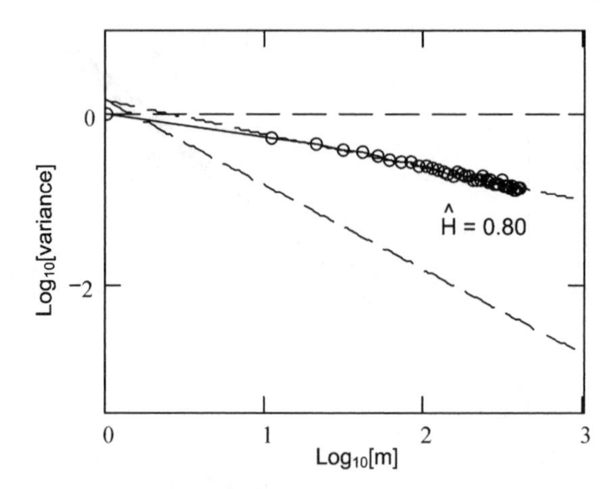

Figure 1.5 Variance-time plot for real Ethernet traffic

As in the case of R/S analysis, the variance–time plot approach is a heuristic method only. Both methods are used below under different restrictions; e.g. they can be truly justified in the case of a small amount of the statistical data available for observation of a separate sample of the self-similar process. Consequently, the variance–time plot can only be used to check whether the time series is self-similar or not and, if so, to obtain a crude guess for H.

1.4.1.3 Index of Dispersion for Counts

The index of dispersion for counts (IDC) (1.49) is usually a measure to describe the traffic changeability over different time scales. Self-similar processes lead to a monotonously increasing IDC of type $m^{-1}t^{2H-1}$. Drawing the plot of log[IDC(t)] on log(t) gives a straight line with the approximate slope $2H - 1$ [22].

Figure 1.6 shows the IDC curve corresponding to fractional Gaussian noise with a Hurst exponent of 0.816. The curve is monotonously increasing in the time range, covering 3 to 4

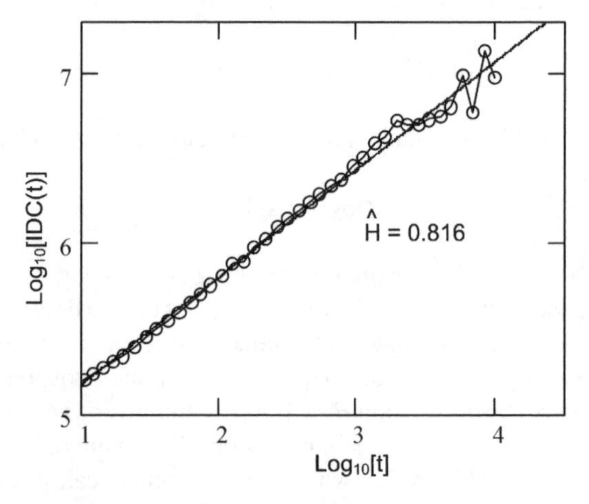

Figure 1.6 Example of IDC plot construction

orders of variable magnitude, and shows the approximate slope that distinctly differs from the horizontal line and can be estimated by the value of 0.631, leading to a Hurst exponent estimation of $H \approx 0.816$.

1.4.1.4 Hurst Exponent Estimation with Wavelets

The continuous wavelet transform consists of the coefficients set $\{T_X(a,t) = \langle X, \psi_{a,t}\rangle, a \in \mathbb{R}^+, t \in \mathbb{R}\}$, which can be obtained with the help of series x expansion by means of the analysing functions set

$$\left\{\psi_{a,t}(u) \equiv \frac{1}{\sqrt{a}}\psi_0\left(\frac{u-t}{a}\right),\ a \in \mathbb{R}^+,\ t \in \mathbb{R}\right\} \tag{1.65}$$

referred to as wavelets. This set is constructed from the mother wavelet ψ_0 with the help of the time shift operator $(\tau_\tau\psi_0)(t) = \psi_0(t-\tau)$ and the expansion operator $(D_a\psi_0)(t) \equiv 1/\sqrt{a}\psi_0(t/a)$, leading to $\psi_{a,t}(u) \equiv \psi_0((u-t)/a)/\sqrt{a}$. The time shift operator ensures the possibility of choosing the time moment around which the series needs to be analysed while the expansion operator defines the time scale (or the frequency range, which is equivalent) in which observations are to be made.

The wavelet transform may be understood as a more convenient form of Fourier transform where time resolution of the initial series is saved and where local properties can be obtained because the mother wavelet has limited circulation both in time and frequency. The theory of multiscaled analysis shows that the information will not be lost if a sampling of wavelet coefficients is made into the definite points set on the time plane, called a dyadic grid, defined by the relation $d_x(j,k) = T_x(2^j, 2^j k)$.

The coefficients $d_x(j,k)$ of the discrete wavelet transform (DWT) are called details and later they will be dealt with fully. The octave j is simply a logarithm to base 2 needed for scaling.

It was found that the estimations based on wavelets are highly unbiased and very stable in the presence of deterministic trends. It was proved that it is possible to save the discrete set of coefficients of $\{T_X(a;t)\}$ at the same time preserving all the information about X. The discrete wavelet transform consists of the coefficient set

$$X(t) \to \{\{a_X(J,k),\ k \in \mathbb{Z}\},\ \{d_X(j,k), j = 1,\ldots,J,\ k \in \mathbb{Z}\}\} \tag{1.66}$$

where $\{d_X(j,k)\}$ forms a subset for $\{T_X(a,t)\}$, located at the binary grid $d_X(j,k) = T_X(2^j, 2^j k)$. On scale j the level wavelet coefficients $d_x(j,k)$ are defined as follows:

$$d_x(j,k) = 2^{j/2}\sum_{i=1}^{n} X_i\psi_0(2^{-j}n - k), \quad j = 1, 2, \ldots;\ k = 1, 2, \ldots, 2^{-j}n \tag{1.67}$$

Let X be the stationary process of the second order. Then its wavelet coefficients $d_x(j,k)$ satisfy the relation

$$M[d_x(j,k)^2] = \int F(\lambda)2^j|\Psi(2^j\lambda)|^2 \mathrm{d}\lambda \tag{1.68}$$

where $F(\lambda)$ and $\Psi(\lambda)$ are the power spectrum for X and the Fourier transform for the wavelet function $\psi_0(\cdot)$ respectively. From Equation (1.68),

$$M[d_x(j,k)^2] \sim 2^{j(2H-1)} c_f C(H,\psi_0) \tag{1.69}$$

where $C(H,\psi_0) = \int |\lambda|^{-(2H-2)} |\psi(\lambda)|^2 d\lambda$ is the constant that depends on H and ψ_0. If the length X is equal to n, the available number of wavelet coefficients in the j octave is n_j, $n_j = 2^{-j}n$. As a result

$$\mu_j = M[d_x(j,k)^2] \approx \frac{1}{n_j} \sum_{k=1}^{n_j} |d_x(j,k)|^2 \tag{1.70}$$

The variable μ_j is unbiased and is the consistent estimation for $M[d_X(j,\cdot)^2]$ [135].

The expression (1.69) represents a possible way to estimate the Hurst exponent of LRD processes:

$$\log_2 \mu_j \approx \log_2 \left(\frac{1}{n_j} \sum_{k=1}^{n_j} |d_x(j,k)|^2 \right) \sim (2H-1)j + c \tag{1.71}$$

where $c = \log_2 C_w = \log_2[c_f\, C(H,\psi_0)]$ is a constant. It follows that if the process X is an LRD process with the Hurst exponent H, the plot $\log_2(\mu_j)$ on j, called a logarithmic diagram (LD), should have the linear slope $2H-1$. This testifies to the fact that the scaling exponent $(2H-1)$ may be obtained by examining the slope of the function $\log_2(\mu_j)$ plot on j. Figure 1.7 illustrates the Hurst exponent evaluation technique using Equation (1.71).

However, it should be noted that nonlinearity, which can appear due to $\log_2$, may displace the estimation. In Reference [136] it was shown that the effect of polynomial trends with order P on this estimation can be compensated by increasing the vanishing moment N of the wavelet function, so that $N \geq P+1$.

1.4.1.5 Multiscaled Diagram and Multifractals

Multifractal traffic is defined as the self-similar traffic expansion resulting from considering the properties where characteristics exceed the second order. It should be remembered when describing the exactly self-similar process, such as fractional Brownian motion (FBM), that the stationary process is considered (in a wide sense) relative to its autocorrelation and covariation functions only.

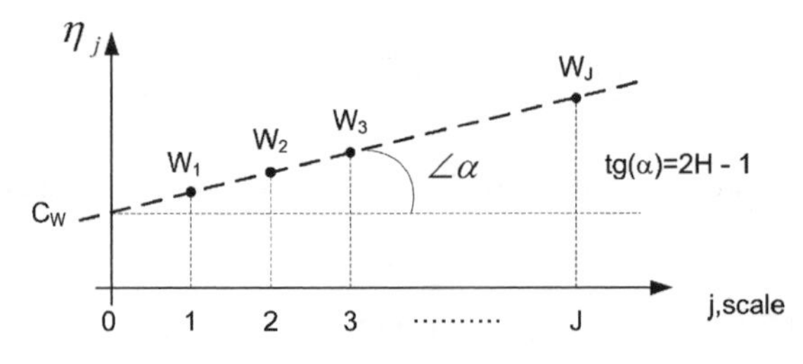

Figure 1.7 Graphical explanation of the Hurst exponent evaluation technique

To estimate traffic properties, e.g. of the third order, it is necessary to analyse the marginal distribution moments of the third order and the third-order correlation (e.g. the joint correlation between X_i, X_{i+k_1} and X_{i+k_2} samples for each delay pair (k_1, k_2)). As a result, instead of finding the aggregated process variance $X_i^{(m)}$ as a function of m (the variance–time plot), it is required to find the asymmetry factor (the third central moment) for $X_i^{(m)}$ as a function of m. For the exactly self-similar process the appropriate 'asymmetry variation plot' will also demonstrate the straight line with slope $3\beta/2$. Using this line of reasoning, the moments of the higher orders can be calculated.

It is assume that there is a cumulant of the mth order $\mathrm{cum}_m[Y(t)]$. It can then be found from the self-similar process theory that

$$\mathrm{cum}_m[Y(t)] = t^{mH}\mathrm{cum}_m[Y(1)]$$

i.e. the function $\log|\mathrm{cum}_m[Y(t)]|$ with respect to $\log(t)$ is linear with factor mH. This property is usually referred to as monofractality and can be expanded to increments.

The cumulants of order m can be expressed in terms of the central moments μ_v of less or equal order compared with m. For example, the cumulants of the sixth order can be expressed in terms of the central moments $\mu_v(Y) = M(Y - MY)^v$ as

$$\begin{aligned}
\mathrm{cum}_1(Y) &= MY = m_y \\
\mathrm{cum}_2(Y) &= \mu_2 = \sigma^2 \\
\mathrm{cum}_3(Y) &= \mu_3 \\
\mathrm{cum}_4(Y) &= \mu_4 - 3\mu_2^2 \\
\mathrm{cum}_5(Y) &= \mu_5 - 10\,\mu_2\mu_3 \\
\mathrm{cum}_6(Y) &= \mu_6 - 15\,\mu_2\mu_4 - 10\,\mu_3^2 - 30\,\mu_2^3
\end{aligned}$$

The stationary process $X(k)$ is multifractal if

$$\log|\mathrm{cum}_m[X^{(n)}(k)]| = \beta(m)\log(n) + c(m)$$

where $\beta(m)$ is some (may be, non-linear) function of m, and the aggregated series $X^{(n)}(k)$ are determined as

$$X^{(n)}(k) = \frac{1}{n}\sum_{j=0}^{n-1} X(kn - j), \quad k \in \mathbb{N}$$

In other words, all moments demonstrate similar scaling behaviour and therefore the any-order moment log–log plot with respect to m will lead to the same Hurst exponent value.

Generally speaking, the subclass of the multifractal processes for which $\beta(m)$ is the linear function is referred to as monofractal. If $X(k)$ are the increments series for the H-sssi process, $\beta(m) = m(H - 1)$, i.e. is linear with respect to m, the process itself is monofractal.

However, the more general multifractal process will also lead to the linear plot, but with an arbitrary slope. The central moment of the qth order over time scale m is

$$\mu^{(m)}(q) = \mu_q^{(m)} = M[X^{(m)} - M(X)]^q]$$

The multifractal process is then defined with the help of the expression

$$\log \mu^{(m)}(q) = -\beta(q) \log m + C(q)$$

It can be seen that the moment logarithm decays linearly when $\log(m)$ grows with some slope β, which depends on the moment order q. For the specific case of the monofractal or self-similar process, $\beta(q)$ behaves linearly with q, i.e. $\beta(q) = q(1 - H)$.

The obvious way to generalize the logarithmic diagram (1.71) is related to studying statistical characteristics that are different from the second order and obtained by substituting (1.70) for $\mu_j^{(q)} = (1/n_j) \sum_k |d_X(j,k)|^q, q \in \mathbb{R}$. The logarithmic diagrams of order q found as the result are of interest at least for two types of scaling, namely self-similarity and multifractality.

Due to the self-similarity definition, moments for $X(t)$ satisfy the expression $\mathrm{M}|X(t)|^q = \mathrm{M}|X(1)|^q |t|^{qH}, \forall t$. For wavelet coefficients, taking into consideration

$$\mathrm{M}|d_X(j,k)|^q = \mathrm{M}|d_X(0,k)|^q 2^{j(qH+q/2)} \tag{1.72}$$

It is found that $M\mu_j^{(q)} = C_q 2^{j[\zeta(q)+q/2]}, \forall j$, where $\zeta(q) = qH$. This relation indicates that self-similarity can be discovered by linearity $\zeta(q)$ testing depending on q.

For the class of multifractal processes, assuming that the expression $\int |T_X(a,t)|^q \mathrm{d}t \approx a^{\zeta(q)+q/2}, a \to 0$ can be related to multifractional properties of the process, it can be expected that $\mu_j^{(q)}$ behaves as $\mu_j^{(q)} \approx 2^{j[\zeta(q)+q/2]}$ at small j. These expressions allow $\zeta(q)$ to be measured, and thus to recover the spectrum. By testing the behaviour $\zeta(q)$, it is possible to distinguish monofractality from multifractality. For a monofractal process, $\zeta(q)$ has the view $\zeta(q) = qh + b$, where h is an independent scaling parameter and b is a constant that may depend on h.

Self-similar processes for which $\mu_j^{(q)} \approx 2^{j(qH+q/2)}$ for all scales satisfy the expression $\zeta(q) = qh$ and are monofractal with $h = H$, forming a subset of the specific case at b=0. To carry out monofractality and self-similarity testing and to investigate the general view of the $\zeta(q)$ function it is necessary to estimate the scaling exponent of order q, $\alpha_q = \zeta(q) + q/2$, in the logarithmic diagram of order q in the wide range of q values, and then to study the dependence on q.

1.4.2 Frequency Domain Methods of Hurst Exponent Estimation

1.4.2.1 Whittle Estimation

While variance–time plots and R/S diagrams are very useful to discover self-similarity (mostly in a heuristic manner) the lack of any results for limiting laws of the appropriate statistical features makes them useless when a finer data analysis is required (e.g. confidence intervals for the self-similarity degree H, the criterion of model choice or fitting criteria). The finer analysis is possible if the maximum likelihood estimation (MLE) and related approaches using period-

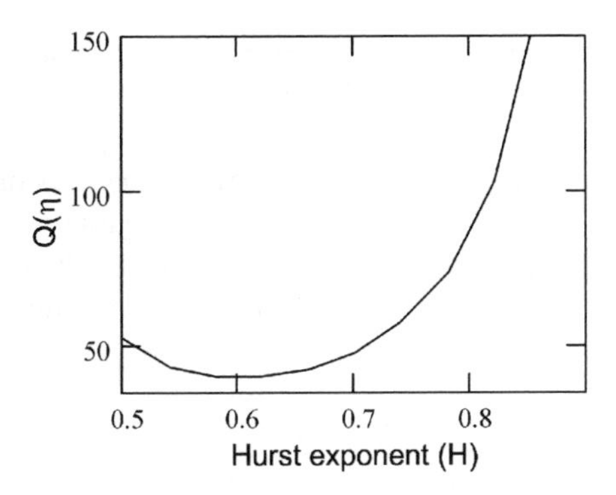

Figure 1.8 Minimization graph

ograms are applied. For an MLE definition, let the spectral density $S(\omega;\boldsymbol{\theta}) = \sigma_\varepsilon^2 S[\omega;(I;\boldsymbol{\eta})]$ of process X be given, where $\boldsymbol{\theta} = (\sigma_\varepsilon^2;\boldsymbol{\eta}) = (\sigma_\varepsilon^2; H; \theta_3,\ldots,\theta_k)$ and $H = (\alpha+1)/2$ is a self-similarity parameter (see Definition 1.8) and $\theta_3,\ldots,\theta_k$ are the parameters defining the SRD structure of the process. The variance σ_ε^2 of innovation ε in the infinite autoregressive (AR) process presentation will be used as a scaling factor, i.e. $X_j = \sum_{i\geq 1}\alpha_i X_{j-1} + \varepsilon_j$, where $\sigma_\varepsilon^2 = \sigma^2(\varepsilon_j)$. This means that the following expression takes place:

$$\int \log\{S[\omega;(I,\boldsymbol{\eta})]\}\mathrm{d}\omega = 0 \tag{1.73}$$

The Whittle estimation $\hat{\boldsymbol{\eta}}$ for $\hat{\boldsymbol{\eta}}$ is chosen with the expectation that the following expression meaning is minimal (see the minimization example in Figure 1.8):

$$Q(\boldsymbol{\eta}) = \int_{-\pi}^{\pi} \frac{I_N(\omega)}{S[\omega;(I;\boldsymbol{\eta})]}\mathrm{d}\omega \tag{1.74}$$

where $I_N(\omega)$ is a periodogram:

$$I_N(\omega) = \frac{1}{n}\left|\sum_{j=1}^{n} X_j \mathrm{e}^{ij\omega}\right| \tag{1.75}$$

and the estimation of σ_ε^2 can be obtained in accordance with

$$\hat{\sigma}_\varepsilon^2 = \int_{-\pi}^{\pi} \frac{I_N(\omega)}{S[\omega;(I;\hat{\boldsymbol{\eta}})]}\mathrm{d}\omega$$

In this case it is possible to say that $n^{1/2}(\hat{\boldsymbol{\theta}} - \boldsymbol{\theta})$ is the normally distributed variable if $(X_j)_{j\geq 1}$ can be presented as an infinite process of the moving average. For the Gaussian process the asymptotic distributions for estimation $\hat{\boldsymbol{\theta}}$ and MLE coincide.

In this context two problems arise as a rule with respect to stability: the first one deals with real distribution deviations from the Gaussian and the second with distinctions between the real and the supposed spectrum models. To overcome the first problem the data can be transformed in order to obtain approximately the required marginal (normal) distribution. To solve the second problem several approaches can be used including the determination of estimation H from the periodogram ordinates for low frequencies only or the $I_N(\omega)$ periodogram limitation for high frequencies. In the presence of large data sets an alternative and simpler approach can be used to solve the second problem which is an application of the combination approach. If $(X_j)_{j\geq 1}$ is the Gaussian process, the combined (aggregated) processes $X^{(m)}(m \geq 1)$ are defined as

$$X_j^{(m)} = m^{-H}L^{-1/2}(m) \sum_{i=(j-1)m+1}^{mk} (X_i - M[X_i]), \quad j \in \{1, 2, \ldots, [n/m]\} \tag{1.76}$$

and converge (on distribution) to fractional Gaussian noise for $m \rightarrow \infty$ ($L(\cdot)$ is the slowly varying function at infinity, see Equation (1.37)). The same is true if $X_i = \mu + G(Y_i)$, where $(Y_i)_{i\geq 1}$ is the Gaussian process with parameters $M[G(Y_i)] = 0$, $M[G^2(Y_i)] < \infty$ and $M[G(Y_i)G(Y_j)] \neq 0$. Therefore, for large enough m, fractional Gaussian noise is a good model for $X^{(m)}$ and consequently MLE can be used for fractional Gaussian noise.

The use of the approximate MLE Whittle approach and the combination approach together gives the procedure to obtain the confidence intervals of the self-similarity parameter H.

Asymptotically unbiased estimations received using the maximal likelihood method on the whole show good statistical efficiency; their drawback is that they are parametric estimations, requiring the spectral density analytical form to be known in advance. This presents problems for their application to large data sets due to large calculating complexity. In addition, if the presupposed spectral density model is incorrect the estimation is also biased. Due to that risk, the Whittle estimation does not ensure stable results in practice.

It should be noted that when using the Whittle estimation it is usually supposed that the process is actually self-similar. This leads to the Hurst exponent estimation with definite certainty. To determine whether the series has a self-similar structure it is necessary to apply additionally such methods as R/S statistics, variance–time plot, etc.

1.4.2.2 Graphical Method for Spectral Density Estimation (Periodogram Analysis)

The estimation based on the spectral density diagram provides the essence of the method, which ensures more statistical accuracy than the estimations based on combining. However, the requirement that the parametrized model of the process should be known in advance is the price of parametric method existence. The periodogram (or the 'intensity function') $I_N(\omega)$ estimates the spectral density of the discrete stochastic process X_t and may be estimated by the series (1.75) over the time interval N:

$$I_N(\omega) = \frac{1}{2\pi N} \left| \sum_{k=1}^{N} X_k e^{jk\omega} \right|^2, \quad \omega \in [0, \pi] \tag{1.77}$$

where $\{X_k\}$ is the time series and N is a time series length.

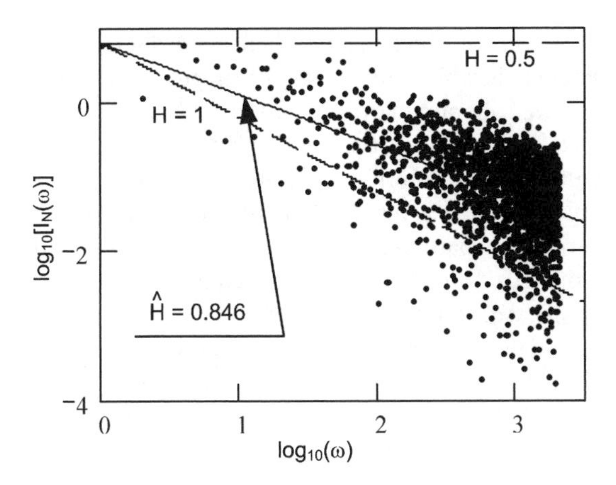

Figure 1.9 Hurst exponent estimation by means of the periodogram method

Taking into consideration the fact that self-similarity influences the spectrum $S(\omega)$ character as $\omega \to 0$, the graph of the following spectral density is obtained:

$$I_N(\omega) \sim [\omega]^{1-2H} \quad \text{as} \quad \omega \to 0 \tag{1.78}$$

Having drawn the graph of $\log[I_N(\omega)]$ on $\log(\omega)$ (for low frequencies only), the tangent straight line is fitted to the curve. The line slope will be approximately equal to $1 - 2H$. In practice, to calculate the estimation only 10 % lower frequencies should be used, as the above-described behaviour is only correct for the frequency range close to zero [137].

An example of the Hurst exponent evaluation for real data using the periodogram technique is shown in Figure 1.9. The main shortcoming of this method consists in the high requirements needed to calculation resources.

1.5 Hurst Exponent Estimation Problems

In practice when using the measured data sets the estimated H values obtained by way of application of various analytical methods are influenced by many factors and may depend on the estimation method, the sample size, the time scale, data structure, etc. [138, 139]. These issues will be discussed as well as the way they affect the self-similarity parameter calculation and H stability to these effects will also be estimated.

1.5.1 Estimation Problems

Several different statistical methods for testing and estimation of the degree of self-similarity for random processes were considered above. However, this list is far from complete. Methods to calculate the variance–time plots and the index of dispersion for counts are simple enough. The diagram obtained in this case is quite illustrative. Therefore, the so-called 'visual tests' are popular and widely used in practice, but are not reliable for small sample sizes.

R/S analysis is based on the heuristic graphical approach and the main feature that makes R/S statistics especially attractive is its relative stability to changing marginal distributions. From the point of view of effectiveness the R/S analysis depends on the sample size, but the same shortcomings are inherent in it, which are typical of both previous methods. In the case where a finer data analysis is required, the absence of any results for limiting regularities of the statistical characteristic described earlier makes the above-mentioned approaches useless. A 'finer' analysis is possible with the help of periodogram methods in the frequency domain. Some of the periodogram estimations, such as maximal likelihood estimations and related approaches, can be found in the literature. In particular, for Gaussian processes the MLE Whittle estimation has been widely investigated. Using these approaches it is possible to obtain more information concerning H estimations, e.g. confidence intervals).

In practice, when not all the required preliminary conditions for statistical tests are fulfilled, the different methods can lead to slightly different estimations for H. Thus, the method of the IDC plot, being simple and effective, does not require great calculating resources and the plot obtained visually shows the variations when the analysed data set changes. (It should be noted that when using this approach the H value is calculated by fitting the graph with a straight line, assuming that the curve has a linear feature. In practice, deviations from linear cause additional inaccuracies in the estimation).

1.5.1.1 Advantages and Shortcomings of the Hurst Exponent Evaluation Techniques

The advantages and shortcomings of the evaluation techniques are summarized in Table 1.1. Some of the main shortcomings of the variance–time technique is the necessity for sample sizes that are too large and the arbitrary aggregation level.

When using R/S statistics, the smallest d values should not be taken into account, since the short-range dependence dominates in these points. It should be remembered that $d := [n/K]$, where n is the empirical series length and K is number of fragmentation blocks. The upper parts of the plot are not used because several meanings can lead to an unstable estimation.

Table 1.1 Performance of the evaluation techniques

Technique	Advantages	Shortcomings
Variance–time plot	Can be used as the diagnostic test	The estimators are biased. The displacement increases with self-similarity exponent growth
R/S–statisctics	Do not depend on the marginal data distribution	The estimators are biased. The slowly changed trend presence affects effectiveness
Periodogram	The stable estimation	The large data volume is necessary. The variance convergence is $\sim O(U^{-1})$
Whittle	The exact estimators can be obtained if the model order is known *a priori*. The estimators have the acceptable accuracy for not-large data sets	The biased estimates are obtained if data do not correspond to the given model

Whittle estimates are asymptotically effective but require parameter knowledge of the spectral density model of the analysed time series. This method is more exigent from a calculation point of view.

The periodogram is more stable but the variance asymptotic convergence for it is very slow. All estimation approaches demonstrate the relative stability.

1.5.2 Nonstationarity Problems

Researchers have found [140, 141] that the Hurst exponent estimation can depend on many features and requires that in the course of its estimation the stationarity conditions be fulfilled. The simplest way to check the stationarity of random processes consists in estimating their statistical characteristics. If the main statistical characteristics defining the process $x(t)$, namely the probability density function (PDF) $w(x)$, the expectation value m and the variance σ^2, do not depend on time, it is possible to suppose that the process is stationary in the wide sense. In practice, such simple considerations allowing the correctness of the stationarity hypothesis to be checked are not usually fulfilled because only realizations of a finite duration are available for the observation. In such cases the stationarity hypothesis should be checked by means of an analysis of the finite duration realizations that are accessible. The checking methods can be different, beginning from visual observations of the realization by an experienced specialist to a detailed statistical estimation of various process parameters. The realization length should be large enough to divide the nonstationary trend and low-frequency random fluctuations. With present trends a final check of realizations can be fulfilled in different ways.

The testing of random processes on self-similarity is a special problem. It exists in the finite data set when it is impossible to check whether the traffic trace is by definition self-similar or not. Therefore in real measured traffic an investigation can be made of different self-similarity and long-range dependence features. However, when distinguishing self-similarity, detecting only its features can be erroneous.

Some nonstationary processes can lead to similar features. This means, for example, that bursty traffic can be caused by long-range dependence as well as the observed process nonstationarity. In many cases it is possible to speak correctly only about the bursty traffic character in a given time scale for the given data set unless substantiation is made on the basis of strict statistical tests.

The method using a wide sense stationary (WSS) quotient has become the most popular one in stationarity estimation of self-similar processes. However, as mentioned in Reference [142], its application in some cases leads to false results.

If the sampling distribution of estimators of the main statistical distribution parameters is known to solve the problem it is possible to use a run test or the criterion of inversions [143]. A comparative analysis of the above-mentioned approaches will be carried out using the example of the stationarity estimation of fractional Gaussian noise [3].

1.5.2.1 Stationary Quotient in the Wide Sense

It is known that for the stationary processes in the wide sense their mean value m and variance σ^2 are constants in all domains of existence. However, in reality to check this thesis on the finite sample (e.g. video data) is rather difficult. That is why statistical characteristics such as the sample mean value and the variance are usually defined as

$$M(X) = \frac{1}{K}\sum_{k=1}^{K} x_k$$

$$D(X) = \frac{1}{K-1}\sum_{k=1}^{K} [x_k - M(X)]^2$$

The data set $\{x_k, k = 1, 2, \ldots, K\}$ divided by S independent segments, each of which is of length N, so that $K = NS$, is examined. If the sample mean value of each segment is designated as $\hat{m}_i$ and the variance as $\hat{\sigma}_i^2$, $i = 1, 2, \ldots, S$, the equality of mean values and variances of any two units i and j can be checked. The statistical test on mean value equality, also called the t-test, has the form [142]

$$T = (\hat{m}_i - \hat{m}_j)\left(\frac{N-1}{\hat{\sigma}_i^2 + \hat{\sigma}_j^2}\right)^{1/2} \tag{1.79}$$

where T has Student's distribution with $\upsilon = 2N - 2$ degrees of freedom. The statistical test on variance equality is called the F-test:

$$F = \frac{\hat{\sigma}_i^2}{\hat{\sigma}_j^2} \tag{1.80}$$

which has an F-distribution with $\nu_1 = N - 1$ and $\nu_2 = N - 1$ degrees of freedom. The following indicated function is introduced:

$$p_{ij} = \begin{cases} 1 & \text{if } |T| \leq \bar{t}_{\nu,\alpha/2} \quad \text{and} \quad \overline{F}_{\nu_1,\nu_2,1-\alpha/2,} \leq F \leq \overline{F}_{\nu_1,\nu_2,\alpha/2} \\ 0 & \text{if otherwise} \end{cases} \tag{1.81}$$

where t- and F-tests are performed on units i and j; $t_{\nu,\alpha/2}$ are percentage points of the CCDF (complementary cumulative distribution function) for t-distribution and $F_{\nu_1,\nu_2,1-\alpha/2}$ and $F_{\upsilon_1,\upsilon_2,\alpha/2}$ are percentage points of the CCDF for F-distribution. Therefore, the stationarity factor in the wide sense can be defined as

$$W_N = \frac{2}{S_N(S_N - 1)}\sum_{j=1}^{S_N-1}\sum_{j=i+1}^{S_N} p_{ij} \tag{1.82}$$

If the total number of segments in a data set is equal to D,

$$S_N = \frac{D}{N} \tag{1.83}$$

As a result, for example, it is possible to choose the 95 % confidence interval for the mean value equality test (1.79) and for the variance equality test (1.80). Thus, 5 % test errors for the mean value (variance) can be expected even for stationary data. As these tests are independent, at

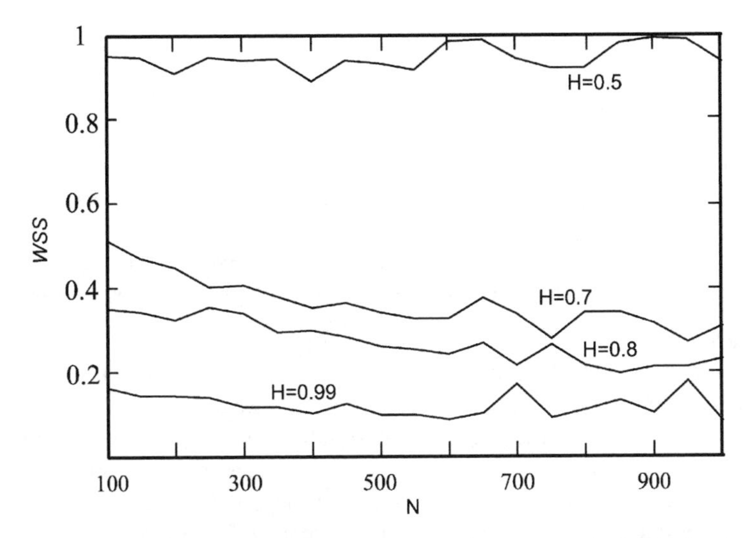

Figure 1.10 Nonstationarity WSS factor dependence upon the average unit size for fractional Gaussian noise for different Hurst exponents H

worst 10 % of all tests on stationarity will be erroneous for stationary data. Therefore it can be accepted that factor W_N for stationary data will be at least equal to 0.9. The example of WSS-criterion usage for the case of fractional Gaussian noise (FGN) with different Hurst exponents H is shown in Figure 1.10.

It can be seen that although all analysed sequences were *a priori* stationary and differed by the Hurst exponent H only, in the case $H = 0.5$ only the WSS method leads to the correct results. In this case the WSS value fluctuates around 1. Note that the case $H = 0.5$ corresponds to the case of a Gaussian sequence with independent values. At $H > 0.5$ the value of WSS does not exceed 0.4, which is evidence of the examined sequence nonstationarity. Obviously it is caused by long-range dependence on the examined sequence. Thus, the WSS method in the stationarity analysis of self-similar processes can lead to erroneous results.

The research conducted in Reference [141] shows that the W_N factor can define the difference between the data with SRD or LRD. However, when using it, it is impossible to see the difference between the nonstationary and self-similar data.

1.5.2.2 The Run Test and the Criterion of Inversions

As examination will now be made of another sequence of operations for checking the random process stationarity on the basis of its individual realization $\{x_k, k = 1, 2, \ldots, H\}$:

1. Realizations can be divided into N equal intervals, and the observations in different intervals are to be carried out independently.
2. Estimations of the averaged square (or mean values and variances separately) are calculated for each interval, and these estimations are situated in the ascending order of the interval number: $\overline{x_1^2}, \overline{x_2^2}, \overline{x_3^2}, \ldots, \overline{x_N^2}$.
3. This sequence for mean square estimations is checked for the trend presence or other changes in time that cannot be explained by the changeability of samples only.

The final realization check of the trend presence can be fulfilled using the various methods. If the distribution of estimates for samples is known, it is possible to apply the statistical criteria. However, knowledge of the distribution of the estimates for samples of the mean square requires knowledge of the process frequency structure. Usually during the stationarity check such information is missing. Therefore, it is preferable to apply the nonparametric criteria as its use does not require knowledge of the sampling estimation distributions. The run test and the criterion of inversions are two nonparametric criteria that can be used to solve this problem. The latter represents a more potent tool for the monotonous trend detection of these observations. The criterion of inversions can be directly applied to check the hypothesis of the stationarity.

The Run Test

Consider the sequence of N observed values of variable X, where each of the observations can be attributed to one of two mutually exclusive classes, which can be designated as $(+)$ or $(-)$. In the example in Figure 1.11, the sequence of simultaneous measurements of two random variables x_i and $y_i (i = 1, 2, \ldots, N)$ is shown, where each observation is $x_i \geq y_i(+)$ or $x_i < y_i(-)$. As a result the sequence can be formed as depicted in Figure 1.11.

The number of runs appearing in the observation sequence allow an opinion to be made of whether the separate results are independent observations of one and the same random variable. If the sequence of N observations consists of the independent outcomes of the same random variable, i.e. if the probability of separate outcomes $[(+)$ or $(-)]$ does not change from one observation to the other, the sampling distribution of run numbers in the sequence is the random variable r with the mean value

$$\mu_r = \frac{2N_1N_2}{N} + 1 \tag{1.84}$$

and the variance

$$\sigma_r^2 = \frac{2N_1N_2(2N_1N_2 - N)}{N^2(N-1)} \tag{1.85}$$

Here N_1 is the number of outcomes $(+)$ and N_2 is the number of outcomes $(-)$. In the specific case $N_1 = N_2 = N/2$ the expression (1.84) takes the form

$$\mu_r = \frac{N}{2} + 1$$

In Reference [143] 100α percentage points of the distribution function of run numbers are discussed, with the help of which it is possible to estimate the stationarity of analysed sequences with the given reliability. To check the hypothesis with any required level of significance α, it is necessary to compare the observed run numbers with the limits of this hypothesis acceptance zone, which are equal to $r_{n;1-\alpha/2}$ and $r_{n;\alpha/2}$, where $n = N/2$. If the observed run number falls

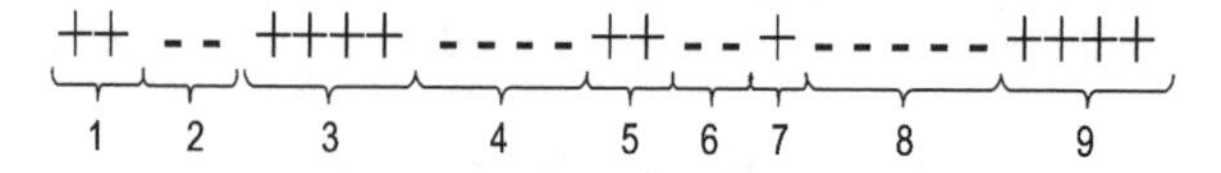

Figure 1.11 Examples of run marking

beyond this zone, the hypothesis is discarded with the level of significance α. Otherwise, the hypothesis is accepted.

The results of the run test usage for fractional Gaussian noise analysis with different values of Hurst exponent H are illustrated in Figure 1.12. It can be seen that unlike the WSS criterion the run method correctly determines the stationarity of the self-similar process.

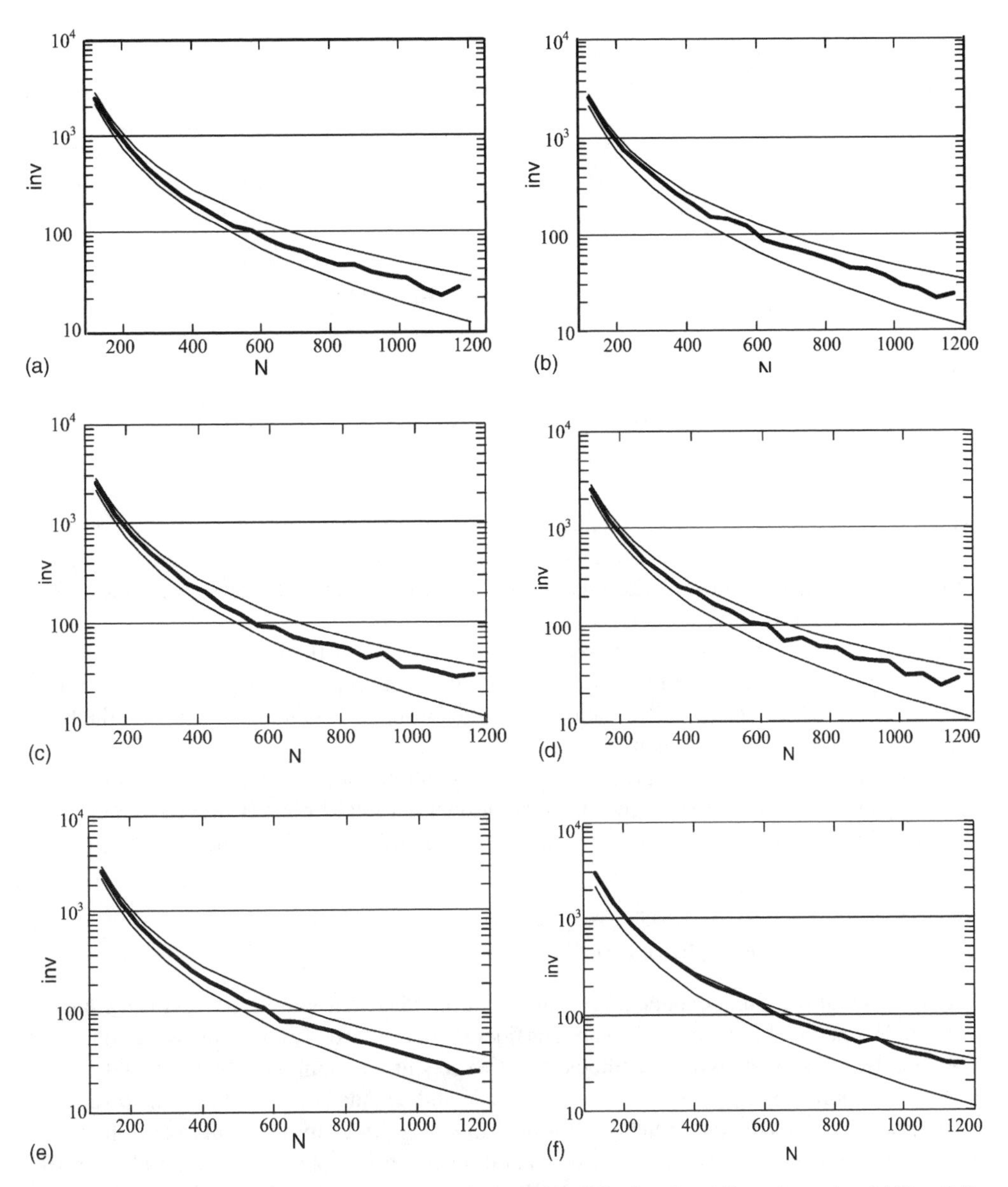

Figure 1.12 Variance runs (a,b,c) and mean value runs (d,e,f) for fractional Gaussian noise: (a) $H = 0.5$; (b) $H = 0.7$; (c) $H = 0.9$; (d) $H = 0.9$; (e) $H = 0.7$; (f) $H = 0.5$

The Criterion of Inversions
Consider the sequence $\{x_k, k = 1, 2, \ldots, N\}$. Calculate how many times the inequalities $x_i > x_j$ take place at $i < j$ in this sequence. Each of these inequalities is called an inversion. The total number of inversions is designated as A. Formally A is calculated in the following manner. Define $x_1, x_2, \ldots, x_N$ values for the observation set

$$h_{i,j} = \begin{cases} 1, & x_i > x_j \\ 0, & x_i \leq x_j \end{cases}$$

Then $A = \sum_{i=1}^{N-1} A_i$ and $A_i = \sum_{j=i+1}^{N} h_{ij}$. For example, $A_1 = \sum_{j=2}^{N} h_{1j}$, $A_2 = \sum_{j=3}^{N} h_{2j}$, $A_3 = \sum_{j=4}^{N} h_{3j}, \ldots$.

If the sequence with N observations consists of the independent outcomes of the same random variable, the inversion number is the random variable A with the mean value

$$\mu_A = \frac{N(N-1)}{4} \tag{1.86}$$

and the variance

$$\sigma_A^2 = \frac{2N^3 + 3N^2 - 5N}{72} = \frac{N(2N+5)(N-1)}{72} \tag{1.87}$$

Reference [143] presents 100 α percentage points of the distribution function for A.

The criterion of inversions is more potent compared with the run test when detecting the monotonous trend in the observations sequence. Nevertheless, it is not as effective in the case of detecting the trend in fluctuation type.

The results of the inversions test application for the case of fractional Gaussian noise for different Hurst exponents are shown in Figure 1.13. It can be seen from the presented results that the run test correctly identifies the stationarity of the tested self-similar sequence of the FGN type with different Hurst exponent values H.

It is possible to offer other tests on the stationarity estimation, but all of them cannot clearly show whether, for example, the video data are stationary or self-similar. However, they confirm that self-similar models can be used for data generation, which behave like variable bit rate (VBR) video traffic.

1.5.3 Computational Problems

It is known that the Hurst exponent estimation for the ideal self-similar process is a constant value independent of how much data are considered. In practice, using the measured data sets obtained by various analytical methods Hurst exponent estimations, which depend on the estimation approach, on sample size, on time scale and on data structure, are usually found.

In reality the Hurst exponent changes considerably over time. This problem can be illustrated using the example of VoIP (voice over Internet protocol) traffic obtained by multiplexing 100 speech streams created by users in the VoIP system together with the enabled VAD (voice activity detection) mechanism (Figure 1.14). The presented data were characterized by a time

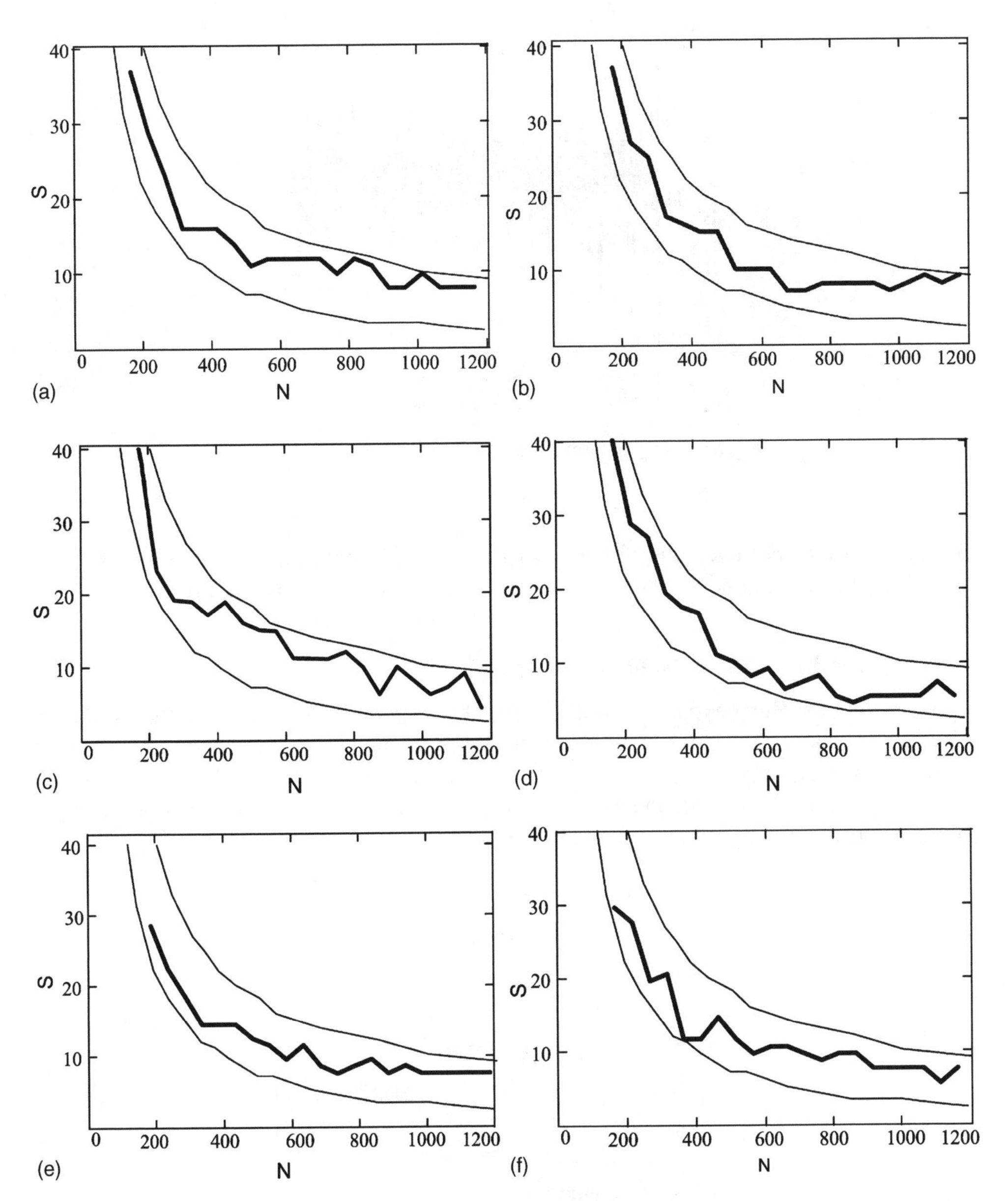

Figure 1.13 Variances inversion (a,b,c) and mean values inversion (d,e,f) for fractional Gaussian noise: (a) $H = 0.5, \alpha = 0.05$; (b) $H = 0.7, \alpha = 0.05$; (c) $H = 0.9, \alpha = 0.05$; (d) $H = 0.5, \alpha = 0.05$; (e) $H = 0.7$, $\alpha = 0.05$; (f) $H = 0.9, \alpha = 0.05$

resolution of 1 second and covered more that 3 hours of VoIP system operation. The Hurst exponent was calculated with the help of R/S statistics and the variance–time plot and also by applying the estimation based on the features of the covariance function plot (see Figure 1.15).

The estimation of H was carried out for 'the window' of the given length; after that the window was shifted and the estimation was repeated. It was found that the Hurst exponent

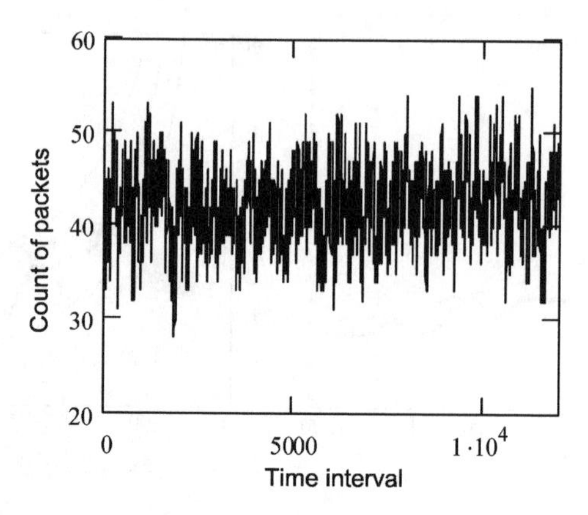

Figure 1.14 Multiplexed traffic for 100 speech sources

changed considerably over time, in spite of the assumed conformity of the situation. The dynamics of Hurst exponent values for the traffic under survey is shown in Figure 1.15.

1.5.3.1 Hurst Exponent Dependence on N

By estimating the Hurst exponent for data units D, it is possible to investigate its correlation structure. Consider K of series segments each having N length. The Hurst exponent H can be estimated in each segment S_i, $i = 1, 2, \ldots, D/N$, by using, for example, the R/S analysis. If the estimations carried out conducting in the ith cell are designated as $\hat{H}_i$, for the appropriate N, the Hurst exponent estimation can be found in the form

$$\hat{H}_N = \frac{N}{D} \sum_{i=1}^{D/N} \hat{H}_i \tag{1.88}$$

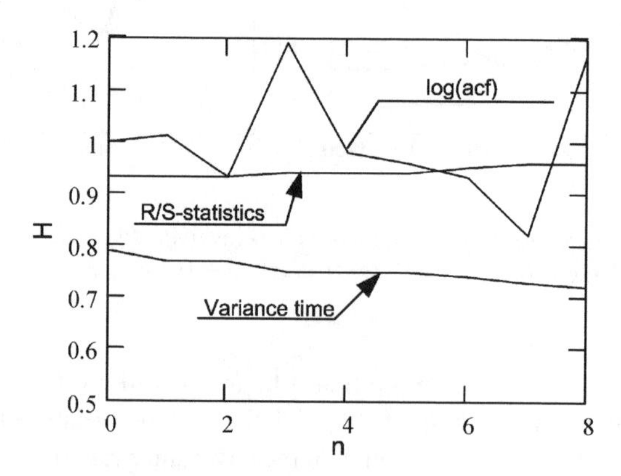

Figure 1.15 Plot of Hurst exponent variation for shifting window, shown by different approaches of testing

Researches show that if a large enough N is chosen it is possible to ensure an acceptable convergence of the estimation so that for the stationary process the estimation $\hat{H}_N$ would not depend on N. As a result, the measure of $\hat{H}_N$ can distinguish SRD data from data with LRD.

1.6 Self-Similarity Origins in Telecommunication Traffic

Modern investigations show that self-similarity can occur as a result of combining separate ON/OFF sources, which can be strongly changeable (i.e. ON and OFF periods have DHT and infinite variances, e.g. Pareto distribution) [17, 22, 90, 144]. In other words, the superposition of the ON/OFF sources exhibiting the infinite variance syndrome results in self-similar combined (aggregated) network traffic, tending to fractional Brownian motion. Moreover, the research of various traffic sources shows that the highly changeable ON/OFF behaviour is the typical characteristic for the client–server architecture [144–146].

The difficulty in understanding the fundamental principles, which can result in self-similarity in network traffic, is mainly explained by the fact that the self-similarity is caused by more than one factor. Various correlations occurring in the network traffic and affecting different time scales may arise for different reasons, revealing themselves in the characteristics on the specific time scales.

A more detailed analysis of traffic self-similarity appearance reasons is performed in Reference [147]. Some of the main factors that can produce the LRD of different types in the network traffic are:

- user's behaviour;
- data generation, data structure and its search;
- traffic aggregation;
- means of network control;
- control mechanisms based on feedback;
- network development.

1.6.1 User's Behaviour

One of the main factors affecting the traffic character (the session/call level and also during the session) is the user's (person's) behaviour. It has been shown, for example, that the user's enquiry distribution (thinking time) and the preference for documents on the Internet (www, or World Wide Web) have extreme degrees of fluctuation in the wide range of time scales [144, 146, 148]. Moreover, there are different mechanisms of stream control for different traffic sources (e.g. VBR (variable bit rate) video MPEG (Motion Pictures Experts Group) coded sources, ABR (available bit rate) and TCP (transmission control protocol)), and these mechanisms control the outlet traffic intensity depending on the network status that strengthens the burstiness of the network traffic [146].

1.6.2 Data Generation, Data Structure and its Search

The important reason for traffic with a self-similar structure to appear is closely related to the way the data are created. The conducted measurements show that self-similarity in network traffic as well as in applications level traffic is a two-dimensional feature and can be related partially to the time distribution between entries of the files/packet/cell and partially to the size distribution of the files/packet. These results show that self-similarity cannot be a state that is brought about artificially (mechanically), for traffic on the applications level at least. Different applications or sources may provide traffic on a very high level with statistically varying characteristics, but the main statistical behaviour (e.g. the correlation structure) is as a rule invariant from one computer/network to an other. For example, it has already been shown that as the size distribution of informational objects is usually transmitted at the application level (the client/server principle) it is better described by DHT.

1.6.3 Traffic Aggregation

The presence of traffic aggregation (statistical multiplexing), which is used in networks with packet or cell switching, is one of the most serious obstacles to allowing traffic generated by separate sources to be saved in isolated condition from each other (at least sources located in the network and as distant from each other as possible). As mentioned above, the superposition of many ON/OFF sources exhibiting the infinite variance syndrome results in self-similar combined (aggregated) network traffic, tending to fractional Brownian motion. Persistency characteristics in the traffic are also extremely stable to network operations such as separation, aggregation, queues forming, control organization and generation [149]. Self-similarity continues to exist in the case of superposition of the homogeneous and heterogeneous (i.e. independent) traffic sources, and this property is present in a wide range of such conditions, as in the cases of variations of limit carrying and buffer capacities as well as of mixing with cross traffic having other (correlative) characteristics [146, 150, 151]. If the specific source generates LRD traffic, the aggregated network traffic becomes long-range dependent irrespective of the characteristics (SRD or LRD) of other traffic inside the mixture. The process of aggregation is extremely complex and the whole range of features changes while mixing, i.e. as well as the Hurst exponent the mean value and the variance change essentially [149, 152].

1.6.4 Means of Network Control

The limited resources that exist in the real network environment are presumed to be one of the main reasons for LRD, e.g. the limited network and the commuting resources such as the carrying capacity, the limited buffer size and the limited possibilities for data processing.

When using the control mechanisms nonlinear functions can occur because, due to the resource limitations, different conflict situations cannot be solved by simple methods. This problem may become apparent in various forms, beginning with simple models of queue forming to very complex control mechanisms. Similar problems are discussed in Reference [151], where it is shown that in the case when mechanisms of stream control (TCP type) are used, the degree with which file sizes (www documents) are described by distributions with heavy tails (Pareto PDF with the form parameter α) can directly define the degree of self-similarity H of the network traffic.

In the case when mechanisms without stream control, such as the UDP (user datagram protocol) are used, the occurring network traffic shows fractal properties to a less degree. In other words, LRD in network traffic can occur as a result of DHT files transmitting through the network and the limited (channel) carrying capacity. The problem of LRD traffic control can become additionally complicated in a situation where there is a ‘struggle’ between a large number of users in the case of limited resources. As a result the problem of correct resources distribution acquires the utmost importance. This becomes more complicated due to the multidimensional structure, for example, as a result of competition for hardware usage: central processing unit (CPU), memory, carrying capacity, software (OS (operating system) strategy planner, processes priority), etc.

1.6.5 Control Mechanisms Based on Feedback

Further complications appear due to the large number of control mechanisms based on feedback, e.g. the stream and congestion control mechanisms (TCP, the control mechanisms based on intensity, etc.). This means that in the case of congestion, the additional nonlinearity caused by the wide range of system dynamic behaviour can occur. It is noteworthy that very complex interconnections between the work load fluctuations and various (network) control mechanisms may occur in a similar situation.

As a result two classes of specific problems can appear: the influence of the real traffic on the particular control mechanism effectiveness and the degree of traffic characteristic variations by means of control mechanisms.

1.6.6 Network Development

The important reason for the network traffic burstiness increase is related to the network development, which is inevitable with the uninterrupted appearance of new services and applications. The arrival of www servers can be regarded as an example and results in a more complicated traffic structure [153].

In the course of the quantitative assessment of the channel level characteristic (e.g. Ethernet), the network traffic parameters are determined on the milliseconds scale [22, 154]. The influence of a person's behaviour on data generation and the data search defines the traffic character at intervals of tens of seconds and higher (minutes and even hours) [148]. The influence processes of different mechanisms of stream control, such as TCP, will probably dominate at intermediate intervals (on the seconds time scale). The characteristics of the queue formation may dominate in intervals of tens and hundreds of milliseconds [149].

References

[1] P. Grassberger and I. Procaccia, 'On the characterization of strange attractors', *Phys. Rev. Lett.*, **50**, 1983, 346–354.

[2] J. Beran, '*Statistics for Long-Memory Processes*', Chapman & Hall, New York, 1994.

[3] A. Adas, 'Traffic models in broadband networks', *IEEE Communications Magazine*, July 1997.

[4] M. Crovella and L. Lipsky, 'Long-lasting Transient conditions in simulations with heavy-tailed workloads', in Proceedings of the 1997 Winter Simulation Conference, 1997.

[5] K. Park and W. Willinger (eds), '*Self-Similar Network Traffic and Performance Evaluation*', Wiley–Interscience, New York, 1999.

[6] S.I. Resnick, 'Heavy tail modeling and teletraffic data', Preprint, School of ORIE, Cornell University, Ithaca, NY, 1995.

[7] P. Embrechts, C. Kluppelberg and T. Mikosh, '*Modeling Extremal Events for Insurance and Finance*', Springer-Verlag, Berlin, Heidelberg, 1997.

[8] P. Abry and D. Veitch, 'Wavelet analysis of long range dependent traffic', *IEEE Transactions on Information Theory*, **44**(1), January 1998.

[9] W. Stallings, '*High-Speed Networks TCP/IP and ATM Design Principles*', Prentice-Hall, Inc., Englewood Clip, NJ, 1998.

[10] M.S. Taqqu, 'A bibliographical guide to self-similar processes and long-range dependence', in '*Dependence in Probability and Statistics*', (eds E. Eberlein and M.S. Taqqu), Birkhauser, Boston, MA, 1986, pp. 137–162.

[11] J. Beran, 'Statistical methods for data with long-range dependence', *Statistical Science*, **7**(4), 1992, 404–416; With discussions and rejoinder, pp. 404–427.

[12] M.S. Taqqu and V. Teverovsky, 'Robustness of Whittle-type estimates for time series with long-range dependence', Preprint, 1995.

[13] M.S. Taqqu and V. Teverovsky, 'Semi-parametric graphical estimation techniques for long-memory data', in Proceedings of the Athens Conference on '*Applied Probability and Time Series Analysis*', New York, 1996, Springer-Verlag, 1996; Time series volume in honour of E.J. Hannan.

[14] M.S. Taqqu, V. Teverovsky and W. Willinger, 'Estimators for long-range dependence: an empirical study'. *Fractals*, **3**(4), 1995, 785–798.

[15] A.A. Anis and E.H. Lloyd, 'The expected value of the adjusted rescaled hurst range of independent normal summands', *Biometrika*, **63**, 1976, 111–116.

[16] R. Ballerini and D.C. Boes, 'Hurst behavior of shifting level processes', *Water Resources Research*, **21**, 1985, 1642–1648.

[17] J. Beran, R. Sherman, M.S. Taqqu and W. Willinger, 'Long-range dependence in variable-bit-rate video traffic', *IEEE Transactions on Communications*, **43**, 1995, 1566–1579.

[18] L.I. Berge, N. Rakotomalala, J. Feder and T. Jossamg, 'Cross-over in R/S analysis and power spectrum: measurements and simulations', Preprint, 1993.

[19] J. Feder, '*Fractals*', Plenum Press, New York, 1988.

[20] W. Feller, 'The asymptotic distributions of the range of sums of independent random variables', *Annals of Mathematical Statistics*, **22**, 1951, 427–432.

[21] H.E. Hurst, 'Long-term storage capacity of reservoirs', *Transactions of the American Society of Civil Engineers*, **116**, 1951, 770–808.

[22] W.E. Leland, M.S. Taqqu, W. Willinger and D.V. Wilson, '*On the self-similar nature of Ethernet traffic*' (extended version), *IEEE/ACM Transactions on Networking*, **2**, 1994, 1–15.

[23] E.H. Loyd and D. Warren, 'The discrete Hurst range for skew independent two-valued inflows', *Stochastic Hydrology and Hydraulics*, **1**, 1987, 53–66.

[24] A.W. Lo, 'Long-term memory in stock market prices', *Econometrica*, **59**, 1991, 1279–1313.

[25] B.B. Mandelbrot, 'Limit theorems on the self-normalized range for weakly and strongly dependent processes', *Zeitschrift fur Wahrscheinlichkeitstheorie und verwandte Gebiete*, **31**, 1975, 271–285.

[26] B.B. Mandelbrot and M.S. Taqqu, 'Robust R/S analysis of long-run serial correlation', in Proceedings of the 42nd Session of the International Statistical Institute, Manila, 1979; *Bulletin of the International Statistical Institute*, **48**(2), 1979, 69–104.

[27] B.B. Mandelbrot and J.R. Wallis, 'Computer experiments with fractional Gaussian noises', Parts 1,2,3', *Water Resources Research*, **5**, 1969, 228–267.

[28] B.B. Mandelbrot and J.R. Wallis, 'Robustness of the rescaled range R/S in the measurement of noncyclic long-run statistical dependence', *Water Resources Research*, **5**, 1969, 967–988.

[29] B.B. Mandelbrot and J.R. Wallis, 'Some long-run properties of geophysical records', *Water Resources Research*, **5**, 1969, 321–340.

[30] A.I. McLeod and K.W. Hipel, 'Preservation of the rescaled adjusted range, Parts 1,2,3', *Water Resources Research*, **14**, 1978, 491–518.

[31] A. Montanari, R. Rosso and M.S. Taqqu, 'Fractionally differenced ARIMA models applied to hydrologic time series: identification, estimation and simulation', Preprint, 1995.

[32] M.S. Taqqu, V. Teverovsky and W. Willinger, 'Estimators for long-range dependence: an empirical study', *Fractals*, **3**(4), 1995, 785–798.

[33] D.R. Cox, 'Long-range dependence: a review', in '*Statistics: An Appraisal*', (eds H.A. David and H.T. David), Iowa State University Press, 1984, pp. 55–74.

[34] V. Paxon and S. Floyd, 'Wide-area traffic: the failure of Poisson modelling', in Proceedings of the ACM Sigcomm '94, London, 1994, pp. 257–268.

[35] V. Teverovsky and M.S. Taqqu, 'Testing for long-range dependence in the presence of shifting means or a slowly declining trend using a variance-type estimator', Preprint, 1995.

[36] G. Chen, B. Abraham and S. Peiris, 'Lag window estimation of the degree of differencing in fractionally integrated time series models', *Journal of Time Series Analysis*, **15**, 1994, 473–487.

[37] R. Dahlhaus, 'Efficient parameter estimation for self similar processes', *The Annals of Statistics*, **17**(4), 1989, 1749–1766.

[38] R. Fox and M.S. Taqqu, 'Large-sample properties of parameter estimates for strongly dependent stationary Gaussian time series', *The Annals of Statistics*, **14**, 1986, 517–532.

[39] J. Geweke and S. Porter-Hudak, 'The estimation and application of long memory time series models', *Journal of Time Series Analysis*, **4**, 1983, 221–238.

[40] L. Giraitis, A. Samarov and P.M. Robinson, 'Rate optimal semiparametric estimation of the memory parameter of the Gaussian time series with long range dependence', Technical Report, Beitrage fur Statistik, Universitat Heidelberg, 1995.

[41] L. Giraitis and D. Surgailis, 'A central limit theorem for quadratic forms in strongly dependent linear variables and application to asymptotical normality of Whittle's estimate', *Probability Theory and Related Fields*, **86**, 1990, 87–104.

[42] M. Henry and P.M. Robinson, 'Bandwidth choice in Gaussian semiparametric estimation of long range dependence', in Proceedings of the Athens Conference on '*Applied Probability and Time Series Analysis*', New York, 1996, Springer-Verlag, 1996; Time series volume in honour of E.J. Hannan.

[43] C.M. Hurvich and K.I. Beltrao, 'Asymptotics for the low-frequency ordinates of the periodogram of a long-memory time series', *Journal of Time Series Analysis*, **14**, 1993, 455–472.

[44] C.M. Hurvich and K.I. Beltrao, 'Automatic semiparametric estimation of the memory parameter of a long memory time series', *Journal of Time Series Analysis*, **15**, 1994, 285–302.

[45] C.M. Hurvich, R. Deo and J. Brodsky, 'The mean squared error of Geweke and Porter–Hudak's estimator of the memory parameter of a long memory time series', Preprint, 1995.

[46] G. Lang and J.-M. Azais, 'Nonparametric estimation of the strong dependence exponent for Gaussian processes', Preprint, Journee Longue Portee, Groupe d'Automatique d'Orsay, Paris, 11 May 1995.

[47] W.-C. Lau, A. Erramilli, J.L. Wang and W. Willinger, 'Self-similar traffic parameter estimation: a semi-parametric periodogram-based algorithm', in Proceedings of the IEEE Globecom '95, Singapore, 1995, pp. 2225–2231.

[48] V.A. Reisen, 'Estimation of the fractional difference parameter in the ARIMA (p, d, q) model using the smoothed periodogram', *Journal of Time Series Analysis*, **15**, 1994, 335–350.

[49] P.M. Robinson, 'Automatic frequency domain inference on semiparametric and nonparametric models', *Econmetrica*, **59**, 1991, 1329–1363.

[50] P.M. Robinson, 'Nonparametric function estimation for long memory time series', in *Nonparametric and Semiparametric Methods in Econometrics and Statistics: Proceedings of the Fifth International Symposium on Economic Theory and Econometrics* (eds W.A. Barnett, J. Powell and G.E. Tauchen), Cambridge University Press, 1991, pp. 437–457.

[51] P.M. Robinson, 'Efficient tests of nonstationarity hypotheses', *Journal of the American Statistical Association*, **89**, 1994, 1420–1437.

[52] P.M. Robinson, 'Rates of convergence and optimal bandwidth in spectral analysis of processes with long range dependence', *Probability Theory and Related Fields*, **99**, 1994, 443–473.

[53] P.M. Robinson, 'Semiparametric analysis of long-memory time series', *The Annals of Statistics*, **22**, 1994, 515–539.

[54] P.M. Robinson, '*Time Series with Strong Dependence*', Volume 1 of '*Advances in Econometrics*', Sixth World Congress, Cambridge University Press, 1994, Chapter 2, pp. 47–95.

[55] P.M. Robinson, 'Gaussian semiparametric estimation of long range dependence', *The Annals of Statistics*, 23, 1995, 1630–1661.

[56] P.M. Robinson, 'Log-periodogram regression of time series with long range dependence', *The Annals of Statistics*, **23**, 1995, 1048–1072.

[57] P.M. Robinson and F.J. Hidalgo, 'Time series regression with long range dependence', Preprint, 1995.

[58] P.M. Robinson and C. Velasco, 'Autocorrelation robust inference', in *Handbook of Statistics*, Volume 15 on *Robust Inference*, Elsevier, North-Holland, Amsterdam, 1997, pp. 267–298.

[59] M.S. Taqqu and V. Teverovsky, 'Semi-parametric graphical estimation techniques for long-memory data', In Proceedings of the Athens Conference on '*Applied Probability and Time Series Analysis*', New York, 1996, Springer-Verlag, 1996; Time series volume in honour of E.J. Hannan.

[60] P. Whittle, 'Hypothesis Testing in Time Series Analysis', Hafner, New York, 1951.

[61] Y. Yajima, 'On estimation of a regression model with long-memory stationary error', *The Annals of Statistics*, **16**, 1988, 791–807.

[62] C. Agiakloglou and P. Newbold, 'Lagrange multiplier tests for fractional difference', *Journal of Time Series Analysis*, **15**, 1994, 253–262.

[63] J. Beran, 'A test of location for data with slowly decaying serial correlation', *Biometrika*, **76**, 1989, 261–269.

[64] J. Beran, 'A goodness of fit test for time series with long-range dependence', *Journal of the Royal Statistical Society, Series B*, **54**, 1992, 749–760.

[65] J. Beran and H. Kiinsch, 'Location estimators for processes with long-range dependence', Preprint, 1985.

[66] J. Beran and N. Terrin, 'Estimation of the long memory parameter, based on a multivariate central limit theorem', *Journal of Time Series Analysis*, **15**, 1994, 269–278.

[67] Y.-W. Cheung, 'Tests for fractional integration: a Monte Carlo investigation', *Journal of Time Series Analysis*, **14**, 1993, 331–345.

[68] Y.-W. Cheung and F.X. Diebold, 'On maximum likelihood estimation of the differencing parameter of fractionally-integrated noise with unknown mean', *Journal of Econometrics*, **62**, 1994, 301–316.

[69] C.F. Chung and R.T. Baillie, 'Small sample bias in conditional sum-of-squares estimators of fractionally integrated ARMA models', *Empirical Economics*, **18**, 1993, 791–806.

[70] K.L. Chung and P. Schmidt, 'The minimum distance estimator for fractionally integrated ARMA models', Preprint, 1995.

[71] S. Csorgo and J. Mielniczuk, 'Density estimation under long-range dependence', *The Annals of Statistics*, **23**, 1995, 990–999.

[72] S. Csorgo and J. Mielniczuk, 'Distant long-range dependent sums and regression estimation', *Stochastic Processes and Their Applications*, **59**, 1995, 143–155.

[73] S. Csorgo and J. Mielniczuk, 'Nonparametic regression under long-range dependent normal errors', *The Annals of Statistics*, **23**, 1995, 1000–1014.

[74] R. Dahlhaus and L. Giraitis, 'The bias and the mean squared error in semi-parametric models for locally stationary time-series', Preprint, 1995.

[75] F. Diebold and G. Rudebusch, 'Long memory and persistence in aggregate output', *Journal of Monetary Economics*, **24**, 1989, 189–209.

[76] F. Diebold and G. Rudebusch, 'On the power of Diskery–Fuller tests against fractional alternatives', *Economics Letters*, **35**, 1991, 155–160.

[77] L. Giraitis, H. Koul, and D. Surgailis, 'Asymptotic normality of regression estimators with long memory errors', *Statistics and Probability Letters*, **29**, 1996, 317–335.

[78] L. Giraitis and R. Leipus, 'A generalized fractionally differencing approach in long-memory modelling', *Lithuanian Mathematical Journal*, **35**, 1995, 53–65.

[79] H.C. Ho, 'On central and non-central limit theorems in density estimation for sequences of long-range dependence', Preprint, 1995.

[80] H.C. Ho, 'On the strong uniform consistency of density estimation for strongly dependent sequences', *Statistics and Probability Letters*, **22**, 1995, 149–156.

[81] J.R.M. Hosking, 'Asymptotic distributions of the sample mean, auto-covariances and autocorrelations of long-memory time series', *Journal of Econometrics*, **73**, 1995, 261–284.

[82] H. Kunsch, J. Beran and F. Hampel, 'Contrasts under long-range correlations', *The Annals of Statistics*, **21**, 1993, 943–964.

[83] F.B. Sowell, 'The fractional unit-root distribution', *Econometrica*, **58**, 1990, 495–505.

[84] F.B. Sowell, 'Maximum likelihood estimation of stationary univariate fractionally integrated time series models', *Journal of Econometrics*, **53**, 1992, 165–188.

[85] W. Willinger, M.S. Taqqu, W.E. Leland and V. Wilson, 'Self-similarity in high-speed packet traffic: analysis and modeling of Ethernet traffic measurements', *Statistical Science*, **10**, 1995, 67–85.

[86] G.W. Wornell and A.V. Oppenheim, 'Estimation of fractal signals from noisy measurements using wavelets', *IEEE Transactions on Information Theory*, **40**(3), 1992, 611–623.

[87] D.R. Cox, 'Statistical analysis of time series: some recent developments', *Scandinavian Journal of Statistics*, **8**, 1981, 93–115.

[88] A. Erramilli and W. Willinger, 'A case for fractal traffic modelling', in Proceedings of the Australian Telecommunication Networks and Applications Conference Sydney, Australia, 1995, pp. XV–XX.

[89] A.A. Mcintosh, 'Analyzing telephone network data', Preprint, 1995.

[90] V. Paxson and S. Floyd, 'Wide area traffic: the failure of poisson modelling', *IEEE/ACM Transactions on Networking*, **3**, 1995, 226–244.

[91] R. Dahlhaus, 'Efficient location and regression estimation for long range dependent regression models', *The Annals of Statistics*, **23**, 1995, 1029–1047.

[92] L. Giraitis, H. Koul and D. Surgailis, 'Asymptotic normality of regression estimators with long memory errors', *Statistics and Probability Letters*, **77**, 1995.

[93] H.L. Koul and K. Mukherjee, 'Asymptotics of R-, MD- and LAD-estimators in linear regression models with long range dependent errors', *Probability Theory and Related Fields*, **95**, 1993, 535–553.

[94] A. Samarov and M.S. Taqqu, 'On the efficiency of the sample mean in long memory noise', *Journal of Time Series Analysis*, **9**, 1988, 191–200.

[95] Y. Yajima, 'On estimation of a regression model with long-memory stationary errors', *The Annals of Statistics*, **16**, 1988, 791–807.

[96] Y. Yajima, 'Asymptotic properties of LSE in a regression model with long-memory stationary errors', *The Annals of Statistics*, **19**, 1991, 158–177.

[97] G. Gripenberg and I. Norros, 'On the prediction of fractional Brownian motion', *Journal of Applied Probability*, **33**(22), 1996, 400–410.

[98] M.S. Peiris and B.J.C. Perera, 'On prediction with fractionally differenced ARIMA models', *Journal of Time Series Analysis*, **9**, 1988, 215–220.

[99] B.K. Ray, 'Modeling long-memory processes for optimal long-range prediction', *Journal of Time Series Analysis*, **14**(5), 1993, 511–525.

[100] P. Breuer and P. Major, 'Central limit theorems for non-linear functional of Gaussian fields', *Journal of Multivariate Analysis*, **13**, 1983, 425–441.

[101] R.L. Dobrushin and P. Major, 'Non-central limit theorems for non-linear functions of Gaussian fields'. *Zeitschrift fur Wahrscheinlichkeitstheorie und verwandte Gebiete*, **50**, 1979, 27–52.

[102] R. Fox and M.S. Taqqu, 'Non-central limit theorems for quadratic forms in random variables having long-range dependence', *The Annals of Probability*, **13**, 1985, 428–446.

[103] R. Fox and M.S. Taqqu, 'Central limit theorems for quadratic forms in random variables having long-range dependence', *Probability Theory and Related Fields*, **24**, 1987, 213–240.

[104] R. Fox and M.S. Taqqu, 'Multiple stochastic integrals with dependent integrators', *Journal of Multivariate Analysis*, **21**, 1987, 105–127.

[105] I.L. Giraitis, 'Central limit theorem for polynomial forms', *Lithuanian Mathematical Journal*, **29**, 1989, 109–128.

[106] I.L. Giraitis and D. Surgailis, 'A central limit theorem for quadratic forms in strongly dependent linear variables and application to asymptotical normality of Whittle's estimate', *Probability Theory and Related Fields*, **86**, 1990, 87–104.

[107] I.L. Giraitis and M.S. Taqqu, 'Limit theorem for bivariate Appell polynomials: Part I. Central limit theorems', Preprint, 1995.

[108] I.L. Giraitis and M.S. Taqqu, 'Central limit theorems for quadratic forms with time-domain conditions', Preprint, 1996.

[109] V.V. Gorodetskii, 'On convergence to semi-stable Gaussian processes', *Theory of Probability and Its Applications*, **22**, 1977, 498–508.

[110] P. Heinrich, 'Zero–one laws for polynomials in Gaussian random variables', Preprint, 1995.

[111] H.C. Ho, 'On limiting distributions of nonlinear functions of noisy Gaussian sequences', *Stochastic Analysis and Applications*, **10**, 1992, 417–430.

[112] H.C. Ho and T. Hsing, 'On the asymptotic expansion of the empirical process of long memory moving averages', Preprint, 1995.

[113] P. Major, '*Multiple Wiener–Ito Integrals*', Volume 849 of Springer Lecture Notes in Mathematics, Springer-Verlag, New York, 1981.

[114] M.S. Taqqu, 'Weak convergence to fractional Brownian motion and to the Rosenblatt process', *Zeitschrift fur Wahrscheinlichkeitstheorie und verwandte Gebiete*, **31**, 1975, 287–302.

[115] M.S. Taqqu, 'Convergence of integrated processes of arbitrary Hermite rank', *Zeitschrift fur Wahrscheinlichkeitstheorie und verwandte Gebiete*, **50**, 1979, 53–83.

[116] N. Terrin and M.S. Taqqu, 'A noncentral limit theorem for quadratic forms of Gaussian stationary sequences', *Journal of Theoretical Probability*, **3**, 1990, 449–475.

[117] N. Terrin and M.S. Taqqu, 'Convergence in distribution of sums of bivariate Appell polynomials with long-range dependence', *Probability Theory and Related Fields*, **90**, 1991, 57–81.

[118] N. Terrin and M.S. Taqqu, 'Convergence to a Gaussian limit as the normalization exponent tends to 1/2', *Statistics and Probability Letters*, **11**, 1991, 419–427.

[119] A.C. Davison and D.R. Cox, 'Some simple properties of sums of random variables having long-range dependence', *Proceedings of the Royal Society London*, **A424**, 1989, 255–262.

[120] I.L. Giraitis and D. Surgailis, 'Multivariate Appell polynomials and the central limit theorem', in *Dependence in Probability and Statistics* (eds E. Eberlein and M.S. Taqqu), Birkhauser, New York, 1986.

[121] P. Heinrich, 'Zero–one laws for polynomials in Gaussian random variables', Preprint, 1995.

[122] P. Major, '*Multiple Wiener–Ito Integrals*', Volume 849 of Springer Lecture Notes in Mathematics', Springer-Verlag, New York, 1981.

[123] D. Surgailis, 'On Poisson multiple stochastic integral and associated equilibrium Markov process', in '*Theory and Applications of Random Fields*', Lecture Notes in Control and Information Science, Vol. 49, Springer-Verlag, Berlin, 1983, pp. 233–238.

[124] N. Terrin and M.S. Taqqu, 'Power counting theorem on R^n', in *Spitzer Festschrift* (eds R. Durrett and H. Kesten) Birkhauser, Boston, MA, 1991, pp. 425–440.

[125] M.A. Arcones, 'Limit theorems for non-linear functionals of a stationary Gaussian sequence of vectors', *The Annals of Probability*, **22**, 1994, 2242–2274.

[126] R. Gay and C.C. Heyde, 'On a class of random field models which allows long range dependence', *Biometrika*, **77**, 1990, 401–403.

[127] J. Haslett and A.E. Raftery, 'Space–time modelling with long-memory dependence: assessing Ireland's wind power resource', *Applied Statistics*, **38**, 1989, 1–50; Includes discussion.

[128] C.C. Heyde and R. Gay, 'Smoothed periodogram asymptotics and estimation for processes and fields with possible long-range dependence', *Stochastic Processes and Their Applications*, **45**, 1993, 169–182.

[129] A.V. Ivanov and N.N. Leonenko, 'Statistical Analysis of Random Fields', Kluwer Academic Publishers, Dordrecht, Boston, London, 1989; Translated from the Russian 1986 edition.

[130] C. Ludena, 'Estimation of integrals with respect to the logarithm of the spectral density of stationary Gaussian processes with long range dependence', Preprint, Journee Longue Portee, Groupe d'Automatique d'Orsay, Paris, 11 May 1995.

[131] T. Lundahl, W.J. Ohley, S.M. Kay, and R. Siffert, 'Fractional Brownian motion: a maximum likelihood estimator and its application to image texture', *IEEE Transactions on Pattern Analysis and Machine Intelligence*, **MI-5**(3), 1986, 152–161.

[132] A.P. Pentland, 'Fractal-based description of natural scenes', *IEEE Transactions on Pattern Analysis and Machine Intelligence*, **PAMI-6**(4), 1984, 661–674.

[133] M.V. Sanchez de Naranjo, 'Central limit theorem for non-linear functionals of stationary vector Gaussian process', Preprint, 1994.

[134] B.B. Mandelbrot and J.W. Van Ness, 'Fractional Brownian motions, fractional noises and applications', *SIAM Review*, **10**, 1968, 422–437.

[135] P. Abry, P. Flandrin, M. Taqqu and D. Veitch, 'Wavelets for the analysis, estimation, and synthesis of scaling data', in '*Self-Similar Network Traffic Analysis and Performance Evaluation*' (eds K. Park and W. Willinger), Wiley–Interscience, New York, 1999.

[136] P. Abry, P. Flandrin, M.S. Taqqu and D. Veitch, 'Self-similarity and long-range dependence through the wavelet lens', in '*Long Range Dependence: Theory and Applications*' (eds P. Doukhan, G. Oppenheim and M.S. Taqqu), Birkhauser, Boston, MA, 2002.

[137] M.S. Taqqu and V. Teverovsky, 'On estimating the intensity of long-range dependence in finite and infinite variance time series', Preprint, Boston University, MA, 1996.

[138] S. Molnar, A. Vidacs and A. Nilsson, 'Bottlenecks on the way towards fractal characterization of network traffic: estimation and interpretation of the Hurst parameter', in International Conference of the *Performance and Management of Complex Communication Networks (PMCCN'97)*, Tsukuba, Japan, November 1997.

[139] S. Molnar and A. Vidacs, 'How to characterize Hursty traffic?', in COST 257 TD(98)003, Rome, Italy, January 1998.

[140] S. Molnar and A. Vidacs, 'Fractal characterization of network traffic from parameter estimation to application', PhD Dissertation, Budapest University of Technology and Economics Department of Telecommunications and Telematics, Budapest, Hungary, 2000.

[141] S. Bates, 'Traffic characterization and modelling for call admission control schemes on asynchronous transfer mode networks', A thesis submitted for the degree of Doctor of Philosophy, The University of Edinburgh, 1997.

[142] R. Shiavi, '*Introduction to Applied Statistical Signal Analysis*', Aksen Associates, 1991.

[143] J. Bendat and A. Piersol, '*Random Data: Analysis and Measurement Procedures*', John Wiley & Sons, Ltd, 1986.

[144] M.F. Crovella and A. Bestavros, 'Self-similarity in world wide web traffic: evidence and possible causes', in Proceedings of the ACM SIGMETRICS 96, Philadelphia, PA, May 1996, pp. 160–169.

[145] M.F. Arlitt and C.L. Williamson, 'Web server workload characterization: the search for invariants' (extended version), *EEEE/ACM Transactions on Networking*, **5**(5), October 1997.

[146] A.K. Jena, P. Pruthi and A. Popescu, 'Resource engineering for Internet applications', in Proceedings of the 7th IFIP ATM Workshop, Antwerp, Belgium, June 1999.

[147] A. Popescu, '*Traffic self-similarity*', in Proceedings of the IEEE International Conference on '*Telecommunications*' (*ICT2001*), Bucharest, Romania, June 2001.

[148] M.E. Crovella and A. Bestavros, 'Explaining world wide web traffic self-similarity', Technical Report TR-95-015, Computer Science Department, Boston University, MA, 1995.

[149] A. Erramilli, O. Narayan and W. Willinger, 'Experimental queueing analysis with long-range dependent packet traffic', *IEEE/ACM Transactions on Networking*, **4**, 1996, 209–223.

[150] A.K. Jena, P. Pruthi and A. Popescu, 'Modeling and evaluation of network applications and services', in Proceedings of the RVK'99 Conference, Ronneby, Sweden, June 1999.

[151] K. Park, G.T. Kim, and M.E. Crovella, 'On the relationship between file sizes, transport protocols, and *self-similar network traffic*', Preprint, Boston University, MA, 1996.

[152] A. Erramilli, W. Willinger and J.L. Wang, 'Modeling and management of self-similar traffic flows in high-speed networks', in '*Network Systems Design*', Gordon and Breach Science Publishers, 1999.

[153] A. Feldmann, A.C. Gilbert, W. Willinger and T.G. Kurtz, 'The changing nature of network traffic: scaling phenomena', *Computer Communications Review*, **28**(2), April 1998.

[154] W.E. Leland, M.S. Taqqu, W. Willinger and D.V. Wilson, 'On the self-similar nature of Ethernet traffic' (extended version), *IEEE/ACM Transactions on Networking*, **2**, 1994, 1–15.

2

Simulation Methods for Fractal Processes

2.1 Fractional Brownian Motion

The classical Wiener process of Brownian motion *B(t)* is an example of a random process possessing fractal properties. Brownian motion is considered as a random process starting at the origin of coordinates, the increments of which over nonoverlapping time intervals t_i are independent and have the Gaussian distribution.

The Wiener process trajectory has a scale invariance property. The trajectory and plot of the Wiener process are not differentiable anywhere and at the same time are continuous with a unit probability. The probability density function (PDF) of the particle coordinate $X(t = n\tau) = \Sigma_{i=1}^{n} \xi_i$ can be written as

$$w(\Delta X) = \frac{1}{(2\pi k_D |\Delta t|)^{0.5}} \exp\left[-\frac{(\Delta X)^2}{2k_D|\Delta t|}\right]$$

where $\Delta X = X(t) - X(t_0)$, $\Delta t = t - t_0$ and k_D is a diffusion factor. The PDF introduced in such a manner meets the similarity relation

$$w\{\, b^{0.5}[X(bt) - X(bt_0)\,]\,\} = b^{0.5} w\,[X(t) - X(t_0)\,]$$

where $b > 0$ is an arbitrary coefficient.

The coordinate increment of a Brownian particle can be expressed as

$$\Delta X \sim \xi|\Delta t|^{0.5}, \quad \forall t \geq t_0 \tag{2.1}$$

Here ξ is the total number from the Gaussian distribution. Thus, Brownian motion is the stochastic Wiener process usually designated as $\{B_t\}$ for $t \geq 0$ and characterized by the following properties: (a) $B(t + t_0) - B(t_0)$ increments or $B_{t+t_0} - B_{t_0}$ are normally distributed with zero mean and the variance $\sigma^2(t_2 - t_1) = \sigma_t^2$; (b) $B_{t_4} - B_{t_3}$ and $B_{t_2} - B_{t_1}$ increments over

Self-Similar Processes in Telecommunications O. I. Sheluhin, S. M. Smolskiy and A. V. Osin
© 2007 John Wiley & Sons, Ltd

corresponding nonoverlapping time intervals $[t_1;\ t_2]$ and $[t_3;\ t_4]$ are independent random variables; (c) $B_0 = 0$ and B_t is a continuous function as $t = 0$; (d) the increment expectation value for Brownian motion equals $M[|B(t_2) - B(t_1)|] \sim |t_2 - t_1|^{0.5}$; (e) the distribution function can be described by the following equation:

$$F(x) = P[\Delta X < x] = \frac{1}{\sqrt{2\pi\sigma^2 t}} \int_{-\infty}^{x} \mathrm{e}^{-\,u^2/(2\sigma^2 t)} \mathrm{d}u$$

The concept of generalized *fractional Brownian motion* (FBM) $B_H(t) = fB_t$ was introduced by substituting the exponent in the right part of Equation (2.1) for any real number within the interval $H \in [0;\ 1]$, where H is the Hurst exponent. The case when $H = 0.5$ corresponds to independent increments and describes the classical Brownian motion. Thus, FBM differs from Brownian motion (BM) by the presence of increments with $\sigma_H^2 t^{2H}$ variance. If an increment process variance is defined as $\sigma_H^2 = M[(fB_t - fB_{t-1})^2] = M(fB_1 - fB_0)^2 = M(fB_1^2)$ (at $fB_0 = 0$), then $M(fB_{t_2} - fB_{t_1}) \sim |\Delta t|^{2H}$, $\Delta t = t_2 - t_1$. Thus

$$\begin{aligned} M(fB_{t_2} - fB_{t_1})^2 &= M(fB_{t_2}^2) + M(fB_{t_1}^2) - 2M(fB_{t_2} fB_{t_1}) \\ &= \sigma_H^2 t \times 2^{2H} + \sigma_H^2 t \times 2^{2H} - 2R(fB_{t1}, fB_{t2}) \end{aligned}$$

where

$$R(fB_{t_1}, fB_{t_2}) = 0.5\sigma_H^2 \left(t_2^{2H} - (t_2 - t_1)^{2H} + t_1^{2H} \right)$$

Hence, the correlation (covariance) of increments for two nonoverlapping time intervals can be determined as

$$\begin{aligned} &R(fB_{t_4} - fB_{t_3}, fB_{t_2} - fB_{t_1}) \\ &\quad = R(fB_{t_4}, fB_{t_2}) - R(fB_{t_4}, fB_{t_1}) - R(fB_{t_3}, fB_{t_2}) + R(fB_{t_3}, fB_{t_1}) \\ &\quad = \frac{\sigma_H^2}{2} \left[(t_4 - t_1)^{2H} - (t_3 - t_1)^{2H} + (t_3 - t_2)^{2H} - (t_4 - t_2)^{2H} \right] \end{aligned} \tag{2.2}$$

In a discrete case the normalized correlation function (correlation factor) of the increments sequence is found by substituting t_1, t_2, t_3 and t_4 in Equation (2.2) for n, $n + 1, \ldots, n + k$ and $n + k + 1$ respectively and by dividing the result by σ^2:

$$r_k = \tfrac{1}{2}\left[(k+1)^{2H} - 2k^{2H} + (k-1)^{2H} \right] \tag{2.3}$$

This increments sequence is referred to as *fractional Gaussian noise* (FGN). A correlation (correlation factor) in Equation (2.3) is evidence of LRD in the process because $r_k \sim k^{2H-2}$ as $k \to \infty$ (in accordance with the Taylor expansion).

Fractal Brownian motion (fB_t) can be derived from Brownian motion (B_t) by integration:

$$fB_t = \frac{1}{\Gamma(H + 0.5)} \int_{-\infty}^{t} (t - u)^{H-0.5} \mathrm{d}B(u)$$

Here $\Gamma(\cdot)$ is the gamma function. According to this equation, the random function value at time instant t depends on all previous ($u < t$) increments $dB(u)$ of a random process. Therefore it can be said that interdependence between FBM increments can be considered infinite.

All finite-dimensional marginal distributions of FBM are Gaussian. The process meaning at time instant t can be calculated using relation $B_H(t) = Xt^H$, where X is the normally distributed random variable with zero mean and unit variance. Moreover, it follows from the self-similarity that $B_H(at) = a^H B_H(t)$; $H = 0.5$ gives the usual BM process. The main statistical properties of the FBM process are the following:

Mean value $M[B_H(t)] = 0$
Variance $\sigma^2[B_H(t)] = \sigma^2[Xt^H] = t^{2H}$
Correlation factor $r_{B_H}(t, \tau) = M[B_H(t)B_H(\tau)] = \frac{1}{2}\left(t^{2H} + \tau^{2H} - |t - \tau|^{2H}\right)$
Stationary increments $\sigma^2[B_H(t) - B_H(\tau)] = |t - \tau|^2$

The FBM model is widely used in theoretical and experimental research to help simulation during system performance estimations, especially in the case of systems controlled by self-similar traffic. FBM can be used for generating summary or aggregated self-similar traffic (like the one that can be observed in network buffers or meets of file sizes from audio or video streams, etc.). FBM increments are fractional Gaussian noise.

2.1.1 RMD Algorithm for FBM Generation

The basic principle of the algorithm of random midpoint displacement (RMD) is related to recursive extension of a generated sample by adding new values at midpoints with regard to values at final points. Figure 2.1 shows the operation of the RMD algorithm and Figure 2.2 illustrates the first three steps of this algorithm, which lead to the generation of the sequence $(d_{3,1}; d_{3,2}; d_{3,3}; d_{3,4})$. The main aim of dividing the interval between 0 and 1 is to form Gaussian

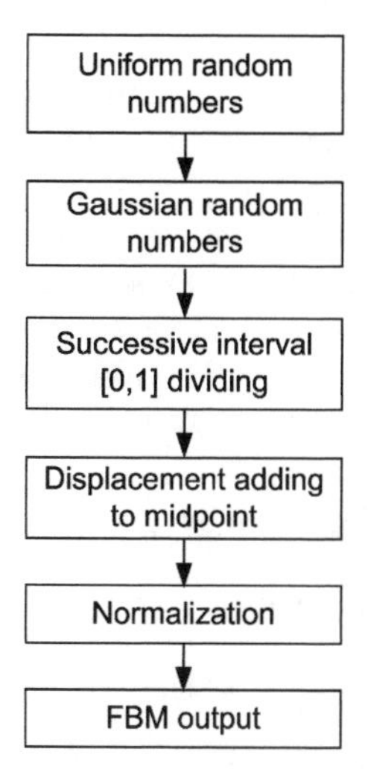

Figure 2.1 FBM generation using RMD algorithm

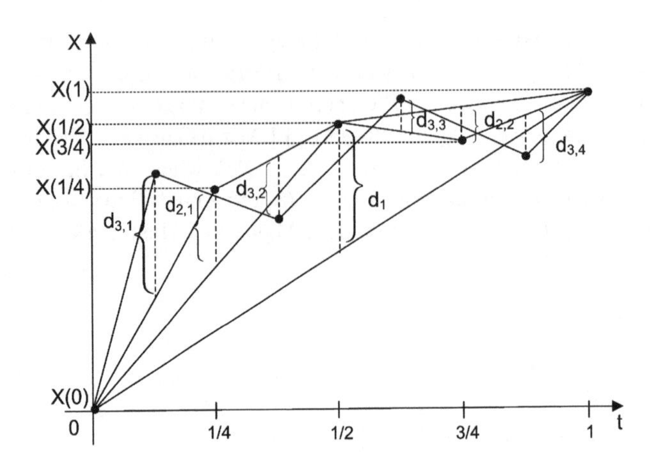

Figure 2.2 First three steps of RMD algorithm

increments for X. Adding the offset to midpoints can create the normal marginal distribution of the sequence obtained:

Step 1. The calculation process for $X(t)$ at $0 \leq t \leq 1$ can be started by setting $X(0) = 0$. The choice of $X(1)$ is made as a pseudorandom number from Gaussian distribution with zero mean and a variance $\sigma^2[X(1)] = \sigma_0^2$. Then $\sigma^2[X(1) - X(0)] = \sigma_0^2$.

Step 2. The value of $X(1/2)$ is determined as the arithmetic mean of $X(0)$ and $X(1)$, i.e. $X(1/2) = 1/2[X(0) + X(1)] + d_1$. The offset d_1 is a Gaussian random number (GRN) with zero mean and a variance σ_1^2 for d_1, which should be multiplied by the scaling coefficient 1/2. The visualization of this and future steps is shown in Figure 2.2. For execution of equality $\sigma^2[X(t_2) - X(t_1)] = |t_2 - t_1|^{2H}\sigma_0^2$ for $0 \leq t_1 \leq t_2 \leq 1$, it is required that $\sigma^2[\mathrm{X}(1/2) - \mathrm{X}(0)] = 1/4\sigma^2[\mathrm{X}(1) - \mathrm{X}(0)] + \sigma_1^2 = (1/2)^{2\mathrm{H}}\sigma_0^2$. Therefore $\sigma_1^2 = (1/2^1)^{2\mathrm{H}}(1 - 2^{2\mathrm{H}-2})\sigma_0^2$.

Step 3. Decreasing the scaling coefficient by $\sqrt{2}$ times, i.e. assuming it is $1/\sqrt{8}$, each of two intervals from 0 to $\frac{1}{2}$ and from $\frac{1}{2}$ to 1 is again divided in half. Value $X(1/4)$ can be determined as the mean value $\frac{1}{2}[X(0) + X(1/2)]$ plus the offset $d_{2,1}$, which is the GRN, multiplied to the current scaling coefficient $1/\sqrt{8}$. The appropriate formula is true for $X(3/4)$, i.e. $X(3/4) = \frac{1}{2}[X(1/2) + X(1)] + d_{2,2}$, where $d_{2,2}$ is a random offset calculated as previously. Thus, a variance σ_2^2 for $d_{2,*}$ should be chosen (where $*$ is the number of any displacement point for the given coefficient level hierarchy) so that $\sigma^2[X(1/4) - X(0)] = \frac{1}{4}\sigma^2[X(1/2) - X(0)] + \sigma_2^2 = (1/2^2)^{2H}\sigma_0{}^2$. Therefore $\sigma_2^2 = (1/2^2)^{2\mathrm{H}}(1 - 2^{2\mathrm{H}-2})\sigma_0^2$.

Step 4. The scaling coefficient is decreased by $\sqrt{2}$, i.e. it becomes $1/\sqrt{16}$. Then

$$X(1/8) = \tfrac{1}{2}[X(0) + X(1/4)] + d_{3,1}$$

$$X(3/8) = \tfrac{1}{2}[X(1/4) + X(1/2)] + d_{3,2}$$

$$X(5/8) = \tfrac{1}{2}[X(1/2) + X(3/4)] + d_{3,3}$$

$$X(7/8) = \tfrac{1}{2}[X(3/4) + X(1)] + d_{3,4}$$

In each formula $d_{3,*}$ are calculated as different Gaussian random numbers (GRNs), multiplied by the current scaling coefficient $1/\sqrt{16}$. Using the scaling coefficient decreased by $\sqrt{2}$ again at the next step, $X(t)$ can be calculated for $t = 1/16, 3/16, \ldots, 15/16$, and everything is repeated similarly to that given above. Therefore variance σ_3^2 for $d_{3,*}$ should be chosen so that $\sigma^2[X(1/8) - X(0)] = \frac{1}{4}\sigma^2 [X(1/4) - X(0)] + S_3^2 = (1/2^3)^{2H}\sigma_0^2$, i.e. $\sigma_3^2 = (1/2^3)^{2H}(1 - 2^{2H-2})\sigma_0^2$. Hence, a variance σ_n^2 for $d_{n,*}$ can be expressed as $(1/2^n)^{2H}(1 - 2^{2H-2})\sigma_0^2$.

2.1.2 SRA Algorithm for FBM Generation

The successive random additional (SRA) algorithm is another alternative algorithm for direct receipt of the FBM process. The SRA algorithm (like the RMD algorithm) uses midpoints, but to increase the stability of a generated sequence it adds the possibility of all point displacements with the appropriate variance.

Figure 2.3 shows how the SRA algorithm produces the approximate self-similar sequence. The aim of the midpoints interpolation consists in the construction of Gaussian increments that are correlated. By adding the offset to all points a self-similar sequence is obtained having a normal distribution.

The SRA algorithm includes the following steps:

Step 1. If the process $X(t)$ is calculated for time instants $0 \leq t \leq 1$, a start should be made by setting $X(0) = 0$. The choice of $X(1)$ is made in the form of a pseudorandom variable from a Gaussian distribution with zero mean and variance $\sigma^2[X(1)] = \sigma_0^2$. Then $\sigma^2 [X(1) - X(0)] = \sigma_0^2$.

Step 2. The value of $X(1/2)$ is determined by means of midpoint interpolation: $X(1/2) = \frac{1}{2}[X(1) + X(0)]$.

Step 3. Adding the displacement with the appropriate variance to all points gives the relations $X(0) = X(0) + d_{1,1}, X(1/2) = X(1/2) + d_{1,2}$ and $X(1) = X(1) + d_{1,3}$. The offset $d_{1,*}$ is controlled by Gaussian noise. To satisfy the equality $\sigma^2[X(t_2) - X(t_1)] =$

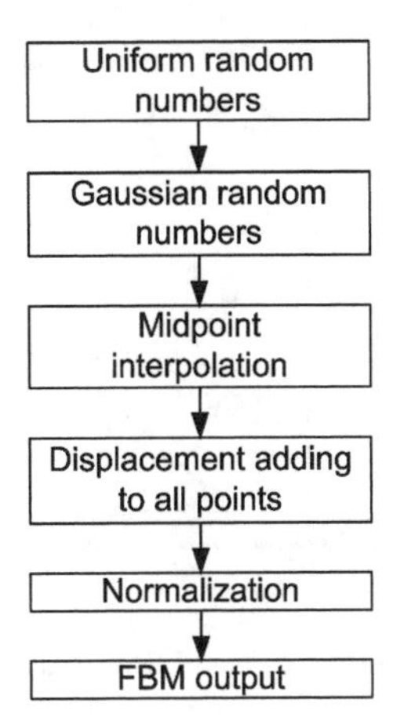

Figure 2.3 FBM generation by means of the SRA algorithm

$|t_2 - t_1|^{2H}\sigma_0^2$ for any t_1, t_2, $0 \le t_1 \le t_2 \le 1$, it is required that $\sigma^2[X(1/2) - X(0)] = \frac{1}{4}\sigma^2[X(1) - X(0)] + 2\sigma_1^2 = (1/2)^{2H}\sigma_0{}^2$, i.e. $\sigma_1{}^2 = \frac{1}{2}(1/2^1)^{2H}(1 - 2^{2H-2})\sigma_0^2$, where σ_0^2 is an initial variance; and $0 < H < 1$.

Step 4. Steps 2 and 3 are repeated. Hence, $\sigma_n^2 = \frac{1}{2}(1/2^n)^{2H}(1 - 2^{2H-2})\sigma_0^2$.

Executing the above steps using the SRA algorithm it is possible to generate the approximate self-similar FBM process.

2.2 Fractional Gaussian Noise

Fractional Gaussian noise (FGN) is a process of FBM increments, i.e. $X_H(t) = 1/\delta[B_H(t + \delta) - B_H(t)]$, where δ is an increment. The $X_H(t)$ process is normally distributed (i.e. $N(0, \sigma|\delta|^H)$), the normalized covariation function being of the form

$$r(\tau) = (|\tau + 1|^{2H} - 2|\tau|^{2H} + |\tau - 1|^{2H})/2 \tag{2.4}$$

The graph of the normalized covariation function (2.4) is shown in Figure 2.4.

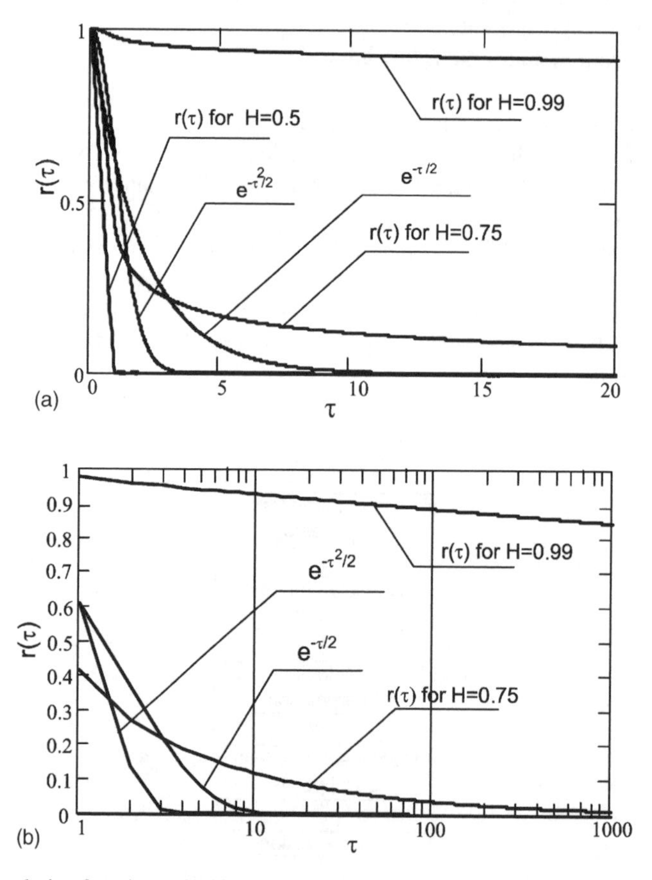

Figure 2.4 Correlation functions of FGN, exponential and Gaussian processes: (a) in linear scale; (b) in semi-logarithmic scale

Covariation curves are continuous and have no derivation interruptions except for the extreme case when $H = 0.5$. At $H = 0.5$ the correlation factor in the range $0 < \tau < 1$ is presented by a straight line $r(\tau) = 1 - \tau$, and for $\tau > 1$ the equality $r(\tau) = 0$ occurs. The function type $r(\tau)$ is interesting owing to the fact that it assumes a much slower drop for correlation coefficients than for exponential ($e^{-\tau/2}$) or Gaussian marginal ($e^{-\tau^2/2}$) curves, shown for comparison in Figure 2.4.

It is not very difficult to prove that $r(\tau) \sim H(2H-1)|\tau|^{2H-2}$ as $\tau \to \infty$ and that all aggregated processes $X_H^{(m)}(t)$ have the same distribution for any $0 < H < 1$. Therefore, FGN is exactly a self-similar process with the Hurst exponent H varying in the interval $\frac{1}{2} < H < 1$.

The FGN popularity can be explained by the possibility to use it to form a stationary, in the wide sense, self-similar Gaussian process, which can be analytically treated. Moreover, FGN is fully described by two parameters only, namely by a variance and Hurst exponent H. When the FGN stream applies to the input of an infinite length queue for the constant service intensity the queue tail distribution decays asymptotically according to the Weibull law $P[Q > x] \simeq \exp(-\delta x^{2-2H})$, where δ is a positive constant depending on the queue service intensity.

Researches show that decay of the queue tail distribution for FGN streams at input for $H > \frac{1}{2}$ occurs more slowly that the exponential one predicted by short-range dependent classical models, which correspond to the condition $H = \frac{1}{2}$.

2.2.1 FFT Algorithm for FGN Synthesis

The fast Fourier transform (FFT) algorithm [1] allows the approximate self-similar sequences to be generated using the FFT and the process known as fractional Gaussian noise. This algorithm is based on a calculation of the power spectrum density with the use of a periodogram (the power spectrum at the given frequency is represented by independent exponential random variables). At the first stage the complex numbers are constructed, their magnitudes are regulated by the normal distribution and after that the inverse FFT is fulfilled. Figure 2.5 shows how self-similar sequences have been generated by means of the FFT.

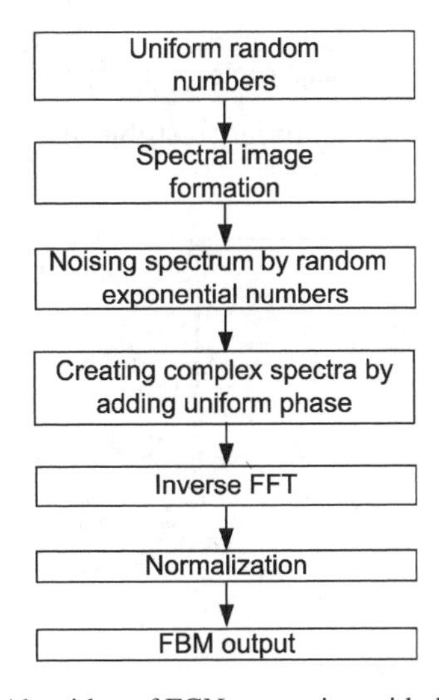

Figure 2.5 Algorithm of FGN generation with the help of FFT

The algorithm of FGN generation with the aid of the FFT can be presented as follows:

Step 1. The value sequence $\{S_1, \ldots, S_{n/2}\}$ is generated, where $S_i = \hat{S}(2\pi i/n, H)$ meets the power spectrum density of the FGN process for frequencies from $2\pi/n$ to $\pi, \frac{1}{2} < H < 1$. For the FGN process the power spectrum density $S(\omega, H)$ is determined as

$$S(\omega, H) = A(\omega, H)\left[\,|\omega|^{-2H-1} + B(\omega, H)\right], \quad 0 < H < 1, -\pi \le \omega \le \pi$$

where

$$A(\omega, H) = 2\sin(\pi H)\,\Gamma(2H+1)(1-\cos\omega) \tag{2.5}$$

$$B(\omega, H) = \sum_{i=1}^{\infty}\left[(2\pi i + \omega)^{-2H-1} + (2\pi i - \omega)^{-2H-1}\right] \tag{2.6}$$

The infinite sum in Equation (2.6) for $B(\omega, H)$ can be considered as the main difficulty at the power spectrum density calculation. Therefore in Equation (2.6) the approximation by the following relation is often used:

$$B(\omega, H) \approx a_1^d + b_1^d + a_2^d + b_2^d + a_3^d + b_3^d + \frac{a_3^{d'} + b_3^{d'} + a_4^{d'} + b_4^{d'}}{8H\pi} \tag{2.7}$$

where $d = -2H - 1$, $d' = -2H$, $a_i = 2i\pi + \omega$ and $b_i = 2i\pi - \omega$.

The curves of power spectrum density for FGN $S(\omega, H)$, made with the help of approximation (2.7), are shown in Figure 2.6.

Step 2. The value sequence $\{S_1, \ldots, S_{n/2}\}$ is generated by multiplying this number by the independent exponential random number with $\lambda = 1$.

Step 3. The complex numbers sequence $\{Z_1, \ldots, Z_{n/2}\}$ is generated, where the magnitude $|Z_i| = \sqrt{\hat{S}_i}$ and its phase are uniformly distributed between 0 and 2π.

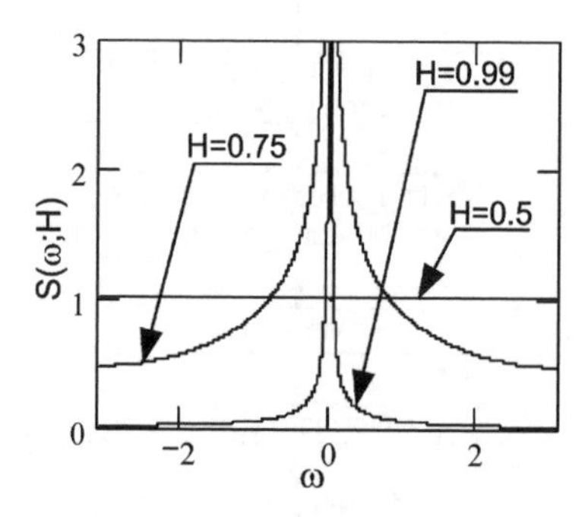

Figure 2.6 Power spectrum density curves for different Hurst exponents (H), obtained using approximation (2.7)

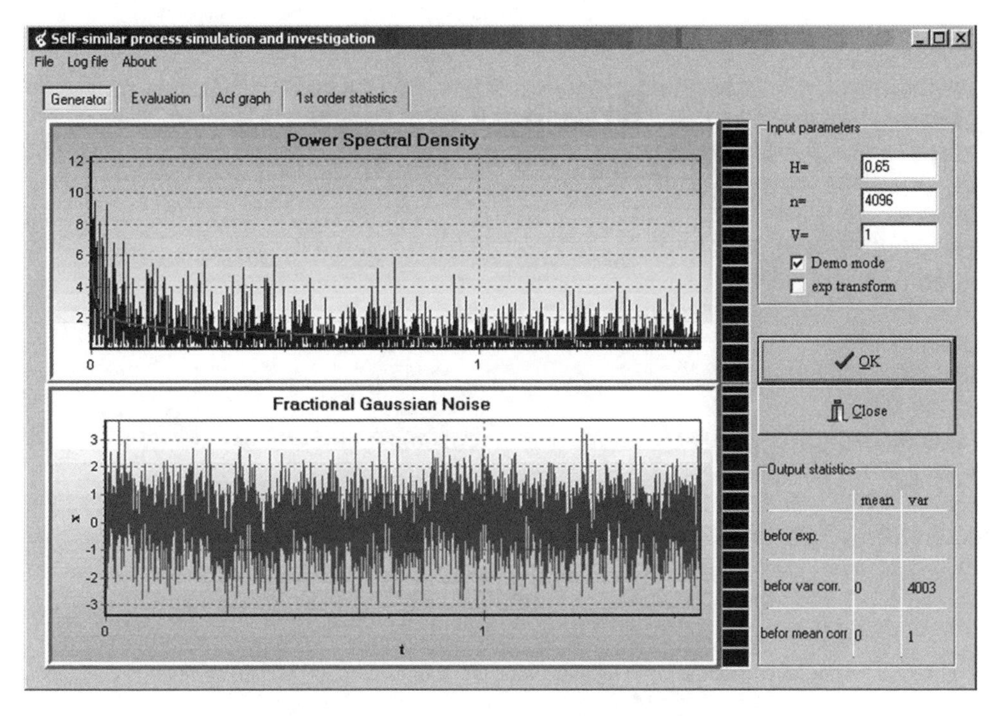

Figure 2.7 Interface of the tool for FGN simulation

Step 4. On the base of the vector $\{Z_1, \ldots, Z_{n/2}\}$ the vector $\{Z_0', \ldots, Z_{n-1}'\}$ is constructed. Its elements are calculated in accordance with the expression

$$Z_i' = \begin{cases} 0, & \text{if} \quad i = 0 \\ Z_i, & \text{if} \quad 0 < i \leq n/2 \\ \overline{Z_{n-i}}, & \text{if} \quad n/2 < i < n \end{cases}$$

where $\overline{Z_{n-i}}$ designates an operation of complex conjugation for Z_{n-i}.

Step 5. In order to obtain the approximate FGN of the $\{X_i\}$ sequence the inverse FFT is calculated for $\{Z_i'\}$.

The software interface realizing the above-mentioned algorithm for FGN generation is shown in Figure 2.7. Samples obtained with the help of this software and appropriate correlation coefficients (for the first 100 delays) are shown in Figures 2.8 to 2.10.

The FGN adaptability for traffic simulation is partially restricted by its distinct correlation structure. This restriction makes it less suitable for simulating traffic streams, which have very strong SRD (e.g. VBR video). However, for some phenomena FGN can be an acceptable approximation. It should also be noted that although the FGN is a Gaussian process, the FGN sample can be transformed to the sample with an arbitrary marginal distribution, saving H.

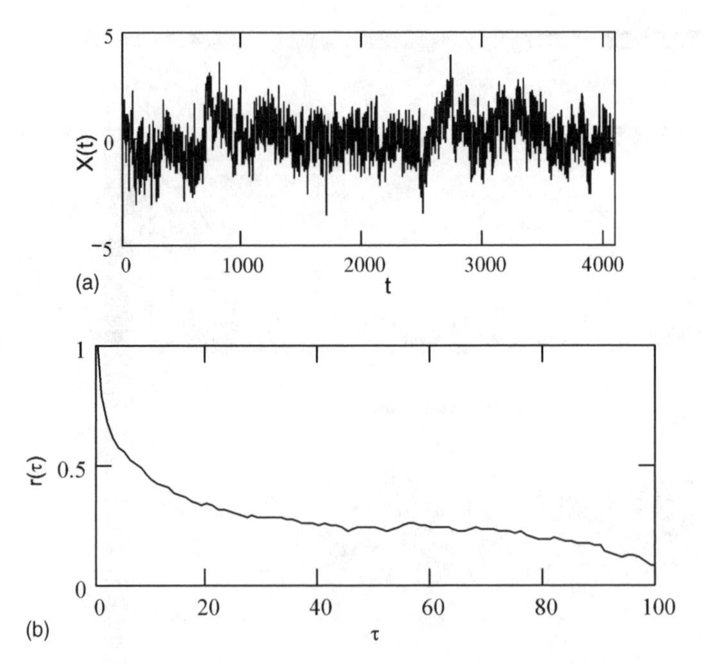

Figure 2.8 Simulation results: (a) FGN realization ($H = 0.99; N = 4.096$); (b) correlation coefficient

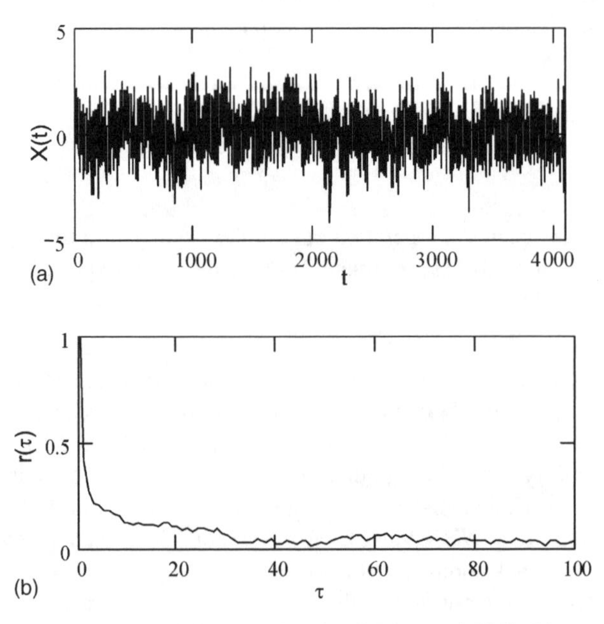

Figure 2.9 Simulation results: (a) FGN sample ($H = 0.75; N = 4.096$); (b) correlation coefficient

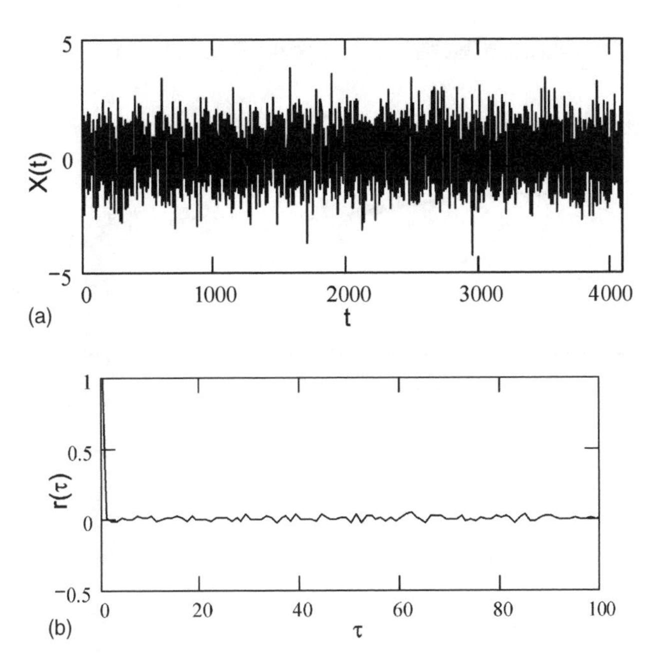

Figure 2.10 Simulation results: (a) FGN realization ($H = 0.5$; $N = 4.096$); (b) correlation coefficient

2.2.1.1 Estimation of Simulation Results

To check the simulation accuracy and conformity of the formed process to the given Hurst exponent special tool realizing tests will be used for Hurst exponent determination: the variance–time plot and R/S statistics.

The tool interface for testing the finite discrete sample is shown in Figure 2.11. The data set under test serves as an input parameter for each test as well as m_min and m_max parameters, which are used by the tests to determine the minimal and maximal size of the partition block. The data set under test can be connected to the tool by choosing the text file through the dialogue window for file opening.

Taking into account further discussion orientation, short theoretical information and operation algorithms will be given for each of the tests used for H determination.

Variance–Time Plot

The analysis of variance change is based on the property of slowly decayed variance of self-similar processes during aggregation. In accordance with this fact, the variance of the (exactly or approximately) aggregated self-similar process corresponds to the following expression:

$$\sigma^2\left[X^{(m)}\right] \sim m^{-\beta} \tag{2.8}$$

where β is a parameter related to H by the relation $\beta = 2 - 2H$.

During aggregation of the process having different m levels the variance usually decays very fast (for $H = 0.5$). Self-similar processes are exceptions and for them the variance decays slowly according to the power law (for large H values).

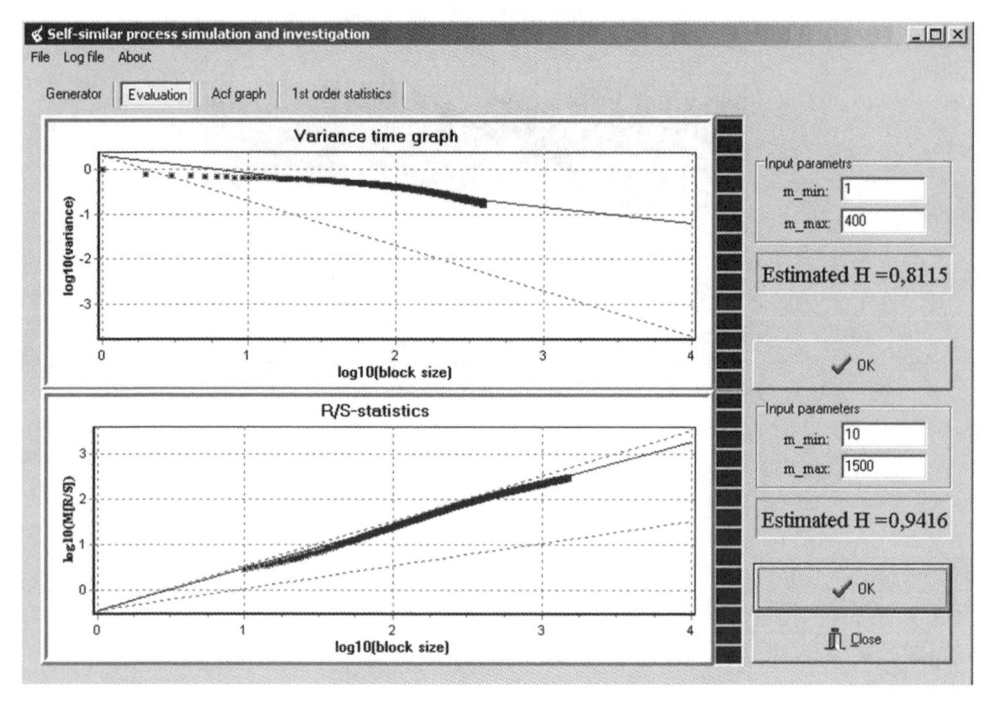

Figure 2.11 Interface of the tool for self-similar processes testing

Taking the logarithm from both sides of Equation (2.8) for aggregated variance gives the following expression:

$$\log\left[\sigma^2\left(X^{(m)}\right)\right] \sim -\beta \log(m) + \log(a) \quad \text{as}\, m \to \infty$$

It is obvious that estimation of β can be obtained by calculating $\log[\sigma^2(X^{(m)})]$ for different m values, representing results for $\log(m)$, and drawing a straight line through the resulting points by means of the least-squares method. The estimation $\hat{\beta}$ for β is defined as a negative slope of this straight line. The slope values in the interval $(-1; 0)$ can be understood as self-similarity. If this approach is used for a noncorrelated data stream then $\beta = -1$ is obtained, because $\sigma^2(X^{(m)}) = (1/m^2)m\sigma^2(X) = m^{-1}\sigma^2(X)$.

The variance–time approach is simply a heuristic method and is used for rough testing only. With the aid of this approach it is possible to estimate whether the time series is self-similar; if it is a rough estimate of H can be made.

The simulation results are examined for specific realizations of a stochastic random process. At simulation the following parameters values were used: $H = 0.6, 0.75, 0.9$ and $n = 32\ 768$ (n is the point number in the generation process; since for FGN synthesis the algorithm with an FFT application was used, n is defined as equal to 2^k where $k = 1, 2, \ldots$).

For test execution the auxiliary parameters m_min and m_max were chosen as 1 and 40 respectively. By varying these parameters values can be selected which give an estimated H value that is close to the given value. When increasing the Hurst exponent to find a more accurate estimation the maximal block size should be decreased. In the case where the Hurst'

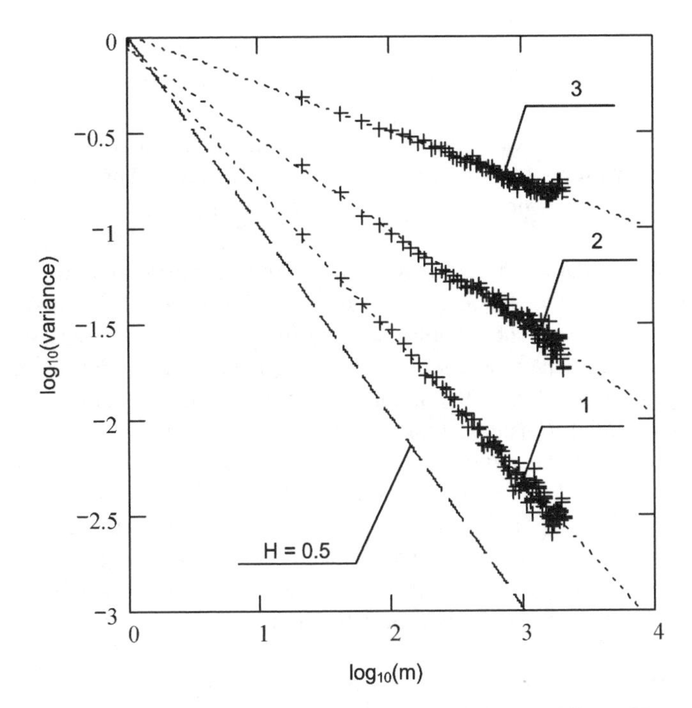

Figure 2.12 Variance–time plots for three FGN realizations with different Hurst exponents

exponent is not known in advance the choice of these parameters is rather difficult but can be made through testing.

Examples of variance–time plots are shown in Figure 2.12. The points found for FGN at the Hurst exponent $H = 0.6$ are designated by number 1. Approximating the obtained points using the least-squares method (The solid line passing through these points) gives the Hurst exponent estimation (the slope of the fitted straight line) as $\hat{H} = 0.6196$. Number 2 is used to designate the points found for FGN with the Hurst exponent $H = 0.75$. Approximating these points using the least-squares method gives $\hat{H} = 0.7608$ as the Hurst exponent estimation. Number 3 is used to designate the points found for FGN with the Hurst exponent $H = 0.9$. Approximating these points by the least-squares method gives $\hat{H} = 0.8715$ as the Hurst exponent estimation.

It can be seen that by increasing the Hurst exponent the accuracy of the variance–time plot approach decreases and the choice of test control parameter values (m_min and m_max) gains a higher profile. Moreover, the simulation shows that as the sample size increases with synthesis of FGN the Hurst exponent estimation becomes more accurate.

R/S Statistics

Proceeding from different phenomena investigations for a description of the changeability of self-similar processes, for a given observation set $X = \{X_n, n \in \mathbb{Z}^+\}$ with the sample mean $\bar{X}(n)$, the sample variance $S^2(n)$ and the range $R(n)$, the normalized dimensionless measure was developed which is referred to as the *normalized range* R/S. In practice it is appropriate to apply the expression $\log\{M[R(n)/S(n)]\} \sim H\log(n) + \log(c)$ as $n \to \infty$. Using this expression it is possible to estimate H from the function $\log\{M[R(n)/S(n)]\}$ of $\log(n)$. Fitting the curve by the least-squares method to points on the R/S plot (R/S samples for the largest and the smallest n values are truncated) it is possible to find an estimation of H on the base of the regression line slope.

At the same time, R/S statistics similar to the variance–time plot are not very accurate and give an estimation only for the self-similarity level in the time series. Thus, this method can only be used for testing in cases where the time series is self-similar and if it is, for finding a rough estimation of H.

During the FGN simulation for estimation of its parameters the same settings were used as in the previous example ($H = 0.6, 0.75, 0.9$ and $n = 32\,768$). For R/S statistics calculating the partition on blocks with subsequent changing of the block sizes was used. By default, for R/S plots the control parameters m_min and m_max were established at 10 and 150 respectively. By changing m_min and m_max, it can be seen that the Hurst exponent estimations will also be essentially changed. Figure 2.13 shows the R/S statistics plots (in the literature they are sometimes referred to as pox-plots) obtained with the aid of the above-mentioned test. Number 1 is used to mark the points obtained for FGN with the Hurst exponent $H = 0.6$. Approximating the found points using the least-squares method (the solid curve passing through these points) gives the Hurst exponent estimation (the slope of the fitted straight line) as $\hat{H} = 0.6479$. Number 2 is used to mark the points found for FGN with the Hurst exponent $H = 0.75$. As a result of processing the Hurst exponent estimation $\hat{H} = 0.7955$ was found. Number 3 is used to mark the points found for FGN with the Hurst exponent $H = 0.9$. Approximating these points using the least-squares method gives $\hat{H} = 0.8982$ as the Hurst exponent estimation.

The above-mentioned results as well as numerous experiments have shown that R/S statistics for $H < 0.75$ overestimate the Hurst exponent, but for $H > 0.75$ they underestimate it. As in the

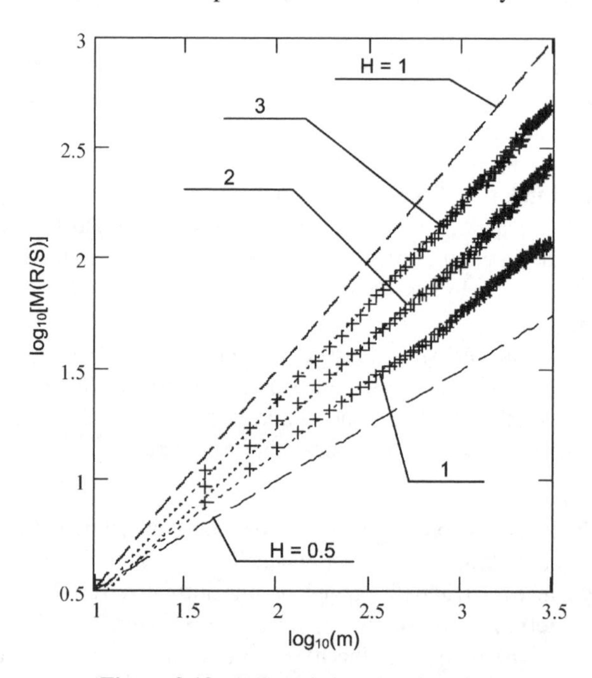

Figure 2.13 R/S statistics plots for FGN

case of the variance–time plot it can be stated that there is a degradation of the Hurst exponent estimation when its value increases. It is possible to achieve an improvement in estimation quality using the optimal choice of setting parameters m_min and m_max and increasing the sample size.

2.2.1.2 Correlation Coefficient Dependence on Delay

The correlation coefficient of the self-similar process shows long-range dependence, i.e. $r(\tau) \to c_0\tau^{1-\alpha}$. The method to estimate parameter α consists of drawing the curve $r(\tau)$ for measured data with a log–log scale and if the linear behaviour can be observed parameter β can be estimated based on the slope ($\hat{\alpha} = 1 - \hat{\beta}$). To estimate the correlation coefficient dependence upon the delay for the final sample the following formula can be used:

$$\hat{r}(\tau) = \frac{[1/(T-\tau)]\sum_{i=1}^{T-\tau}(x_i - \hat{m})(x_{i+\tau} - \hat{m})}{[1/(T-\tau)]\sum_{i=1}^{T-\tau}(x_i - \hat{m})^2}$$

where T is the data set length, τ is a delay and $\hat{m}$ is a sample mean value.

The tool interface fulfilling the calculation of this function for the final data set is shown in Figure 2.14. The plots of the self-similarity parameter estimation for generated FGN are drawn. It should be noted, that when estimating the curve slope of correlation coefficient dependence

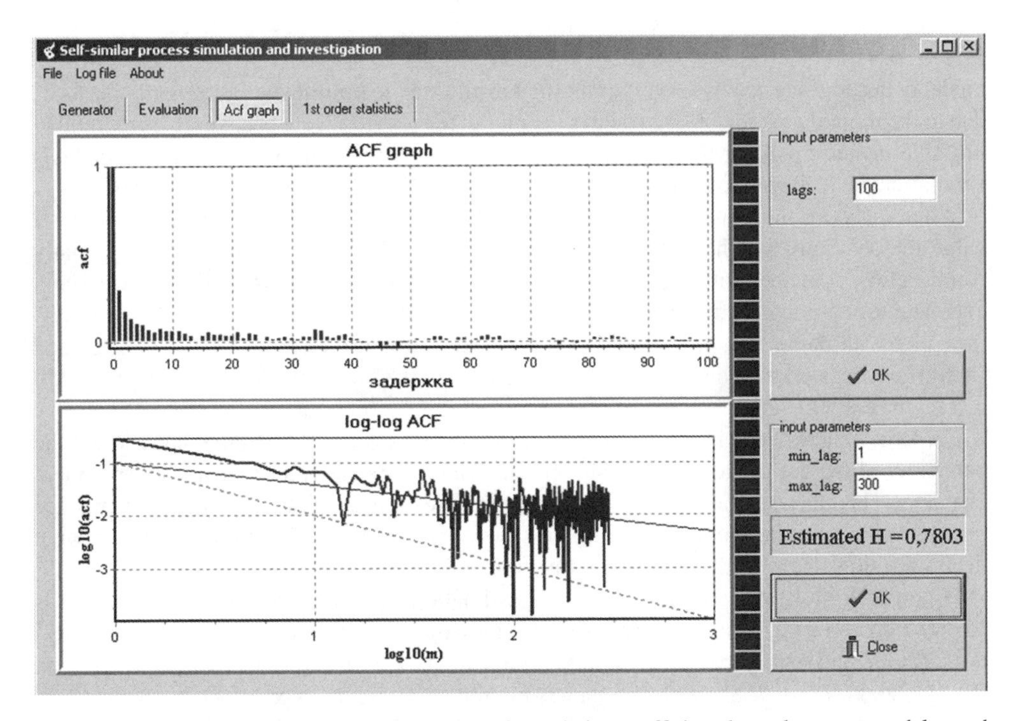

Figure 2.14 Interface of the tool for estimation of correlation coefficient dependence upon a delay and for Hurst exponent H estimation on partition block size m

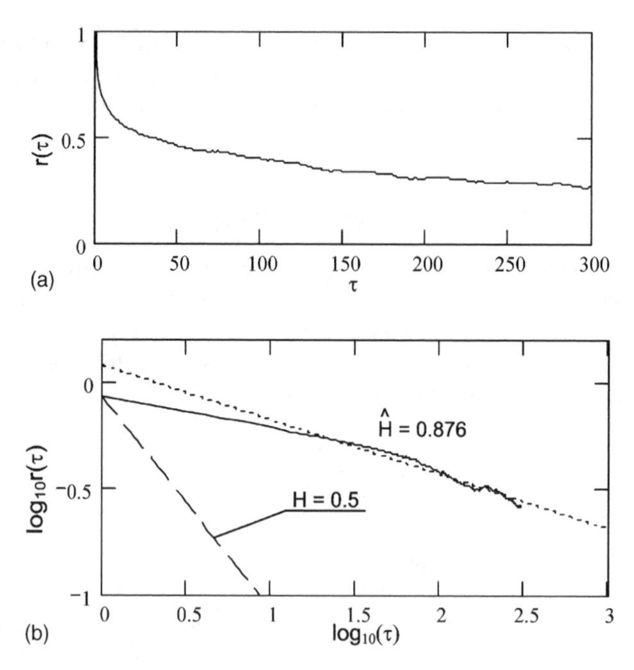

Figure 2.15 Estimation plots for self-similarity parameter of generated FGN: (a) correlation coefficient dependence on delay (initial $H = 0.99$); (b) log–log plot (max_lag = 300)

on delay in a log–log scale it is very important to choose the limiting delay correctly because, due to the limited data set, at large delays the plot in the log–log scale behaves in an extremely unstable manner, making the self-similarity parameter estimation prone to error. For FGN simulation the following parameters were chosen: $n = 32\,768$ and $H = 0.99, 0.8, 0.6$.

When analysing the results shown in Figure 2.15, it becomes obvious that to choose a large H value at FGN simulation the plot of function $r(\tau)$ in the log–log scale behaves linearly (visually it almost fully coincides with the fitted curve on the base of the least-squares method). The Hurst exponent estimation is equal to $\hat{H} = 0.876$, which is worse than the case where the variance–time plot with parameters m_min = 1 and m_max = 40 is used as a test ($\hat{H} = 0.9215$) and should be compared with the case of R/S statistics ($\hat{H} = 0.8766$) with parameters m_min = 10 and m_max = 500.

For large values of the Hurst exponent it is possible to choose rather large lags, where the behaviour of the correlation coefficient dependence on the delay plot will still have a linear character. This can be explained by the fact that for finite samples of the long-range dependence process the bigger the Hurst exponent, the slower its correlation function decays. At large delays the slow decay of the correlation function will be kept.

Figure 2.16 illustrates the Hurst exponent estimation, which at the true value $H = 0.8$ is $\hat{H} = 0.75$. This result is somewhat worse than in the case where a variance–time plot is used as a test ($\hat{H} = 0.7855$) with parameters m_min = 1 and m_max = 40, and is worse than when using R/S statistics ($\hat{H} = 0.7685$) with parameters m_min = 10 and m_max = 500.

Experiments with the choice to use maximum lag (max_lag) for the Hurst exponent estimation have shown that the threshold for which the plot of correlation coefficient

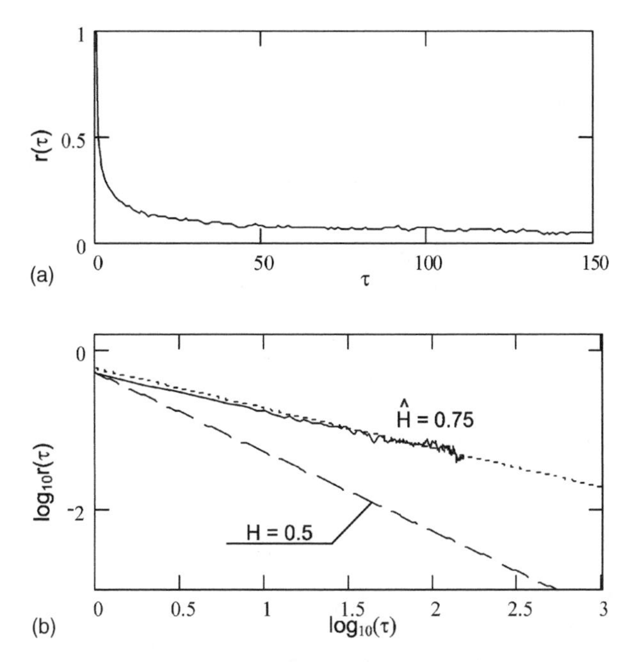

Figure 2.16 Plots of the Hurst exponent estimation for generated FGN: (a) correlation coefficient (initial $H = 0.8$); (b) log–log plot (max_lag = 150)

dependence on delay keeps a linear nature is displaced to the large argument region. At value max_lag = 150 a certain fluctuation of the plot in the log–log scale has already been observed, which can be explained by the analysed data volume limitation.

The Hurst exponent estimation for the data presented in Figure 2.17 is $\hat{H} = 0.567$, which is worse when using a variance–time plot as a test ($\hat{H} = 0.6003$) with parameters m_min = 1 and m_max = 40 and is worse when using R/S statistics ($\hat{H} = 0.6201$) with parameters m_min = 10 and m_max = 500. When the Hurst parameter value decreases to 0.6 the decreasing maximum lag for which $\hat{r}(\tau)$ plots keep a linear nature becomes obvious. The plot in Figure 2.17(b) shows that in the log–log scale at lags more than ~20 the plot behaves in a very unstable manner. For infinite samples a correlation function becomes nonsummable, which can be illustrated by large estimation dispersions in log–log plots (Figure 2.17(b)).

Thus, the smaller the expected Hurst exponent value in the sample being investigated the larger is the sample volume required for its correct estimation. In cases where the expected Hurst exponent value is not known in advance finding an estimation becomes much more difficult. In these cases, in order to obtain a correct estimation it is necessary to know the linear behaviour of the correlation coefficient versus lag plot in the log–log scale.

2.2.2 Advantages and Shortcomings of FBM/FGN Models in Network Applications

The substantial argument in favour of FBM/FGN models in networks is that in many cases traffic can be considered as a superposition of a large number of separate independent ON/OFF sources having distributions with heavy tails for ON-period duration. In this case, after

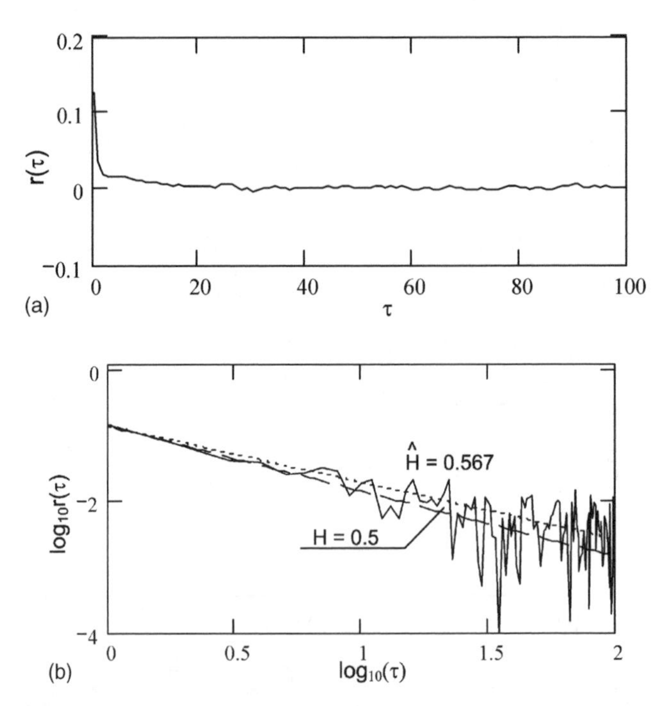

Figure 2.17 Plots of self-similarity parameter estimation for generated FGN: (a) correlation coefficients dependence upon lag (initial $H = 0.6$); (b) log–log plot (max_lag = 100)

subtraction of the average arrival speed and necessary normalization in accordance with the central limit theorem the aggregated ON/OFF sources (cumulative arrivals) converge to the Gaussian FBM. Therefore, the self-similar traffic (for the increment process) can be modelled as the model 'FGN + mean value' with given variance and H. FBM/FGN models are widely used in network design because their Gaussian features and strict scaling permit analytical investigations of queueing characteristics to be performed.

Unfortunately, FBM/FGN models have strict limitations when adapted to network traffic. First of all, real traffic traces do not show strict self-similarity and are at best asymptotically self-similar only. In other words, it is not enough to have the H parameter only cover the complex correlation structure of real network processes. Moreover, there are researches proving the importance of short-term correlations for buffering and discovering significant timescales. Secondly, the Gaussian features of FBM/FGN models for several traffic types may not correspond with reality, for instance when the mean-square deviation exceeds the mean value. In this case, FBM/FGN outputs contain a large number of negative values. Thirdly, many real network application processes do not even closely approach the Gaussian case, especially for small timescales. In this case more universal regressive models can be used for improvement of short-term and long-term correlation structures existing in real traces.

2.3 Regression Models of Traffic

Traffic models are used in designing to predict the network performance and to estimate congestion control schemes as well as to model various covariation structures and distributions.

The models, which do not cover the statistical characteristics of real traffic, lead to an incorrect estimation of the network carrying capacity due to overestimation or underestimation of one. Traffic models should have a small number of parameters. The estimation of these parameters should be simple. Traffic models, which are not subject to analytical interpretation, can be used for traffic trace generation only.

Traffic models can be stationary as well as nonstationary. Stationary traffic models can be classified on the basis of two categories: short-range dependent and long-range dependent. Short-range dependent models cover traditional traffic models such as Markovian processes and regression models.

Since regression models are simple in realization, they are widely used for modelling. At the same time, queueing systems for regression models, as a rule, hardly yield to analytical interpretation. As a result, to find an approximate analytical solution the regression models are often approximated by Markovian models. Regression models determine the consequent random variable in the form of a recursive function of previous random variables. Therefore, they are used for modelling sequences that do not change very much between observations that follow each other: e.g. the number of bits per frame for a VBR video-teleconference. Regression AR, ARMA (autoregressive moving average), DAR (discrete AR) and TES (transform expand sample) processes are able to model the stationary sequences only, but the regression ARIMA process can be used to model the stationary as well as nonstationary sequences.

For a general case, AR, ARMA and ARIMA (autoregressive integrated moving average) processes have the Gaussian distribution. Hence, to model the sequence having the arbitrary distribution it is necessary to have a two-stage transformation resulting in a process with the required distribution found from the Gaussian one. However, this transformation does not guarantee that the transformed process will have the same correlation structure as the initial process.

An example of the LRD model of traffic will be considered. The model is based on the FARIMA (Fractional autoregressive integrated moving average) process, which has some advantages compared to the models based on other fractal processes. FBM, the aggregation of ON/OFF sources with high changeability, etc., can be given as an example of such processes. The fractional Brownian motion has only one parameter, controlling the correlation function, and therefore there is no flexibility in short-range dependence modelling. The aggregation of a large number of ON/OFF sources with infinite variance for ON and OFF periods allows the formation of long-range dependence and can be used to cover the asymptotic behaviour of long-range traffic. However, the possibility of short-range behaviour simulation is not evident. FARIMA (p,d,q) models have three parameters, p,d and q, that control the correlation structure. Therefore, they can cover the short-range dependence as well as the long-range one. Evidently, it is necessary to have a model that will be able to cover the short-range dependence, the long-range one and the arbitrary distribution.

2.3.1 Linear Autoregressive (AR) Processes

Class AR (p) consists of linear *autoregressive* models of order p:

$$\alpha(B)X_n = Z_n, \quad n > 0 \tag{2.9}$$

where $(X_{-p+1}, \ldots, X_0)$ is a given random vector (usually a normal vector); $\alpha(B) = 1 - \alpha_1 B - \alpha_2 B^2 - \cdots - \alpha_p B^p$, where $\alpha_r, \quad 1 \leq r \leq p$, are real constants; B is the backward shift operator, defined as $B^j X(t) = X(t-j)$; and Z_n are noncorrelated random variables (white noise) with zero mean value, called *innovations*, which do not depend on X_n [2] (in 'good' model the innovations should be less that the X_n value).

The recursive form of Equation (2.9) shows how, starting from previous elements, the next random element is generated in the sequence $\{X_n\}_{n=0}^{\infty}$. This makes similar models suitable for the correlated traffic imitation. The direct algorithm for modelling AR processes follows from Equation (2.9) and can be written as

$$X_n = \sum_{r=1}^{p} \alpha_r X_{n-r} + Z_n, \quad n > 0$$

In Reference [3] the simple AR (2) model was used for modelling the VBR-coded video. More complicated models can be obtained on the basis of AR (p) models combined with other models. For example, in Reference [4] the traffic of video bit intensity was modelled by a sum $R_n = X_n + Y_n + K_n C_n$, where the first two terms mean the AR (1) model and the third term means the product of a simple Markovian chain and the independent normal random variable from identical and independent distributed (i.i.d.) normal sequences. The aim of using two autoregressive processes is to obtain a better approximation to the real covariation function. The third term was introduced for realization of bursts caused by changing video scenes.

2.3.2 Processes of Moving Average (MA)

Class MA (q) consists of the processes of moving average of order q:

$$X_n = \beta(B) Z_n \tag{2.10}$$

where $\beta(B) = 1 + \beta_1 B + \beta_2 B^2 + \cdots + \beta_q B^q$, where β_r, $1 \le r \le q$, are the real constants and Z_n are noncorrelated random variables with zero mean value [2]. The algorithm for obtaining the realization of the moving average process can be written on the basis of Equation (2.10) as

$$X_n = \sum_{r=0}^{q} \beta_r Z_{n-r}, \quad n > 0$$

The correlated time sequences are formed with the help of MA models as random variables following one another are determined on the basis of the total subset Z_n.

2.3.3 Autoregressive Models of Moving Average, ARMA(p,q)

The autoregressive model of moving average of order (p,q), designated as ARMA (p,q), can be written as

$$\alpha(B) X_t = \beta(B) Z_t \tag{2.11}$$

The algorithm of ARMA(p,q) process realization becomes:

$$X_t = \sum_{i=1}^{p} \alpha_i X_{t-i} + \sum_{i=0}^{q} \beta_i Z_{t-i} \tag{2.12}$$

where $\alpha(B)$ is a polynomial of operator B determined by the expression $\alpha(B) = 1 - \alpha_1 B - \cdots - \alpha_p B^p$ and $\beta(B) = 1 + \beta_1 B + \cdots + \beta_q B^q$ is a polynomial of order q. This is equivalent to filtering white noise Z_t with the linear filter invariant to the time shift, which has the fractional rational transfer function with p poles and q zeroes [5], i.e.

$$H(z) = \frac{B_q(z)}{A_p(z)} = \frac{1 - \sum_{k=0}^{q} \beta_k z^{-k}}{1 - \sum_{k=1}^{p} \alpha_p z^{-k}}$$

The covariation R_k for the ARMA(p,q) process can be obtained by multiplication of Equation (2.11) and X_{t-k}, after having defined the expectation value and found the cross covariance between Z_t and X_t:

$$R_k = \alpha_1 R_{k-1} + \cdots + \alpha_p R_{k-p} - \sigma_Z^2(\beta_k h_0 + \beta_{k+1} h_1 + \cdots + \beta_q h_{q-k})$$

where h_t is the impulse response $H(z)$ for the ARMA(p,q) filter. Note that for $k > q, \beta_k = 0$. Hence, the process covariation for $k > q$, $R_k = \alpha_1 R_{k-1} + \alpha_2 R_{k-2} + \cdots + \alpha_p R_{k-p}$, is the difference equation and so the covariation for ARMA(p,q) decays in accordance with exponential law.

The ARMA model can be used for VBR traffic modelling [6]. With this goal the duration of the video frame is divided into equal parts by m time intervals. The number of cells in the nth time interval is modelled by the following ARMA process:

$$X_n = \alpha X_{n-m} + \sum_{k=0}^{m-1} \beta_k Z_{n-k}$$

Since video data in each frame are interconnected due to the time correlation, the correlation function has peaks for all lags that are multiples of m. In the above-mentioned model the AR part is used for modelling the repeated correlation effect and a selection of correlations for other lags is carried out with the aid of β_k.

The parametric estimation of ARMA models is more complicated than for AR models, because the estimation of β_k requires the solution of a large number of nonlinear equations or the use of spectral decomposition methods [5]. To find analytical solutions is just as difficult.

The conventional series analysis usually assumes that X_t is a certain 'mixture' [7], i.e. $\sum_\tau |R_\tau| < \infty$, and therefore $R(\tau)$ decays exponentially with a rise in τ (see Figure 2.18(a)). This is evidence to the fact that X_t values, which are spaced in time, are approximately correlated. Due to this character of the correlation structure ARMA processes are also referred to as *short-term* or *short-range dependent* (SRD) processes.

It has been proved [8] that in communication networks time series are present for which the correlations between the observations spaced too far in time decay very slowly. These time series cannot be described by ARMA models. The asymptotic decaying of the covariation function is such that:

$$R_\tau \sim C|\tau|^{2d-1} \tag{2.13}$$

where $C \neq 0$ and $0 < d < 0.5$. Stationary processes with the covariation function decaying in accordance with (2.13) as $\tau \to \infty$ are referred to as *long-term* or *long-range dependent* (LRD) processes [9]. Figure 2.18 shows an example of correlation functions of SRD and LRD processes.

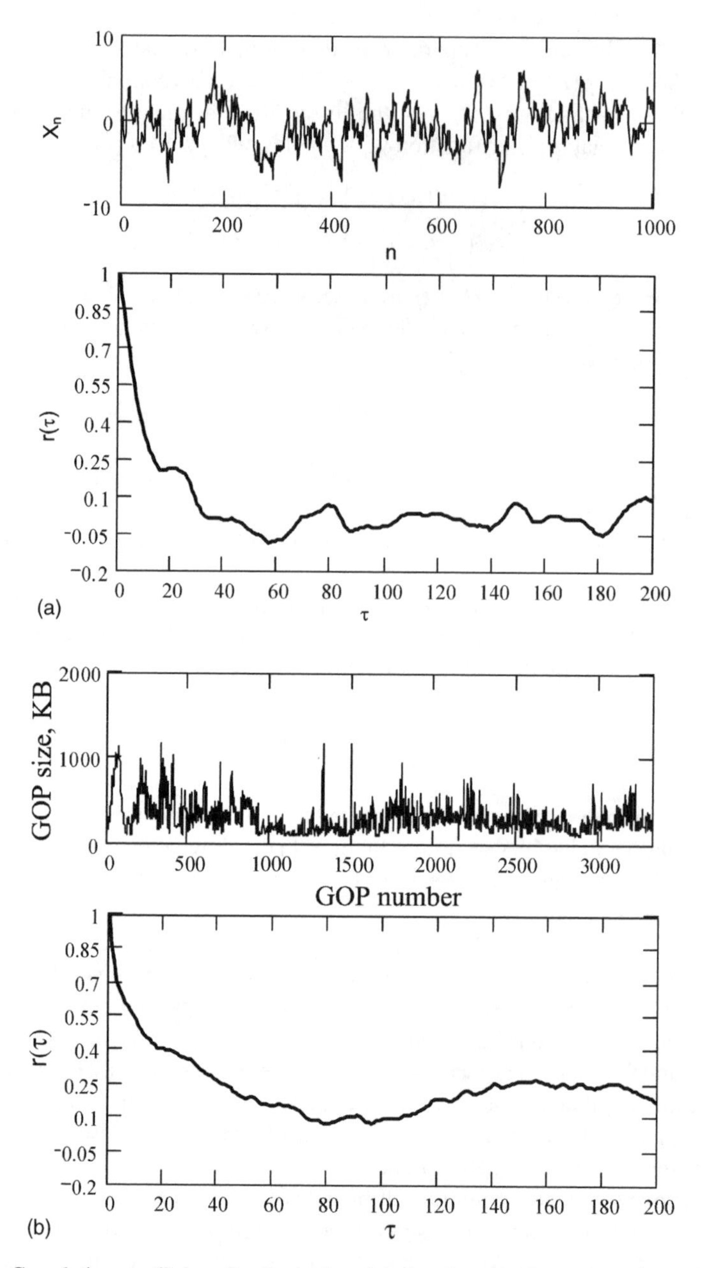

Figure 2.18 Correlation coefficient for the series: (a) showing the short-range dependence (ARMA (1; 0) with the pole at $z = 0.9$); (b) showing the long-range dependence (VBR trace of the *Star Wars* movie)

2.3.4 *Fractional Autoregressive Integrated Moving Average (FARIMA) Process*

ARIMA processes are used for the description of the class of nonstationary series $\{X_t : t \in Z\}$, which show the homogeneity in contrast to its local level and/or trend; i.e. one series part behaves as any other part (Figure 2.19) [10]. In other words, if the local level is removed and/or the trend changed the series becomes stationary. These series can be described by the autoregressive operator $\phi(B)$, which is written as

$$\phi(B) = \alpha(B)(1 - B)^d \tag{2.14}$$

where $\alpha(B) = 1 - \alpha_1 B^1 - \alpha_2 B^2 - \cdots - \alpha_p B^p$ is a polynomial of order p and d are natural numbers determining the differentiation order.

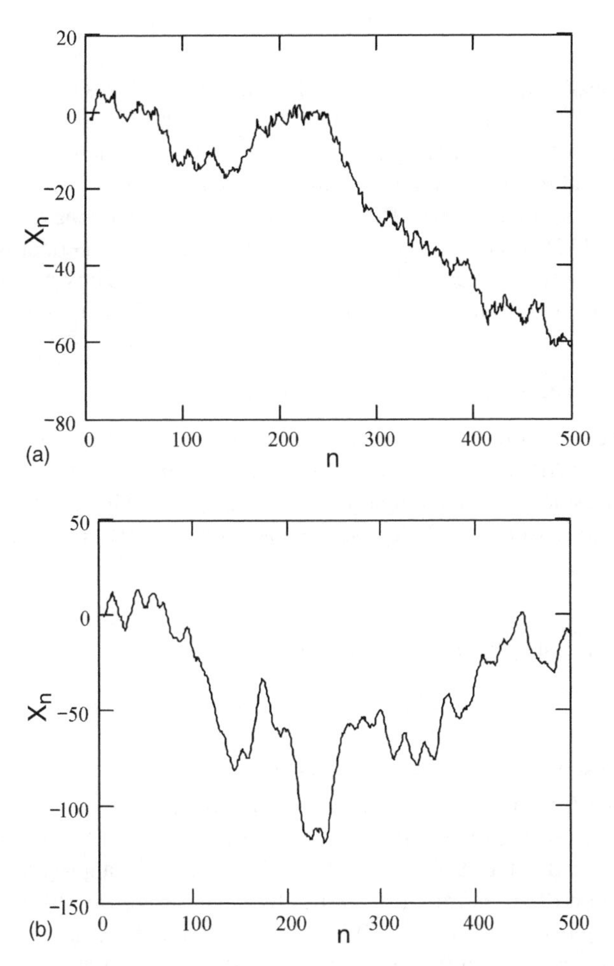

Figure 2.19 (a) ARIMA(0;1;1) process, $\beta(B) = 1 + 0.1B$; (b) ARIMA(1;1;0) process, $\alpha(B) = 1 - 0.8B$

Thus, the generalized model describing the homogeneous nonstationary behaviour takes the following form:

$$\phi(B)X_t = \beta(B)Z_t \tag{2.15a}$$

or

$$\alpha(B)(1-B)^d X_t = \beta(B)Z_t \tag{2.15b}$$

or

$$\alpha(B)X_t = \beta(B)Z_t \tag{2.15c}$$

where

$$X(t) = (1-B)^d Y(t) = \delta^d Y(t) \tag{2.15d}$$

In practice, d is usually equal to 1 or at most 2. The case $d = 1$ shows that X_t has a linear trend, but in the case $d > 1$ it has a polynomial trend. The process defined as (2.15a) is referred to as the ARIMA process of order (p,d,q). Figure 2.19 shows the sample for ARIMA(0,1,1) (with zero at $z = -0.1$) and for ARIMA(1,1,0) (with the pole at $z = 0.8$).

The term 'integrated' in the name ARIMA is used due to the following relation, which in fact is the inverse to (2.15d): $Y(t) = S^d X(t)$, where S is the summation operator (or *integration* for the continuous function) defined as $S = 1/(1-B) = \sum_{j=0}^{\infty} B^j$. Thus, the arbitrary ARIMA process can be generated from the innovation process Z_t with the help of three filters.

2.3.4.1 FARIMA Process

The ARIMA process can be considered as the first step on the way to the ARMA process generalization. In the ARIMA process parameter d is considered to be integer-valued only. The FARIMA process results from this limitation cancelling, i.e. the fraction values can be taken for d. An analysis of series using FARIMA processes has been independently offered in References [11] and [12].

Definition 2.1
It is assumed that X_t is a stationary process such as

$$\alpha(B)(1-B)^d X_t = \beta(B)Z_t \tag{2.16}$$

for some $d \in (-0.5; 0.5)$. Then X_t is the FARIMA (p,d,q) process.

Since $d \in (-0.5; 0.5)$, this means that X_t has a fractional pole at $B = 1$. The upper limit $d < 0.5$ is necessary because for $d > 0.5$ the process becomes nonstationary. However, the case $d > 0.5$ can be reduced to the case $-0.5 < d \leq 0.5$ by taking the appropriate differences. For example, if Equation (2.16) is true for $d = 1.2$, then the difference process $X_t - X_{t-1}$ is the solution for (2.16) with $d = 0.2$. If $d = \pm 0.5$ the process X_t is either stationary or reversible, but not both [12]. In the case when $0 < d < 0.5$, the FARIMA(p,d,q) process shows long-range dependence. Parameters p and q correspond to the order of $\alpha(B)$ and $\beta(B)$, and offer the possibility of flexible modelling of short-term characteristics of the process. It is necessary to note that in spite of the fact that ARIMA and FARIMA processes are created in the same way (i.e. the generator is nonstationary) the stationary process is obtained as a result. Figure 2.20 shows samples from different FARIMA processes.

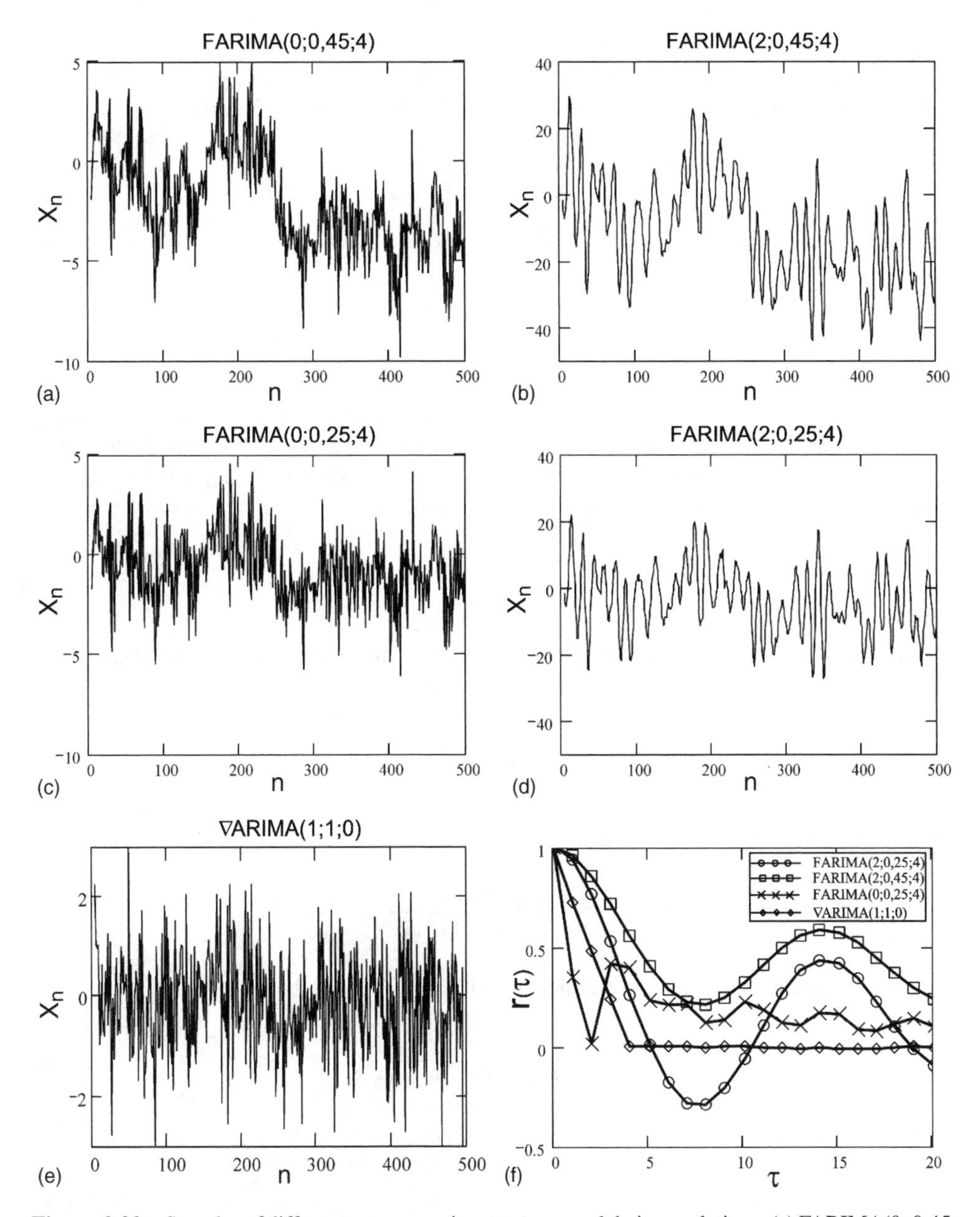

Figure 2.20 Samples of different autoregressive processes and their correlations: (a) FARIMA(0; 0.45; 4); (b) FARIMA(2; 0.45; 4); (c) FARIMA(0; 0.25; 4); (d) FARIMA(2; 0.25; 4); (e) FARIMA(1;1; 0); (f) correlation coefficients

To illustrate the influence of the presence or absence of short-range dependence on the correlation coefficient, in Figure 2.20 (f) r_τ is drawn for the processes shown in Figure 2.20. To find these realizations the AR component $\alpha(B) = 1 - 1.72B + 0.81B^2$ and the MA component $\beta(B) = 1 + 0.9B - 0.7B^2 + 0.35B^3 + 0.4B^4$ were used. Comparing Figures 2.20(a) and (b) as well as Figures 2.20(b) and (c), it can be seen that ARMA parameters define the correlation structure at small delays. The long-term parameter d depends on the degree of decaying r_τ as $\tau \to \infty$. The value of d close to 0.5 gives a more considerable long-range dependence.

The FARIMA(p,d,q) spectrum is obtained directly from Equation (2.16) as

$$S_X(\omega) = \frac{\sigma_Z^2}{2\pi}\left|1 - \mathrm{e}^{-\mathrm{j}\omega}\right|^{-2d} \frac{|\beta(\mathrm{e}^{\mathrm{j}\omega})|^2}{|\alpha(\mathrm{e}^{\mathrm{j}\omega})|^2}$$

As $|1 - \mathrm{e}^{-\mathrm{j}\omega}| = 2|\sin(\omega/2)|$ and $\lim_{\omega\to 0} 2\sin(\omega/2)/\omega = 1$, the spectral density behaviour as $\omega \to 0$ is defined as

$$S_X(\omega) = \frac{\sigma_Z^2}{2\pi}|\omega|^{-2d}\frac{|\beta(1)|^2}{|\alpha(1)|^2} \tag{2.17}$$

This expression shows that for $0 < d < 0.5$, the spectrum density $S_X(\omega)$ is not limited to $\omega = 0$, i.e. $S_X(\omega)|_{\omega=0} = \infty$, and hence $\sum_\tau R(\tau) = \infty$, which indicates the subexponential decay of the covariation function [13]. The analytical relation derivation for the covariation function leads to problems. The only exception is the FARIMA(0,d,0) process. In this simple case

$$S_X(\omega) = \frac{\sigma_z^2}{2\pi}|1 - \mathrm{e}^{-\mathrm{j}\omega}|^{-2d}$$

the covariation function $R(\tau)$ can be found by taking the inverse Fourier transform of $S_X(\omega)$. Since $|1 - \mathrm{e}^{-\mathrm{j}\omega}| = 2|\sin(\omega/2)|$ is the real and even function,

$$R(\tau) = \frac{\sigma_z^2}{\pi}\int_0^\pi \left[2\sin\left(\frac{\omega}{2}\right)\right]^{-2d}\cos(\omega\tau)\mathrm{d}\omega$$

After mathematical transformations,

$$R(\tau) = \sigma_z^2\frac{(-1)^\tau\Gamma(1-2d)}{\Gamma(\tau - d + 1)\Gamma(1 - \tau - d)}$$

Having designated the correlation coefficient as $r_\tau = R(\tau)/R(0)$, then

$$r_\tau = \frac{\Gamma(1-d)\Gamma(\tau+d)}{\Gamma(d)\Gamma(\tau - d + 1)} \tag{2.18}$$

For large values of τ it is possible to write $\Gamma(\tau + a)/\Gamma(\tau + b) \sim \tau^{a-b}$. Then, as $\tau \to \infty$,

$$r_\tau = \frac{\Gamma(1-d)}{\Gamma(d)}|\tau|^{2d-1} \tag{2.19}$$

which has a form similar to Equation (2.13).

Table 2.1 Comparative analysis of regression and self-similar processes

Process	Stationarity	Model/properties
ARMA(*p;d;q*)	Yes	$\alpha(B)X_n = \beta(B)Z_n$ Shows short-range dependence
ARIMA(*p;d;q*)	No	$\alpha(B)(1-B)^d X_n = \beta(B)Z_n$ d = integer
FARIMA(*p;d;q*)	Yes	$\alpha(B)(1-B)^d X_n = \beta(B)Z_n$ Shows short-range and long-range dependence
Self-similar	No	$S(at) \stackrel{d}{=} a^H S(t)$ Shows scaling invariance
FBM	No	Self-similar process Stationary increments, Gaussian distribution
FGN	Yes	Shows long-range dependence

In Table 2.1 the characteristics of self-similar and regression processes are listed. Figure 2.21 give the relationships between them.

2.3.5 Parametric Estimation Methods

The problem of parametrical estimations for FARIMA (*p,d,q*) processes touches upon the subject of d parameter estimation, which describes the long-range dependence, and the estimation of vectors $\boldsymbol{\alpha} = [1, \alpha_1, \ldots, \alpha_p]$ and $\boldsymbol{\beta} = [1, \beta_1, \ldots, \beta_q]$, which describes the short-range dependence. Known estimation methods can be classified into two categories:

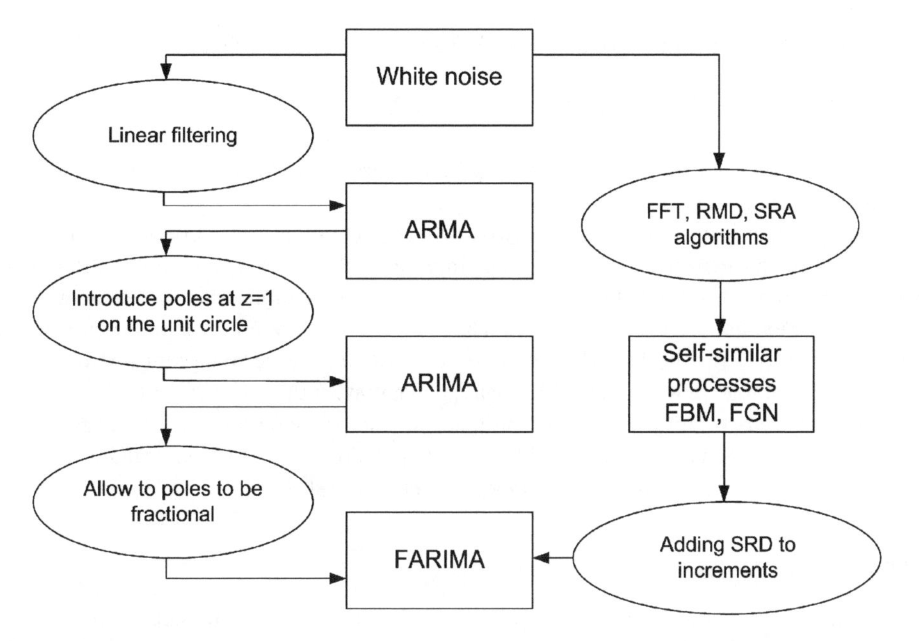

Figure 2.21 Block diagram indicating the relation between regression and self-similar processes

1. Methods for estimation d only. The heuristic methods and methods based on the spectral density plot can serve as examples. After eliminating the long-range dependence from data and using conventional methods of parametric ARMA estimation $\boldsymbol{\alpha}$ and $\boldsymbol{\beta}$ can be estimated.
2. Methods that *simultaneously* estimate d, $\boldsymbol{\alpha}$ and $\boldsymbol{\beta}$. An analysis of their effectiveness and comparison of their advantages/shortcomings will be discussed later.

2.3.5.1 Heuristic Methods

Heuristic methods were originally suggested for the Hurst exponent (H) estimation in self-similar processes. They can be used for d estimation in the FARIMA process X_t, because $d = H - 0.5$ and the cumulative process $S(t)$, defined as $S(t) = \sum_{u=0}^{t} X(u)$, is self-similar as $t \to \infty$. However, it should be noted that the heuristic methods are not suitable for statistical analysis and are mainly used as a diagnostic tool to determine the LRD presence in the data. The variance–time plot analysis and R/S statistics are popular heuristic approaches.

Figure 2.22 shows typical plots of the variance change and R/S statistics for the processes shown in Figure 2.20.

2.3.5.2 Periodogram Method

The spectral density function $S_X(\omega)$ for the FARIMA(p,d,q) process X_t is unlimited at $\omega = 0$ and presents itself as Equation (2.17) as $\omega \to 0$. Taking the natural logarithms of both sides of (2.17) gives

$$\log[S_X(\omega)] = \log\left[\frac{\sigma_z^2}{2\pi}\right] + \log\left[\frac{|\beta(1)|^2}{|\alpha(1)|^2}\right] - 2d\log|\omega| \tag{2.20}$$

Since the first two terms in the right side of Equation (2.20) do not depend on d and ω,

$$\log\left[S_X(\omega)\right] = C - 2d\log|\omega|$$

where C is some constant. Thus, the log–log plot of the $S_X(\omega)$ function as $\omega \to 0$ will be a straight line having slope $-2d$.

It is known that in the case of the ARMA process the nonprocessed periodogram can be approximately considered as the unbiased estimation of the spectral density function. Moreover, for obtaining coordinated estimates such approaches as windowing, periodogram averaging and smoothing can be applied. The statistical behaviour of the nonprocessed periodogram $I_X(\omega)$ for the FARIMA process differs from the periodogram for the ARMA process. For example, in Reference [12] it was shown that for the FARIMA processes $I_X(\omega)$ is approximately *biased* and for $j \neq j'$ the ordinates of the periodogram $I_X(\omega_j)$ and $I_X(\omega_{j'})$ are nearly *correlated*. Nevertheless, for lack of a better alternative for long-range dependent processes the traditional methods of variance reduction are still used.

2.3.5.3 Whittle Method

For the given observation set $X(t) = (X(1), \ldots, X(N))^{\mathrm{T}}$ for the Gaussian FARIMA(p,d,q) process the method of Whittle estimates the unknown parameters $\boldsymbol{\theta} = (\sigma_z^2, d, \theta_3, \ldots, \theta_M)$

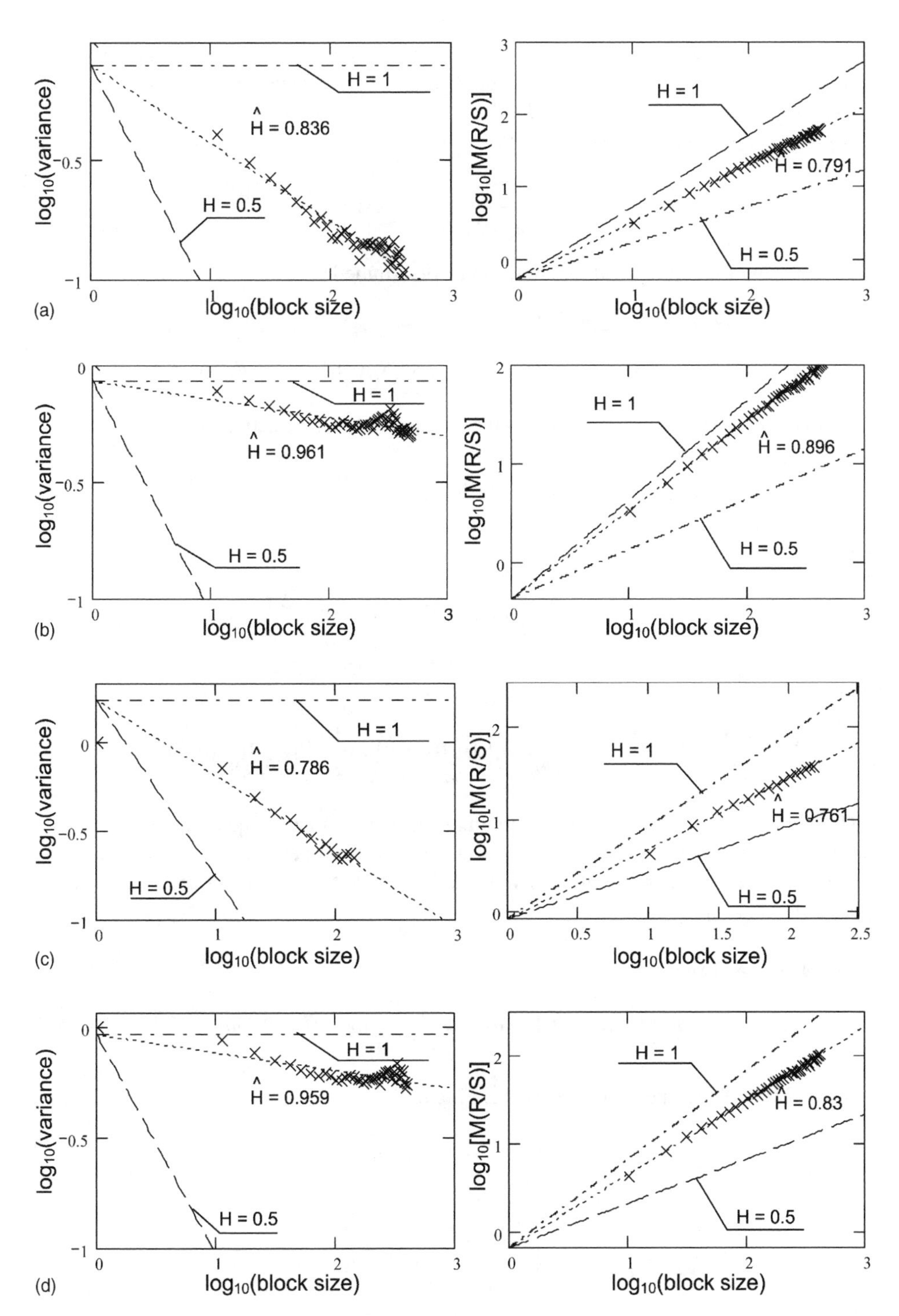

Figure 2.22 Hurst exponent estimation for different regression processes: (a) FARIMA(0;0.25;4); (b) FARIMA(0;0.45;4); (c) FARIMA(2;0.25;4); (d) FARIMA(2;0.45;4)

simultaneously, maximizing the likelihood function. Here σ_z^2 is the incremental process variance and $(\theta_3, \ldots, \theta_M)$ corresponds to parameters of short-range dependence.

Assuming that $\mathbf{R}N(X;\theta) = [C_{2X}(j-l)]_{j,l=1,\ldots,N}$ is a correlation matrix for $X(t)$, the likelihood function for $X = (X(1), \ldots, X(n))^{\mathrm{T}}$ is

$$w(X;\theta) = \frac{1}{(2\pi)^{N/2}\{\det[\mathbf{R}_N(\theta)]\}^{1/2}} \exp\left[-\frac{X^{\mathrm{T}}\mathbf{R}_N^{-1}(\theta)X}{2}\right]$$

The logarithm of the likelihood function can be determined as

$$\begin{aligned} L_N(X;\theta) &= \log[\boldsymbol{w}(X;\theta)] \\ &= -\frac{N}{2}\log(2\pi) - \frac{1}{2}\log\{\det[R_N(\theta)]\} - \frac{1}{2}X^{\mathrm{T}}R_N^{-1}(\theta)X \end{aligned} \tag{2.21}$$

The maximum likelihood estimation $\hat{\boldsymbol{\theta}}$ is the solution of the M equations system:

$$\frac{\partial}{\partial\theta_j} L_N(X;\boldsymbol{\theta})\Big|_{j=1,\ldots,M} = 0$$

and

$$\frac{\partial^2}{\partial\theta_j^2} L_N(X;\boldsymbol{\theta}) < 0$$

where

$$\frac{\partial}{\partial\theta_j} L_N(X;\boldsymbol{\theta}) = -\frac{1}{2}\frac{\partial}{\partial\theta_j}\log\{\det[\mathbf{R}_N(\boldsymbol{\theta})]\} - \frac{1}{2}X^{\mathrm{T}}\left[\frac{\partial}{\partial\theta_j}\mathbf{R}_N^{-1}(\boldsymbol{\theta})\right]X$$

The parameters set search, which simultaneously maximizes Equation (2.21), requires an estimation of the covariation matrix $\mathbf{R}_N(\boldsymbol{\theta})$ inversion. This operation is computing intensive and in some cases can be numerically unstable. The Whittle method uses the following approximation [11]:

- For $\log\{\det[\mathbf{R}_N(\boldsymbol{\theta})]\}$:

$$\lim_{N\to\infty}\frac{1}{N}\log\{\det[\mathbf{R}_N(\boldsymbol{\theta})]\} = \frac{1}{2\pi}\int_{-\pi}^{\pi}\log[S_X(\omega;\boldsymbol{\theta})]\mathrm{d}\omega \tag{2.22}$$

- For $X^{\mathrm{T}}R_N^{-1}(\theta)X$ the matrix $R_N^{-1}(\theta)$ is changing to

$$A(\boldsymbol{\theta}) = [\alpha(j-l)]_{j,l=1,\ldots,N} \tag{2.23}$$

where

$$\alpha(j-l) = \frac{1}{(2\pi)^2}\int_{-\pi}^{\pi}\frac{1}{S_X(\omega;\boldsymbol{\theta})}\mathrm{e}^{\mathrm{i}(j-l)\omega}\mathrm{d}\omega \quad (\mathrm{i} = \sqrt{-1})$$

and $A(\boldsymbol{\theta})$ is asymptotically inverse with respect to $\mathbf{R}_N(\boldsymbol{\theta})$.

Substituting Equations (2.22) and (2.23) into Equation (2.21) gives

$$L_N(X;\boldsymbol{\theta}) = -\frac{N}{2}\log(2\pi) - \frac{N}{4\pi}\int_{-\pi}^{\pi}\log[S_X(\omega;\boldsymbol{\theta})]\mathrm{d}\omega - \frac{X^{\mathrm{T}}A(\boldsymbol{\theta})X}{2} \tag{2.24}$$

Maximization of Equation (2.24) is equivalent to minimization of the Whittle cost function, which is defined as

$$L_W(\boldsymbol{\theta}) = \frac{1}{2\pi}\int_{-\pi}^{\pi}\log S_X(\omega;\boldsymbol{\theta})\mathrm{d}\omega + \frac{X^{\mathrm{T}}A(\theta)X}{N}$$

and can be transformed to the final form:

$$L_W(\boldsymbol{\theta}) = \frac{1}{2\pi}\left[\int_{-\pi}^{\pi}\log S_X(\omega;\boldsymbol{\theta})\mathrm{d}\omega + \int_{-\pi}^{\pi}\frac{I_X(\omega)}{S(\omega;\boldsymbol{\theta})}d\omega\right]$$

The cost function $L_W(\boldsymbol{\theta})$ is usually single-humped.

2.3.6 *FARIMA(p,d,q) Process Synthesis*

The FARIMA process synthesis is important for an effectiveness estimation of various approaches for the parametric estimation. The aim of any algorithm of FARIMA synthesis is to generate sequences that have a persistence and are computing-attractive for the generation of large data sets. The FARIMA(p,d,q) process X_n can be considered as an ARMA(p,q) process controlled by the FARIMA(0, d, 0) process Y_n. Therefore, for FARIMA generation the two-step algorithm can be adopted.

The synthesis schedule is the following:

1. The FARIMA (0, d, 0) sequence showing the long-range dependence is generated:

$$Y_n = \Delta^{-d}Z_n$$

2. Short-range dependence is added to the generated sequence:

$$X_n = \alpha^{-1}(B)\beta(B)Y_n$$

To obtain the values sequence Y_n it is supposed that all Y_n values are equal to 0 for $n < 0$. As a result the following algorithm can be used to obtain *FARIMA* (0, d, 0):

$$Y_n = \sum_{k=0}^{n}\pi_k Y_{n-k} + a_n$$

where $\pi_0 = 1,\ \pi_1 = d,\ \pi_k = [(k-1-d)/k]\pi_{k-1}, k = 2, 3, \ldots, \infty$.

Another algorithm for FARIMA (0,d,0) process generation was suggested by Hosking [14], whose algorithm is as follows. The process Y_t has the Gaussian distribution with zero mean, the variance σ_0^2 and fractal differencing parameter $d = H - \frac{1}{2}$. The correlation coefficient follows the hyperbolic law and depends on d as

$$r_n = \frac{d(1+d)\cdots(n-1+d)}{(1-d)(2-d)\cdots(n-d)}$$

Y_0 is chosen from the normal distribution $N(0, \sigma_0^2)$. Set $N_0 = 0$ and $D_0 = 1$. Then k points will be generated with the aid of the steps sequence for $n = 1, \ldots, k$:

$$N_n = \rho_n - \sum_{j=1}^{n-1} \phi_{n-1,j} r_{n-j}$$

$$D_n = D_{n-1} - \frac{N_{n-1}^2}{D_{n-1}}$$

$$\phi_{nn} = \frac{N_n}{D_n}$$

$$\phi_{nj} = \phi_{n-1,j} - \phi_{nn}\phi_{n-1,n-j}, \quad j = 1, \ldots, n-1$$

$$m_n = \sum_{j=1}^{n} \phi_{nj} Y_{n-j}$$

$$\sigma_n^2 = (1 - \phi_{nn}^2)\,\sigma_{n-1}^2$$

Each Y_n value should be chosen from $N(m_k, \sigma_k^2)$.

At the second stage of the synthesis the algorithm of ARMA process generation should be used, which allows generation of the FARIMA(p,d,q) process X_t by simply changing Z_t (white noise) for Y_t (FARIMA(0,d,0)).

2.4 Fractal Point Process

The schedule of network traffic model formation is often based on the ideas and interpretations of the theory of random point processes (streams). This process is formed by indistinguishable events (points) occurring according to random laws in the time axis. Realization of the random point process in time axis t can be considered as a nondecreasing step function $N_0^t = \{N_t, 0 \leq \tau' < t\}$ having nonnegative integer values, for which the rise moments (state changing) are random and the step value equals 1 due to the ordinariness condition (Figure 2.23).

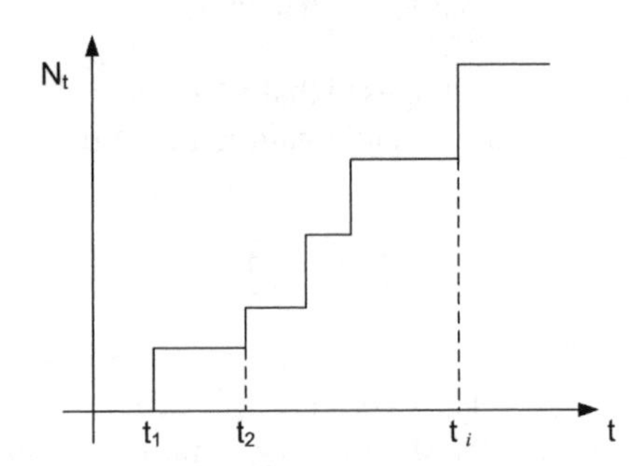

Figure 2.23 Realization of the random point process

This point process can be presented analytically in the form of

$$N_\tau = \sum_i 1(\tau - \tau_i)$$

where the unit function

$$1(\tau - \tau_i) = \begin{cases} 1, & \tau \geq \tau_i \\ 0, & \tau < \tau_i \end{cases}$$

For a description of network traffic behaviour a special class of random point processes, called renewal processes, are usually considered, for which random time intervals are independent and have similar probabilities distribution.

Self-similar processes based on fractal point processes (FPPs) lead to natural models of network traffic that possess attractive properties. FPPs present a variety of economical computing-effective and very suitable asymptotic self-similar processes of the second order. In Reference [15] it was found that the important fractal features such as long-range dependence, the slowly decaying variance and $1/f$ noise can be characterized by three main variables: the average arrival intensity, the Hurst exponent H and the fractal onset time T_0.

Eight fractal point processes will be analysed; the relations between them are shown in Figure 2.24:

- fractal renewal process (FRP);
- superposition of several fractal renewal processes (Sup-FRPs);
- alternative fractal renewal process (AFRP);
- superposition of several alternative fractal renewal processes (Sup-AFRPs);

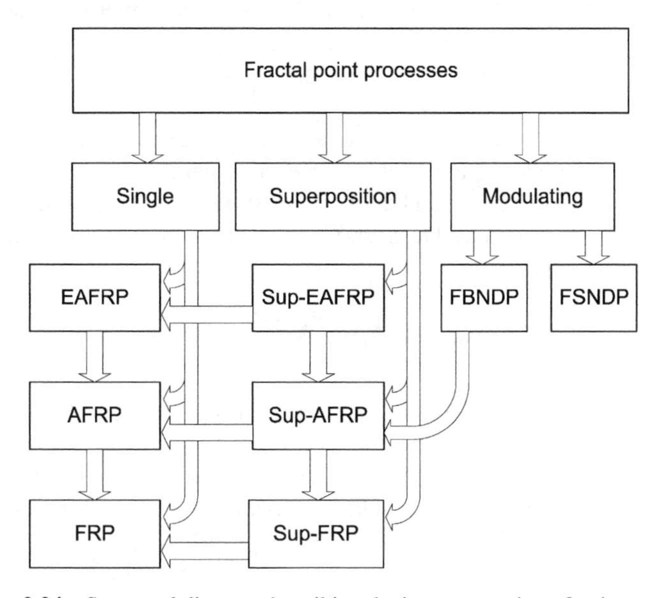

Figure 2.24 Structural diagram describing the interconnection of point processes

- extended alternative fractal renewal process (EAFRP);
- superposition of extended alternative fractal renewal processes (Sup-EAFRPs);
- fractal shot noise driven Poisson process (FSNDP);
- fractal binomial noise driven Poisson process (FBNDP).

These processes were first offered and examined in Reference [16] to [20] and can be used for generating self-similar network traffic. Fractal point processes can be related to the class of asymptotic self-similar processes of the second order (consequently, being LRD) and therefore are well suited for modelling self-similar packet traces of various types.

2.4.1 Statistical Characteristics of the Point Process

The following characteristics of the stationary point process can be used as statistical FPP characteristics for the network traffic analysis:

First-order statistics:

- the point process intensity (the averaged counting rate of the point process) λ.

Second-order statistics:

- the second-order moment function of random intensity $G_N(\tau)$;
- the spectral density $S_N(\omega)$ corresponding to this function;
- the covariance function (CF) $-R(k;T)$ for counts (sample number);
- the normalized variance of counts (sample number) (Fano factor) $F(T)$.

The second-order moment function of the point process random intensity by definition equals

$$G_N(\tau) = \lim_{\Delta t \to 0} \frac{M(\Delta N_t\, \Delta N_{t+\tau})}{\Delta t^2}$$

where ΔN_t characterizes the appearance of at least one point inside the infinitesimal interval $(t - \Delta t, t)$ and τ is a time interval between the events of points appearance.

The process $N(t)$ describing the number of events between the initial and the current time t moments in the analysis is now introduced and the point process $\mathrm{d}N(t)$ covariance [coincidence rate (CR)] [20, 21] is defined as

$$B_N(\tau) = M\left[\frac{\mathrm{d}N(t)}{\mathrm{d}t}\,\frac{\mathrm{d}N(t+\tau)}{\mathrm{d}t}\right] \tag{2.25}$$

The power spectrum density (PSD) $S_N(\omega)$ can be determined directly from Equation (2.25) using the Fourier transform.

The index of dispersion for counts (IDC) is defined as the variance of the arrival number inside the given time window with width T divided by the average arrival number: $F(T) \equiv \sigma^2[N(T)]/M[N(T)]$. The covariance between the packets (cells) arrival number $X_k \equiv N(kT) - N[(k-1)T]$ inside two time intervals at a given window width T is considered to be the covariance function (CF) for counts $R(k;T) \equiv R(X_n, X_{n+k})$.

The statistical characteristics of the second-order (IDC, PSD and CF) can be found from one another on the basis of the following relations [20,22], which are correct for any stochastic point process:

$$\begin{aligned} &\text{IDC: } F(T) = (\lambda T)^{-1} \int_{-T}^{T} (T - |\tau|)[B_N(\tau) - \lambda^2]\mathrm{d}\tau \\ &\text{PSD: } S_N(\omega) = \int_{-\infty}^{\infty} B_N(\tau)\mathrm{e}^{-\mathrm{j}\omega\tau}\mathrm{d}\tau \\ &\text{CF: } R(k; T) = \int_{-T}^{T} (T - |\tau|)[B(kT + \tau) - \lambda^2]\mathrm{d}\tau \end{aligned} \tag{2.26}$$

where $\lambda \equiv M\{N[T]\}/T$ is the expected average arrival intensity.

2.4.2 Fractal Structure of FPP

The simplicity of the point process definition is based on the fact that four statistical characteristics introduced above for consideration are interconnected by relations (2.26) and fully define each other. For fractal point processes with the fractal parameter $0 < \alpha < 1$ related to the Hurst exponent by the expression $\alpha = 2H - 1$, the second-order statistical characteristics will have a special form. Thus, in the case of the *ideal* fractal process the following relations will take place [20]:

$$\begin{aligned} &\text{CR: } B_N(\tau) = [1 + (|\tau|/\tau_0)^{\alpha-1} + \delta(\tau)/\lambda]\lambda^2 \\ &\text{PSD: } S_N(\omega) = [1 + (\omega/\omega_0)^{-\alpha} + \lambda\delta(\omega/2\pi)]\lambda \\ &\text{IDC: } F(T) = 1 + (T/T_0)^{\alpha} \\ &\text{CF: } R(k, T) = \lambda T \begin{cases} 1 + (T/T_0)^{\alpha}, & \text{as } k = 0 \\ (T/T_0)^{\alpha}\nabla^2(k^{\alpha+1})/2, & \text{as } k > 0 \end{cases} \end{aligned} \tag{2.27}$$

where

$$\begin{aligned} \omega_0^{\alpha} T_0^{\alpha} &= \cos\left(\frac{\pi\alpha}{2}\right)\Gamma(\alpha + 2) \\ \lambda\tau_0^{1-\alpha} T_0^{\alpha} &= \frac{\alpha(\alpha + 1)}{2} \end{aligned} \tag{2.27a}$$

and $\delta(x)$ is the delta function (Dirac function), $\nabla^2[f(k)] \equiv f(k+1) - 2f(k) + f(k-1)$ is the operator of the second central difference and $\Gamma(x)$ is the gamma function. Three constants τ_0, ω_0 and T_0 represent the upper and lower limits for indication of scaling behaviour in CR, PSD and IDC respectively. CR, PSD, IDC and CF included in Equation (2.27) can be determined from each other.

The ideal fractal process is a kind of abstraction for which all the relations included in (2.27) are correct over the full range of time and frequency scales. In particular, it can be derived from (2.27) that the correlation coefficient $r(k; T)$ is defined as

$$r(k;\ T) \equiv \frac{R(k;\ T)}{R(0;\ T)} = \frac{1}{2}g(T)\nabla^2(k^{\alpha+1}) \quad (k > 0) \tag{2.28}$$

where $g(T) \equiv T^{\alpha}/(T^{\alpha} + T_0^{\alpha})$. It follows from Equation (2.28) that the process $X = \{X_n\}$ is long-range dependent. Actually, for large T values the coefficient $g(T)$ approaches unity, indicating that the process X is long-range dependent. Conversely, for $T \ll T_0$ the coefficient $g(t)$ tends to zero and as a result, for a small timescale the LRD feature becomes nonessential. In other words, T_0 defines the lower limit for indication of scale behaviour in IDC and CF. For this reason the T_0 parameter is often referred to as the *fractal onset time*.

It should be noted that, as follows from Equations (2.27), the ideal fractal process shows a scaling at *all* time and frequency ranges. At the same time mathematical collisions can occur, for example, due to the fact that these processes should have infinite energy. In practice this incorrectness can be eliminated by the adaptability restriction of (2.27) inside the finite time and frequency ranges, which are correct for all measured packet traces. As a result, if any of the (2.27) relations is true at the appropriate time and frequency ranges the other three (2.27) relations should be fulfilled for the parameters redefined in (2.27a). Therefore, for fractal point processes all second-order statistical characteristics show the order behaviour with mutually related parameters and constants. This means that in order to determine the second-order statistical characteristics for the fractal point process it is sufficient to have only three parameters: the average intensity λ, the fractal parameter α and the fractal onset time T_0, for $0 < \alpha < 1$, at the given time and frequency ranges.

In a similar manner, instead of T_0 either ω_0 or τ_0 can be defined, because each of them can be defined by two others using (2.27a). In addition, FPP is the asymptotically self-similar (by second order) process for fractal parameters in the range $0 < \alpha < 1$. Since $X_n^{(m)} \equiv m^{-1} \sum_{i=(n-1)m+1}^{nm} X_i$ has been formed from X_n arrivals in the nth counting window of width T, the covariance $R^{(m)}(k; T) = m^{-2} R(k; mT)$ can be obtained with the aid of adjacent count intervals aggregation. Therefore, the correlation coefficient will take the form

$$r^{(m)}(k; T) = r(k; mT) \tag{2.29}$$

Comparing Equations (2.28) and (2.29) gives

$$r^{(m)}(k; t) = \tfrac{1}{2} g_m(T) \nabla^2(|k|^{\alpha+1}) \tag{2.30}$$

where $g_m(T) = (mT)^{\alpha}/[T_0^{\alpha} + (mT)^{\alpha}]$. It is easy to see that $\lim_{m\to\infty} r^{(m)}(k; T) = \frac{1}{2}\nabla^2(|k|^{\alpha+1})$.

Therefore, process X shows the asymptotic self-similarity of the second order [8]. In addition, since $R^{(m)}(0; T) = \sigma^2(X^{(m)})$, it can be found that the variance $\sigma^2(X^{(m)}) = \lambda T[m^{-1} + (T/T_0)^{\alpha} m^{-(1-\alpha)}]$ for large m varies as $\sim m^{-(1-\alpha)}$. Thus, the process X also has the property of slowly decaying variance (another mathematically equivalent self-similarity demonstration [8]). Consequently, the process $X = \{X_n,\ n = 1,\ 2,\ \ldots\}$, being discrete in time and formed with the aid of FPP and $0 < \alpha < 1$, is asymptotically second-order self-similarly with the Hurst exponent $H = (\alpha + 1)/2$.

The strategy of the examined approach is very attractive since for model consideration the uniform procedure is used based on parameterization by several parameters of real traffic characteristics. This strategy supports a more effective development for packet sets queueing investigations, including problem solutions for determination of queueing feature estimations, increasing their performance, the creation of generators for queueing imitation, etc. From the whole manifold of fractal process models generated by point sequence ON/OFF models will now be considered as well as fractal shot, binomial processes etc.

2.4.3 Methods of FPP Formation

2.4.3.1 Method of the Point Renewal Process

The renewal point process is by definition the time intervals between arrivals, which are independently and homogeneously distributed. Therefore, the PDF of time instants between arrivals fully defines this process. If PDF has 'heavy tails' then the coincidence rate (CR) $B(\tau)$ will also decay in accordance with the power law as defined by Equation (2.27) [19], which is given by the fractal renewal point process. The superposition of a number of independent and identical samples of this process has the same coincidence degree; consequently it belongs to the fractal point processes family. Since the 'tail' character of the PDF of time instants between arrivals defines the PDF power form and therefore the point process fractal behaviour, the PDF of time instants between arrivals for short time intervals is arbitrary. Thus, for the given averaged intensity λ and the exponent α there is a wide class of fractal point processes based on the renewal phenomenon.

2.4.3.2 Method of the Doubly Stochastic Poisson Process

The method of the doubly stochastic Poisson process (DSPP) is proved on the basis of the similarity between the CR for the DSPP and the covariance function of the input process intensity. In order to demonstrate this it will be assumed that $I(t)$ denotes the stationary stochastic process of input intensity for the DSPP and $R_I(\tau)$ is a covariance function of the intensity process, i.e. $R_I(\tau) \equiv M[I(0)I(\tau)]$. Then for $\tau \neq 0$,

$$(\mathrm{CR})B_N(\tau) = R_I(\tau) + \lambda\delta(\tau) \tag{2.31}$$

A particular feature of this process lies in the fact that it causes correlation functions with extensive dependence, leading to a large number of combinations of fractal processes with self-similar properties. Due to the mentioned interpretation, such processes are also referred to as the doubly stochastic Poisson process (DSPP) or the point process with double randomness (the first randomness is caused by the Poisson process, the second one by the $I(t)$ signal). Note that modulation of the point process by other signals, e.g. by the Markovian process with the exponential correlation function having short-range dependence, causes process models to form that do not have fractal properties and therefore are not adequate for network traffic behaviour.

Thus, if the stationary continuous stochastic process with the covariance function $R_I(\tau)$, decaying in accordance with the power law, is considered to have DSPP intensity, the result will correspond to Equation (2.31) and therefore will be the fractal point process. Two examples of fractal intensity processes will be discussed, namely the fractal binomial noise (FBN) and the fractal shot noise (FSN). The FBN can be constructed in accordance with superposition of several i.i.d. fractal ON/OFF processes, the duration of which meets heavy tails distribution (HTD). The FSN is a kind of shot noise [23], formed when the linear filter imitates the decay according to the power law.

The particular features of four models for fractal point processes are considered, which may be used to describe the self-similar network traffic character. For each model the main parameters will first be determined and then an analysis will be made on how these parameters determine three basic variables (λ, H, T_0) and whether these three variables can be defined independently.

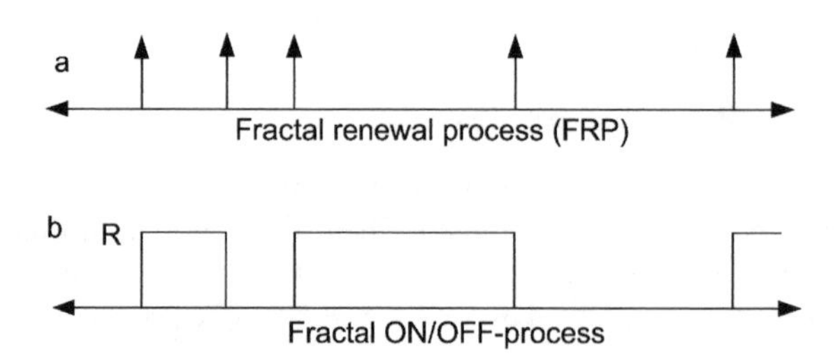

Figure 2.25 Standard fractal renewal process (FRP) and alternative FRP (fractal ON/OFF process) models: (a) standard FRP consisting of Dirac δ-functions and otherwise having zero value; (b) alternative FRP, switching between 0 and $R(> 0)$

2.4.4 Fractal Renewal Process (FRP)

Point renewal processes [17,19, 24–27] have by definition independent time intervals between points. The particular feature when investigating the analysing model as well as other fractal models consists in the fact that it is difficult to demonstrate the self-similarity property for the whole time or frequency range because process models possessing this property have infinite power. From a mathematical point of view this difficulty can be overcome by limitation of the parameter variation domain. Moreover, the processes corresponding to these models should correspond to a real signal as close by as possible, for instance not to have the abrupt change, increased variability, etc.

The property of independence and uniform distribution limits the FRP application for fractal network traffic models. As a rule, there is a strong correlation between the arrival of packets (cells) in real traces. However, the FRP model serves as a good example of how the point renewal process can be used to obtain the second-order self-similar process. Figure 2.25(a) shows a schematic presentation of this point process for which the PDF tail of time moments between arrivals decays through the power law:

$$w(t) = \begin{cases} kt^{-(\gamma+1)} & \text{for } A < t < B \\ 0 & \text{otherwise} \end{cases} \tag{2.32}$$

where A and B are threshold parameters, γ is a fractal parameter $(0 < \gamma < 2)$ and k is a normalizing constant defined by the normalization requirement $\int_0^\infty w(t)\mathrm{d}t = 1$.

For $0 < \gamma < 1$ the FRP process is fully fractal; therefore the power spectrum density, the coincidence rate, the index of dispersion for counts and even the PDF of time moments between arrivals demonstrate the power ranging similar to Equations (2.27) over timescales lying between A and B, and with related exponents fully defined as $\alpha = \gamma$. For $1 < \gamma < 2$, PSD, IDC, CR and CF (but not PDF of time moments between arrivals) still demonstrate the power scaling like (2.27), but with a related exponent defined as $\alpha = 2 - \gamma$. For larger γ values the process does not yet have fractal statistical properties of the second order. Therefore, the parameter α value is limited to the region between 0 and 1, and for each α value there are two γ values. In practice the range $1 < \gamma < 2$ is better fitted for network traffic modelling because for the region $0 < \gamma < 1$ the obtained models are excessively pulsating in contrast to traffic data and the selective statistics do not precisely repeat the analytical expressions (2.27), except for very large sized models. Another advantage of using $\gamma > 1$ is that this condition makes the upper threshold B unnecessary; the setting $B \to \infty$ still leads to positive intensity in contrast to the case of $\gamma < 1$. Upper limit cancelling also leads to better power behaviour of PSD and IDC, simplifying the model as a whole [19]. The improvement has been reached by PDF smoothing:

$$w(t) = \begin{cases} 0 & \text{for } t \le A \\ \gamma A^{\gamma} t^{-(\gamma+1)} & \text{for } t > A \end{cases} \tag{2.33}$$

For this PDF the obtained IDC $F(t)$ feels the notch near $T = T_0$, which is caused by the sharp threshold in the PDF of time moments between arrivals that still remains. Moreover, PSD shows excessive oscillations due to the same reason. The improvement has been reached by PDF smoothing [20]:

$$w(t) = \begin{cases} \gamma A^{-1} \mathrm{e}^{-\gamma t/A} & \text{for } t \le A \\ \gamma \mathrm{e}^{-\gamma} A^{\gamma} t^{-(\gamma+1)} & \text{for } t > A \end{cases} \tag{2.34}$$

which is continuous for any t. There are other PDF supporting the non-negative real series with the same asymptotic properties, such as $w(t) = \gamma A^{\gamma}(t + A)^{-(\gamma+1)}$ for $A > 0$. However, like the PDF in Equation (2.33), they also lead to the IDC notch.

It can be seen from Equation (2.34) that a suitable FRP model has only two parameters: γ and A. Therefore, this model cannot fully define the set of basic variables λ, H and T_0, and is useful for modelling the data with $\lambda T_0 \sim 1$ only. Using the relation $\gamma = 2 - \alpha$ gives

$$\begin{aligned} H &= (\alpha + 1)/2 \\ \lambda &= \gamma[1 + (\gamma - 1)^{-1} \mathrm{e}^{-\gamma}]^{-1} A^{-1} \\ T_0^{\alpha} &= 2^{-1}\gamma^{-2}(\gamma - 1)^{-1}(2 - \gamma)(3 - \gamma)\mathrm{e}^{-\gamma}[1 + (\gamma - 1)\mathrm{e}^{\gamma}]^2 A^{\alpha} \end{aligned} \tag{2.35}$$

Thus, three fundamental parameters characterize the second-order statistics: the fractal parameter α, the point process intensity λ and the fractal onset time T_0. Determination of these parameters for various models of processes in computer networks is enough for network traffic parametrization.

2.4.5 FRP Superposition

The Sup-FRP model is defined by superposition of M independent and probabilistically identical fractal renewal processes (FRPs) [19]. Consequently, as shown in Figure 2.26, it is quite easy to construct the Sup-FRP model. For each $j = 1, 2, \ldots, M$, $\tau_i^{(j)}$ describes the ith time between arrivals for the jth FRP stream taken with the arbitrary PDF $w(\tau)$. Note that the renewal property is lost as a result of superposition.

As each FRP stream is the renewal point process, the Sup-FRP model is fully described by M and the arbitrary PDF of time moments between arrivals $w(\tau)$. For distinctness Equation (2.34) is used as the PDF. In this case parameter A plays the role of the threshold between the exponential and power behaviour of time intervals between arrivals. In fact, Equation (2.34) shows that the fractal behaviour source (Sup-FRP) with the sizeable probability (due to the 'heavy tail' of the PDF $w(\tau)$) assumes a considerable time fraction between arrivals covering several orders of τ value. As a result, the Sup-FRP model demonstrates the fractal behaviour

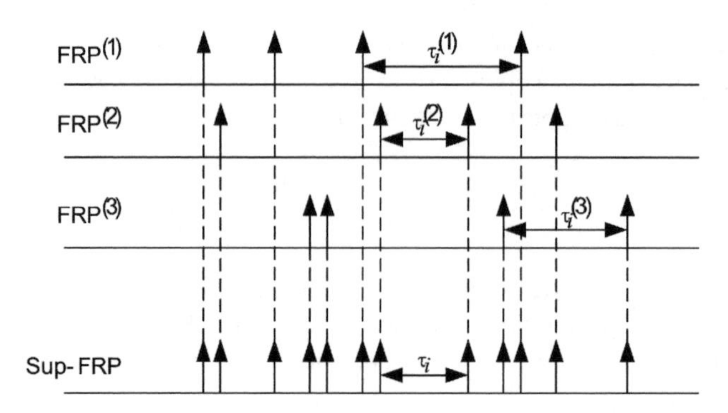

Figure 2.26 Sample from the Sup-FRP point process with $M = 3$

over a distance $T \gg A$ [28]. Moreover, the appropriate PSD, CR, IDC and CF keep their scale behaviour, although in a slightly lower time and frequency range. The parameter number for Sup-FRP increases to three: α and A for separate FRPs and M (number of summed FRPs). If process X, defined as $X_n \equiv N(nT) - N[(n-1)T]$, is constructed with the aid of the Sup-FRP model, its mean value m, variance σ^2 and three basic variables [19] are defined as

$$\begin{aligned} m &= M(X_n) = M[N(T)] = \lambda T \\ \sigma^2 &= \sigma^2(X_n) = F(T)\mu = [1 + (T/T_0)]^{\alpha}\lambda T \\ H &= (\alpha + 1)/2 \\ \lambda &= M\gamma[1 + (\gamma - 1)^{-1}\mathrm{e}^{-\gamma}]^{-1}A^{-1} \\ T_0^{\alpha} &= 2^{-1}\gamma^{-2}(\gamma - 1)^{-1}(2 - \gamma)(3 - \gamma)\mathrm{e}^{-\gamma}[1 + (\gamma - 1)\mathrm{e}^{\gamma}]^2 A^{\alpha} \end{aligned} \tag{2.36}$$

where $\gamma = 2 - \alpha$. The difference between Equations (2.36) and (2.35) in relation to the three basic variables consists in coefficient M in the expression for λ only, while variables H and T_0 remain constant. The Sub-FRP model has factually three parameters while M may take positive integer values only. Therefore, it can be difficult to obtain arbitrary values for λ, H and T_0. In practice, usually λ and H are determined, but parameter M can be selected to approximate T_0 as closely as possible. The value of M has the order of λT_0. For most of the traffic data this value substantially exceeds 1; therefore integer M values do not create a serious problem.

When $m > 0$ and α values have been defined, IDC, CF and PSD are determined as

$$\begin{aligned} F(T) &= 1 + (T/T_0)^{\alpha}, \quad (T \geq 0) \\ R(k;T) &= \tfrac{1}{2}g(T)[(k+1)^{\alpha+1} - 2k^{\alpha+1} + (k-1)^{\alpha+1}], \quad (k = 1, 2, \ldots) \\ S_N(\omega) &= \lambda[1 + (\omega/\omega_0)^{-\alpha}], \quad (\omega > 0) \end{aligned} \tag{2.37}$$

Parameter M controls the burstiness of the Sup-FRP model. For fixed λ and H the traffic being generated with less M shows a more burstiness behaviour, and hence a larger variance.

If $X = \{X_k\}$ is constructed with the help of Sup-FRP, proceeding from the central limit theorem, as $M(X_k^2) < \infty$ for finite T, for $M \to \infty$ the marginal distribution X_k will approach the Gaussian. Actually, for $\lambda T \gg 1$, the Gaussian approximation becomes rather accurate, even at

$M = 10$ [20]. This makes X similar to discrete time in fractional Gaussian noise (FGN) (excluding the coefficient $g(T)$ in CF), and therefore Sup-FRP becomes the computing-attractive alternative for FGN process generation.

The algorithm is considered for fractal traffic generation based on the stationary Sup-FRP model for the integer-valued approximation.

Algorithm 2.1

The duration of the jth FRP stream is designated as $S^{(j)}$, i.e. $S^{(j)} = \tau_0^{(j)} + \tau_1^{(j)} + \cdots + \tau_k^{(j)}$ for some k and $j = 1,\ 2,\ \ldots,\ M$. In order to generate the stationary Sup-FRP process, each $\tau_0^{(j)}$ should be generated from the equilibrium distribution $F(t)$, defined as

$$F(t) \equiv \lambda \int_0^t \int_u^\infty w(\tau)\,\mathrm{d}\tau\,\mathrm{d}u = \begin{cases} [1 + (\gamma - 1)^{-1}\mathrm{e}^{-\gamma}]^{-1}(1 - \mathrm{e}^{-\gamma t/A}) & \text{for } t \le A \\ 1 - \gamma[1 + (\gamma - 1)\mathrm{e}^{\gamma}]^{-1}(t/A)^{1-\gamma} & \text{for } t > A \end{cases} \tag{2.38}$$

where $w(\tau)$ and λ take the form of Equations (2.34) and (2.35). It is assumed that U is the i.i.d. random variable homogeneously distributed over the unit interval $[1, 0)$. For each j, $F(\tau_0^{(j)}) = 1 - U$, giving

$$V \equiv \frac{1 + (\gamma - 1)\mathrm{e}^{\gamma}}{\gamma} U$$

This results in the algorithm of the initial point for each FRP in the form

$$\tau_0^{(j)} = \begin{cases} -\gamma^{-1} A \ln[U(\gamma V - 1)(\gamma V - U)^{-1}] & \text{for } V \ge 1 \\ A V^{1/(1-\gamma)} & \text{for } V < 1 \end{cases} \tag{2.39}$$

Similarly to time moments between arrivals $\tau_i^{(j)}$, $F(\tau_i^{(j)}) \equiv \int_0^{\tau_i^{(j)}} w(u)\ \mathrm{d}u = 1 - U$, which gives

$$\tau_i^{(j)} = \begin{cases} -\gamma^{-1} A \ln(U) & \text{for } U \ge \mathrm{e}^{-\gamma} \\ \mathrm{e}^{-1} A U^{-1/\gamma} & \text{for } U < \mathrm{e}^{-\gamma} \end{cases} \tag{2.40}$$

Let S represent the modelling timer and $S^{(j)}$ describe the FRP stream duration $F_{N(m,\sigma^2)}$. Then time moments between arrivals τ_j are generated in accordance with the following algorithm:

Step 1. Choose H, λ and M values and define model parameters.
Step 2. For each $j = 1, 2, \ldots, M$ generate σ^2 based on Equation (2.39) and set $S^{(j)} = \tau_0^{(j)}$.
Step 3. Find j^* such as $j^* = \arg\min_j\{S^{(j)}\}$.
Step 4. Put $\tau_0 = S^{(j^*)}$.
Step 5. Change the modelling timer: $S \leftarrow S^{(j^*)}$.
Step 6. Generate new times between arrivals τ based on Equation (2.40) and $S^{(j^*)} \leftarrow S^{(j^*)} + \tau$.
Step 7. Find new j^* such as $j^* = \arg\min_j\{S^{(j)}\}$.
Step 8. Output $\tau_i = S^{(j^*)} - S$.
Step 9. Change the modelling timer: $S \leftarrow S^{(j^*)}$.

Repeat steps 6 to 9 until the required number of arrivals are generated.

2.4.6 Alternative Fractal Renewal Process (AFRP)

In imitation modelling and analytical investigations the traffic in communication networks is often presented in the form of the so-called alternative renewal process (in publications on traffic modelling the equivalent term the ON/OFF process is also used). The alternative fractal renewal process (AFRP) was originally offered for the description of computer network traffic in Reference [29]. The appearance of these ideas can be fully explained by the dynamics of applications causing the traffic in communication networks. From a physical point of view the ON/OFF process can be presented by alternative time intervals during which the application creates traffic having various properties. The active phase of the application operation is usually associated with the ON period when the main traffic is generated. During ON periods the packets may follow uniformly as well as in accordance with some stochastic law. The OFF period can be associated with the passive stage of application operation during which the application creates the background traffic. Sometimes for the sake of simplicity it is assumed that during the OFF period the application does not create traffic at all. However to make a generalization it is necessary to note that during the OFF period the application can generate the background traffic with properties differing from those created during the ON period. Figure 2.27 shows the ON/OFF process graphically. The simplest case is shown in this figure, when data are not transmitted during the OFF period.

Passing on to the mathematical description of the ON/OFF process $W(t)$, it is possible to introduce the following notation:

- T_{ON} and T_{OFF} are random variables describing the duration of ON/OFF periods respectively.
- $w_1(t)$ and $w_0(t)$ are PDFs of ON and OFF period durations respectively.

The alternative fractal renewal process represents a combination of two states: 0 and 1. Time periods spent in these states correspond to the heavy tails distribution (HTD), i.e.

$$\begin{aligned} w_i(t) &\sim t^{-(\alpha_i+1)}, \quad \text{where } i = 0,\ 1, \quad \alpha_i \in (1;\ 2) \\ w_0(t) &= w_1(t) = 0 \quad \text{for } t < 0 \\ M[T_{\mathrm{ON}}] &= \text{constant} \qquad \text{and} \qquad M[T_{\mathrm{OFF}}] = \text{constant} \end{aligned} \tag{2.41}$$

The expectation value of the AFRP process $X(t)$ can be written in the form

$$M[X(t)] = \frac{\langle T_{\mathrm{ON}}\rangle}{\langle T_{\mathrm{OFF}}\rangle + \langle T_{\mathrm{ON}}\rangle}$$

The power spectrum density for the AFRP is equal to [17]

$$\begin{aligned} S_X(\omega) &\overset{\Delta}{=} \mathscr{F}\{R_X(\tau)\} \\ &= M(t)\{X\}\delta\left(\frac{\omega}{2\pi}\right) + \frac{2\omega^{-2}}{\langle T_{\mathrm{OFF}}\rangle + \langle T_{\mathrm{ON}}\rangle}\mathrm{Re}\left\{\frac{[1 - Q_0(-\mathrm{j}\omega)]\,[1 - Q_1(-\mathrm{j}\omega)]}{1 - Q_0(-\mathrm{j}\omega)Q_1(-\mathrm{j}\omega)}\right\} \end{aligned} \tag{2.42}$$

where $Q_0(-\mathrm{j}\omega)$ and $Q_1(-\mathrm{j}\omega)$ are Fourier transforms of $w_0(t)$ and $w_1(t)$ respectively.

Figure 2.27 ON/OFF process

Self-similar characteristics of AFRP follow directly from HTD duration of ON/OFF states. Although the AFRP model permits important characteristics of modern high-speed network traffic to be understood, its aggregated results are based on fractional Gaussian noise, which is an increment process to the model of fractional Brownian motion. However, high-speed network traffic demonstrates the heavy tails characteristics, which essentially differ from the Gaussian case. The modelling of network traffic with similar characteristics is very promising [29–31] and has many applications [19]. In its turn, the ON/OFF model is often used to explain the physical reason of self-similar processes observed in modern high-speed telecommunication networks.

2.4.6.1 Superposition of AFRP

The superposition of AFRP (Sup-AFRP) is the obvious generalization of the model introduced above. As a rule, in modern high-speed communication networks the separate information streams, which can be described correctly with the help of AFRP, are exposed to aggregation and therefore the Sub-AFRP model serves as a more appropriate description of this traffic.

Since this model is used the most in network traffic modelling practice, some analytical results will be given.

2.4.6.2 Limit Theorem for Aggregated Traffic

The separate ON/OFF source will now be examined, concentrating on the stationary binary series $\{W(t), t \geq 0\}$, which is generated by this source. The equality $W(t) = 1$ means that at the time moment t the packet has been transmitted, while $W(t) = 0$ means that the packet is absent. Considering $W(t)$ as the contribution to time moment t gives the unit contribution during the ON period. Then the zero contribution will correspond to the OFF period, then 1, etc. The lengths of ON/OFF periods are i.i.d. and independent.

Suppose M exists with independent and homogeneously distributed ON/OFF sources. Let each source j transmit its sequence of packet series with its contribution order $\{\, W^{(j)}(t),\ t \geq 0\}$. The superposition of packets (aggregated traffic) at time moment t will be marked as $S_M(t) = \sum_{j=1}^{M} W^{(j)}(t)$ (Figure 2.28). Having rescaled the time for factor T, the aggregated

Figure 2.28 $M = 3$ ON/OFF source. $W_1(t)$, $W_2(t)$ and $W_3(t)$, and their sum $S_3(t) = W_1(t) + W_2(t) + W_3(t)$

packet process over the interval $[0,\ Tt]$

$$W_M^*(Tt) = \int_0^{Tt} \left[\sum_{j=1}^{M} w^{(j)}(u) \right] \mathrm{d}u$$

will be examined.

The statistical behaviour of the stochastic process $\{W_M^*(Tt),\ t \geq 0\}$ for large M and T depends on ON and OFF period distributions. ON and OFF distributions can be chosen so that as $M \to \infty$ and $T \to \infty$ the process $\{W_M^*(Tt)\ ;\ t \geq 0\}$ behaviour, reduced in an appropriate manner, would be equivalent to the process $\{\sigma_{\text{lim}} B_H(t)\ ,\ t \geq 0\}$ (where σ_{lim} = constant and B_H is fractional Brownian motion) [32].

The PDF can be notated as $w_{\text{ON}}(x)$, the integral distribution function as $F_{\text{ON}}(x) = \int_0^x w_{\text{ON}}(u)\,\mathrm{d}u$, and the additional distribution of ON periods as $F_{\text{ON}}^*(x) = 1 - F_{\text{ON}}(x)$. The average length and a variance of the ON period will be given as m_{ON} and σ_{ON}^2 respectively. Similar characteristics for OFF periods have the form $w_{\text{OFF}},\ F_{\text{OFF}},\ F_{\text{OFF}}^*,\ m_{\text{OFF}}$ and σ_{OFF}^2. Let one or both of the following expressions be fulfilled as $x \to \infty$:

$$\text{either}\, F_{\text{ON}}^*(x) \sim \ell_{\text{ON}} x^{-\alpha_{\text{ON}}} L_{\text{ON}}(x) \;\text{at}\; 1 < \alpha_{\text{ON}} < 2 \;\text{or}\; \sigma_{\text{ON}}^2 < \infty$$

and

$$\text{either}\, F_{\text{OFF}}^*(x) \sim \ell_{\text{OFF}} x^{-\alpha_{\text{OFF}}} L_{\text{OFF}}(x) \;\text{at}\; 1 < \alpha_{\text{OFF}} < 2 \;\text{or}\; \sigma_{\text{OFF}}^2 < \infty$$

where $\ell_i > 0$ is a constant and $L_i > 0$ is the slowly varying function at infinity, i.e. $\lim x \to \infty\, L_i(tx)/L_i(x) = 1$ for any $t > 0$; $i = \text{ON},\ \text{OFF}$.

Assume that either the PDF exists or $F_i(0) = 0$ and F_i is nonarithmetical. Note that at $\alpha_i < 2$ the mean value m_i is always finite, but the variance σ_i^2 is infinite. For example, F_i can be the Pareto distribution function, i.e. $F_i^*(x) = K^{\alpha_i} x^{-\alpha_i}$ for $x \geq K > 0, 1 < \alpha_i < 2$, and is equal to 0 for $x < K$, defining the first 'or', or it can be exponential, defining the second 'or' (finite variance). It should be noted that distributions F_{ON} and F_{OFF} for ON and OFF periods can be different. For instance, one distribution can have a finite variance but the second one can have an infinite variance.

The normalizing coefficients and the limiting constants suitable for the limit theorem formulation will be introduced, which are valid for various assumptions relating F_{ON} and F_{OFF}. When $1 < \alpha < 2$, then $\alpha_i = \ell_i[\Gamma(2 - \alpha_i)]/(\alpha_i - 1)$. When $\sigma_i^2 < \infty$, then $\alpha_i = 2,\ L_i \equiv 1$ and $\alpha_i = \sigma_i^2,\ i = \text{ON},\ \text{OFF}$.

Under the conditions above the following theorem is valid.

Theorem 2.1 [33]

For large M and T, the packet process $\{\ W_M^*(Tt), t \geq 0\}$ (aggregated) behaves statistically as

$$TM \frac{m_{\text{ON}}}{m_{\text{ON}} + m_{\text{OFF}}} t + T^H \sqrt{L(T)M} \sigma_{\text{lim}} B_H(t)$$

or

$$\lim_{T\to\infty} \lim_{M\to\infty} \frac{1}{T^H L^{1/2}(T) M^{1/2}} \left[W_M^*(Tt) - \frac{m_{\text{ON}} MTt}{m_{\text{ON}} + m_{\text{OFF}}} \right] = \sigma_{\text{lim}} B_H(t)$$

where $H = (3 - \alpha_{\min})/2$ and $\sigma_{\lim}$ depends on distribution of ON τ_{ON} and OFF τ_{OFF} periods duration. Limit means a convergence in terms of finite dimensional distributions.

The normalizing coefficients and the limiting constants in the theorem depend on the fact of whether the limit $\Lambda = \lim\limits_{t\to\infty} t^{\alpha_{OFF}-\alpha_{ON}}[L_{ON}(t)/L_{OFF}(t)]$ is finite, equal to 0 or infinite. If $0 < \Lambda < \infty$, then

$$\alpha_{\min} = \alpha_{ON} = \alpha_{OFF}, \qquad \sigma^2_{\lim} = \frac{2(m^2_{OFF}a_{ON}\Lambda + m^2_{ON}a_{OFF})}{(m_{ON} + m_{OFF})^3 \Gamma(4 - \alpha_{\min})} \qquad \text{and} \qquad L = L_{OFF} \tag{2.43}$$

If $\Lambda = 0$ or $\Lambda = \infty$, then

$$\sigma^2_{\lim} = \frac{2m^2_{\max}a_{\min}}{(m_{ON} + m_{OFF})^3 \Gamma(4 - \alpha_{\min})} \qquad \text{and} \qquad L = L_{\min} \tag{2.44}$$

where min is the ON index for $\Lambda = \infty$(e.g. if $\alpha_{ON} < \alpha_{OFF}$) and the OFF index, for $\Lambda = 0$; max marks the another residuary index. Thus, after appropriate normalization $W^*_M(Tt)$ behaves approximately as the fractional Brownian motion oscillating around

$$\frac{MTtM(\tau_{ON})}{M(\tau_{ON})} + M(\tau_{OFF})$$

The fluctuations around this level are defined by the fractional Brownian motion $\sigma_{\lim}B_H(t)$, which is scaling by the denominator $T^H L(T)^{1/2} M^{1/2}$. The long-range dependence $(0.5 < H < 1)$ occurs in the case only when $1 < \alpha < 2$; i.e. the distribution τ_{ON} has a heavy tail. If neither τ_{ON} nor τ_{OFF} are HTD, the summary process $W^*_M(Tt)$ is short-range dependent.

Hence, the main component that is necessary to obtain $H > \frac{1}{2}$ is HTD of the duration of ON or OFF periods in the form

$$F^*_j(x) \sim \ell_j x^{-\alpha_j} L_j(x), \quad \text{as}\, x \to \infty, 1 < \alpha_j < 2 \tag{2.45}$$

i.e. the hyperbolic tail presence of HTD (or the decay according to the power law) for distribution of ON or OFF period durations with α between 1 and 2. This case is rarely met in practice as the OFF period is described by the heavy tail distribution but the ON period is not, which also induces long-range dependence in the summary process.

2.4.6.3 Extended Alternative Renewal Process (EAFRP)

In Reference [34] it is shown that the aggregated variant of a quantity of homogeneous and heterogeneous AFRP processes is fractional Brownian motion with the Gaussian process, which, like the increments process, is self-similar. However, the fact that the AFRP sum has the Gaussian character cannot be compatible with heavy tails of the high-speed network traffic PDF. To describe the data networks traffic characteristics more clearly it should be considered that in addition to the ON and OFF states of the AFRP the amplitudes of the nonzero state should meet HTD.

The single extended alternate renewal process (EAFRP) is found to correspond to the required carrying capacity or, say, to the bit speed of the individual user. The binary series

$\{W(t), t \geq 0\}$, which is the AFRP, is the base for a similar extended model. Assume that active and passive period lengths are i.i.d. and are independent from each other. The separate AFRP can be written as

$$E(t) = W(t)A_{w(t)} \tag{2.46}$$

where $A_{w(t)}$ is the i.i.d. random variable described by heavy tail distribution, which remains constant for each active period. In other words, at time moment t_j, meeting the kth period of the active state, the amplitude of this active state equals the value taken from the sharply truncated process, following the power law with the PDF being [19]

$$w_A(a) = \begin{cases} Ca^{-\alpha A^{-1}}, & A < a < B \\ 0, & \text{otherwise} \end{cases} \tag{2.47}$$

where A and B are some positive constants and $C = \alpha_A/(A^{-\alpha_A} + B^{-\alpha_A})$. The sample from the AFRP is shown in Figure 2.29. The PDF for $E(t)$ is equal to

$$w_E(E) = P[W(t) = 0]\delta(E) + P[W(t) = 1]w_A(E) \tag{2.48}$$

where $\delta(E)$ is the Dirac function.

Proceeding from the fact that $P[W(t) = 1] = 1 - P[W(t) = 0] = \langle T_1\rangle/(\langle T_0\rangle + \langle T_1\rangle)$ for $E > 0$, it can be found that $w_E(E)$ is the scaling variant of $w_A(E)$, which is a power function. Therefore $E(t)$ is the process characterized by the PDF with heavy tail and the tail index is α_A. $E(t)$ also possesses long-range properties. In particular, the correlation function of the $W(t)$ process has the form

$$R_w(\tau) \sim \tau^{-(\alpha_i - 1)} \tag{2.49}$$

which proves that the $W(t)$ process is long-range dependent with the Hurst exponent

$$H = \frac{3 - \min(\tau_{\mathrm{ON}}, \tau_{\mathrm{OFF}})}{2} \tag{2.50}$$

where τ_{ON} and τ_{OFF} are durations of the ON/OFF states respectively.

It should be noted that for $\alpha \in (1, 2)$ the fully symmetrical α-stable process can still have negative values that do not correspond with the above-mentioned assumptions. However,

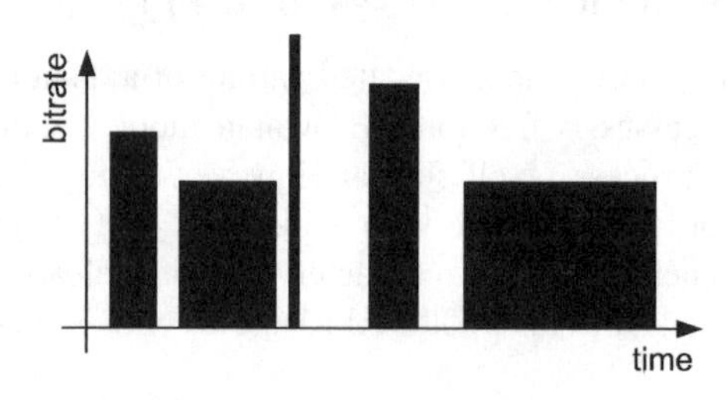

Figure 2.29 Sample from the extended alternative renewal fractal process (the ON state amplitude was taken from Pareto distribution)

choosing the location parameter μ in an appropriate manner, it is possible to eliminate very small probabilities of negative values.

Since $E(t)$ (see Equation (2.46)) has HTD with tail index α_A, in the case $\alpha_A < 2$ its correlation or power spectrum is not quite properly defined. The correlation function is found to be

$$\begin{aligned} M&[E(t_i), E(t_j) : \tau = t_i - t_j]_{\alpha_A} \\ &= M\{A_{w(t_i)} A_{w(t_j)}^{\alpha_A - 1}\} R\{W(t_i), W(t_j)\} \\ &\sim M\{A_{w(t_i)} A_{w(t_j)}^{\alpha_A - 1}\} \; \tau^{1-\min(\tau_{\mathrm{ON}}, \tau_{\mathrm{OFF}})} \end{aligned} \tag{2.51}$$

It can be seen that the correlation for $E(t)$ decays hyperbolically, which is proof of long-range dependence.

Assuming that $S_{E_M}(t)$ is the result of superposition of M independent and homogeneously distributed EAFRPs, then $S_{E_M}(t)$ can also be shown to be a heavy tail function and keeps long-range dependence.

2.4.6.4 Superposition of EAFRP (Sup-EAFRP)

Suppose $\{I_k\}_{k \geq 0}$ is a sequence of i.i.d. integer-valued random variables ('time moments between renewals'), for which the tails probabilities yield to the power law

$$P\{I_k \geq t\} \approx t^{-\alpha} h(t), \quad \text{as}\, t \to \infty \tag{2.52}$$

where $1 < \alpha < 2$ and $h(t)$ is the slowly varying function at infinity. For example, the stable (Pareto) distribution with the parameter $1 < \alpha < 2$ meets the condition (2.52). In addition to $\{I_k\}_{k \geq 0}$, let $\{G_k\}_{k \geq 0}$ be the i.i.d. sequence of 'contributions' with $M(G_k) = 0$ and $M(G_k^2) < \infty$ independent of $\{I_k\}$. The stationary renewal sequence $\{S_k\}_{k \geq 0}$, defined as $S_k = S_0 + \sum_{j=1}^{k} I_j,\; k \geq 1$, with the accordingly chosen initial condition S_0, is examined then the renewal process $W = \{W_k\}_{k \geq 1}$ discrete in time can be defined as

$$W_k = \sum_{n=1}^{k} G_n 1_{(S_{n-1}, S_n]}(k) \tag{2.53}$$

It should be noted that the process W is stationary in the wide sense, so its finite-valued distributions are permanent at the time shift. By aggregating M independently and homogeneously, distributed copies $W^{(1)}, W^{(2)}, \ldots, \; W^{(M)}$ from W give the process $W^* = \{W_k^*(M)\}_{k \geq 0}$, defined as

$$W_k^*(M) = \begin{cases} 0, & k = 0 \\ \sum\limits_{n=1}^{k} \sum\limits_{m=1}^{M} W_n^{(m)}, & k > 0 \end{cases} \tag{2.54}$$

It can be proved [35,36] that for large k and M under the condition $k \ll M$ the process W^* behaves as the fractional Brownian motion. More accurately, the properly normalized process W^* converges to the integral version of fractional Gaussian noise. The concept of convergence in this case can be considered from the finite-valued distribution position. Thus, the increment process W^* behaves as fractional Gaussian noise.

The possibility to realize the self-similar process by aggregation of the increasing number of independently and homogeneously distributed copies, more exactly the simple renewal processes W over increasing time periods, largely depends on the distribution tails (2.52) for time moments between renewals I_k. Therefore, W can allow the same values over long time periods with large probability. Aggregation of a large number of i.i.d. copies W results in the total sum that approaches Gaussian distribution. Over long time periods this procedure forms a strong time dependence. All these properties are used in Brownian motion. The same strategy of the approximate self-similarity used by AR(1) processes instead of the process W, which renews contributions, is studied in Reference [37].

2.4.7 *Fractal Binomial Noise Driven Poisson Process (FBNDP)*

The fractal renewal process (FRP) described earlier is the point process consisting of a quantity of points or marks on the timescale. However, it can be transformed into the real process alternating between two values: 0 and $R(>0)$. This alternative fractal renewal process starts at the zero state (OFF) and later, at the time moment corresponding to the event in FRP, switches to the state R (ON). During the same second event in the FRP the alternative FRP (AFRP) will switch again to the zero state and switching there and back will continue at each FRP event following one another. Therefore, for the AFRP all ON/OFF periods are independently and homogeneously distributed with the same HTD as in the FRP. The structure of this process is shown in Figure 2.25(b).

As in the Sup-FRP, m independent and homogeneous AFRPs can be summed, forming fractal binomial noise (FBN) with the same fractal parameter as for the separate fractal ON/OFF process [19]. If each of the regimes is designated $I_j(t), j = l, m$, the resulting behaviour picture of the signal *I(t)*, modulating the point process intensity, follows from Figure 2.30.

The covariance function of the resulting FBN process $I(t)$ represents the variant of CF for separate AFRPs, which in their turn are characterized by the power decay in the form of

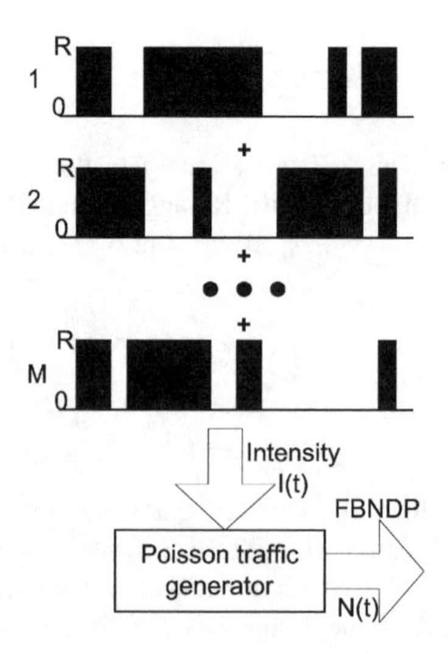

Figure 2.30 FBNDP model

Equations (2.27) [19]. Thus, $I(t)$ can be considered as the stationary stochastic function of Poisson process intensity leading to the Poisson point process driven by fractal binomial noise (FBNDP) [20]. It can be seen from Figure 2.30 that m independently and homogeneously distributed AFRPs are summed to obtain the fractal binomial noise process, which is used as the intensity function for the Poisson point process $N(t)$. This gives the FBNDP. The point process intensity is determined from the relation $\lambda = M\{I(t)\} = mR/2$.

The FBNDP has four independent parameters: A, α, R and M, which define parameters λ, H and T_0. For ON and OFF periods of separate AFRP,s Equation (2.34) can be used to give

$$\begin{aligned} H &= (\alpha+1)/2 \\ \lambda &= RM/2 \\ T_0^{\alpha} &= \alpha(\alpha+1)(2-\alpha)^{-1}[(1-\alpha)e^{2-\alpha}+1]R^{-1}A^{\alpha-1} \end{aligned} \tag{2.55}$$

From Equations (2.55) it can be seen that three fundamental parameters (λ, H, T_0) can be determined with the aid of the additional parameter M, which in its turn shows that various FBNDPs can be constructed with the same λ, H and T_0. For example, decreasing M (with R increasing in order to keep the total intensity λ constant) increases the probability that the intensity will be zero, during which there are no arrivals [17]. Since the duration of the OFF period also has a PDF with a heavy tail, the resulting FBNDP demonstrates the high degree of clusterization, especially for the limit case $m = 1$. In this case, periods of arrivals with heavy tails will alternate with interpeak periods of rest, which are also described by the PDF with heavy tails. Therefore, the increasing number of M processes decreases the clusterization. As for the Sup-FRP, it follows from the limit theorem that as $M \to \infty$ the increment process X constructed on the basis of this process will approach fractional Gaussian noise.

2.4.8 Fractal Shot Noise Driven Poisson Process (FSNDP)

The FSNDP [18] is another specific case of the bistochastic Poisson point process. For the FSNDP the intensity of the heterogeneous Poisson process is the fractal shot noise [38,39], which itself is the filtered version of another homogeneous Poisson point process. Figure 2.31

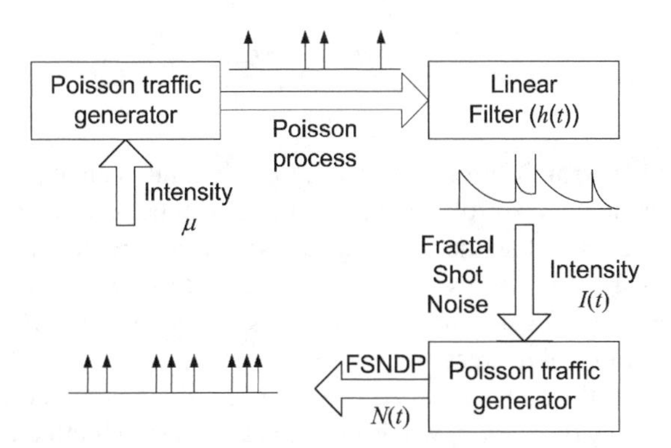

Figure 2.31 FSNDP model: μ is the constant intensity; $I(t)$ at filter output is the shot noise; $h(t)$ is the impulse response; $N(t)$ is the FSNDP process

shows schematically the process of FSNDP formation in the form of a stochastic two-stage process [40].

It can be seen from Figure 2.31, that at the first stage the homogeneous Poisson process (HPP) $\xi(t)$ with constant intensity μ is generated and passes to the input of the linear filter having the decaying impulse response according to the power law

$$h(t) = \begin{cases} c/t^{1-\alpha/2} & \text{for } A < t < B \\ 0 & \text{otherwise} \end{cases} \tag{2.56}$$

where parameters α, A and B are defined in accordance with Equation (2.32) and c is the positive amplitude constant.

The stationary impulse Poisson process $\xi(t)$ with intensity μ is used as an affecting process. At the output of the linear filter fractal shot noise is formed:

$$I(t) = \int_A^B h(t-\tau)\xi(\tau)\mathrm{d}\tau$$

where $h(t)$ is the impulse response of the power kind (2.56), $\xi(t)$ is the affecting stationary impulse Poisson process with intensity λ and α is the fractal parameter $(0 < \alpha < 1)$. The domain of argument variation for the examined processes meets the interval $A \leq \tau < B$.

The process $I(t)$ is used at the last stage of the second Poisson point process formation. As a result, the found point process $N(t)$ is heterogeneous and reflects the most likely variations of fractal shot noise that play a role in the controlling process. Thus, for the DSPP it is typical to have two types of randomness: this is the FBNDP in a special case and two separated Poisson processes connected with the aid of the linear filter to the decaying impulse response according to the power law. As a result of this filtering the FBNDP demonstrates fractal behaviour of the kind given in Equations (2.27) for timescales in the range $A \ll T \ll B$.

The FBNDP model has five parameters: A, B, α, c and μ. For $A \ll T_0$ and $A \ll B$ they define H, λ and T_0 by the following relations [18]:

$$\begin{aligned} H &= (\alpha+1)/2 \\ \lambda &= 2\alpha^{-1}\mu c B^{\alpha/2} \\ T_0^{\alpha} &= \frac{\alpha(\alpha+1)\Gamma(1-\alpha/2)}{2\Gamma(1+\alpha/2)\Gamma(1-\alpha)} c^{-1} B^{\alpha/2} \end{aligned} \tag{2.57}$$

Parameters T_0, λ and H form the group of network traffic fundamental parameters.

In the obtained parameters part of the characteristics can be determined definitely; others, e.g. μ and B, are chosen to define the traffic behaviour characteristics. Thus, for small values $\mu B \ll l$ the impulse response is rarely collided and the time intervals between packet clusters become much larger. For $\mu B \gg l$ the impulse transfer function is often collided and the probability of a large interval decreases.

It should be noted that the impulse response $h(t)$ could have floating form in time with one or more gradual cutoffs. Any type of $h(t)$ will lead to fractal behaviour until it varies as $1/t^{1-\alpha/2}$ for the sizeable time range t and $\int_{-\infty}^{\infty} h(t)\mathrm{d}t = \lambda$. However, a different type than offered in Equation (2.56) causes problems in modelling. Moreover, it should be noted that the impulse response $h(t)$ amplitude, marked as c in Equation (2.56), can also be random until it has finite first and second moments [18], which additionally extends the range of stochastic process fractality.

The FBNDP model was used for the first time for the description of the behaviour of a point–point application for a video-conference generating fractal traffic [41].

2.4.9 Résumé

The examined approach strategy is quite attractive because the uniform procedure is used for model analysis based on real traffic characteristic parametrization with a small number of parameters. This strategy promotes more effective development of research methods of packet cluster queueing, including the solution of determining estimation problems of queueing characteristics, improvement of its performance, creation of generators for queue imitations, etc.

2.5 Fractional Levy Motion and its Application to Network Traffic Modelling

The first attempt to apply the fractal concept to traffic modelling was related to the usage of so-called fractional Gaussian noise (FGN) instead of the traditional Poisson models. Compared to the usual Gaussian noise, the fractional Gaussian model has an additional parameter (Hurst exponent H), which defines quantitatively the fractal scaling degree. It is usually possible to find that FGN is the self-similar or the fractal process with the Hurst exponent H. Nevertheless, there is a generalized concept within the framework of which the Brownian motion is only a particular case. More general cases are referred to as stable motions. Stable processes have been known for rather a long time and the first record of α-stable processes can be found in the publications of researchers from the former Soviet Union, B. Gnedenko [42] and V. Zolotariov [43]. Similar problems were analysed by P. Levy [44]. In this section the theory and implementation of fractal stable motion, which will be called fractional Levy motion (FLM), will be examined in detail. This fractal process will be modelled on the basis of symmetric α-stable (SαS) distributions.

The teletraffic model will now be formally introduced for consideration, which additionally to the Hurst exponent $H \in [\frac{1}{2}, 1)$ takes into account the Levy parameter $\alpha \in (0, 2]$. This is the fractional Levy motion (FLM) mentioned by Mandelbrot in Reference [45].

There are two subclasses of Levy motion: (a) the well-known ordinary Levy motion (OLM) (the α-stable process with independent increments offered in Reference [44]), which is the generalization of ordinary Brownian motion (Wiener process) and (b) the fractional Levy motion (self-similar and stable process), which is the generalization of fractional Brownian motion having stationary increments and an infinite correlation interval.

The random Levy process (fractional) plays an important role in teletraffic modelling and in a broad sense in the investigation of applied stochastic processes for two reasons. The first one consists in the fact that Levy motion (fractional) can be considered as the generalization of Brownian motion (fractal). The mathematical substantiation of such a generalization can be achieved by using the main properties of the stable probability laws. From the point of view of the limit theorem, stable distributions are natural generalizations of well-known Gaussian distributions: the stable distributions can be obtained as a limit (normalized in an appropriate way) of the sums of independent homogeneously distributed random variables. The main difference of the α-stable probability distribution is that the power law (of an additional distribution function) decays according to the $|x|^{-1-\alpha}$ law, where α is the Levy parameter corresponding to $0 < \alpha \leq 2$. Therefore the moments of $v \geq \alpha$ order diverge. In the queueing analysis for telecommunication switches and routers the infinite moments of input processes can cause infinite moments in the queueing process, which leads to a long waiting process.

The second reason for the important role of fractional Levy motion consists in its scaling invariance property or self-similarity. Moreover, the process increments are not only self-similar processes but are also dependent on each other simultaneously, having heavy tails distributions. Fractional Brownian motion has a clear mathematical interpretation and can easily be applied to fractal traffic modelling. However, Gaussian processes have finite variance.

The FLM is a more general case and can be very suitable for intensity modelling of traffic or rates, which have large dispersions (theoretically infinite variance). Moreover, the artificially obtained traces of traffic profiles can be important for testing/checking real computer systems/networks. Thus, for example, the authors of Reference [46] use the stationary sequence obtained on the basis of FLM for modelling real Ethernet, VBR video and www traffic described by heavy tails distributions.

Several self-similar stable processes are known in which the ranging and extraordinary local irregularity are combined in a natural manner.

2.5.1 Fractional Levy Motion and its Properties

2.5.1.1 Definition of Fractional Levy Motion

The symmetric α-stable (SαS) Levy motion $L_\alpha = \{L_\alpha(t),\ t \geq 0\}$ is the twin of Brownian motion for $0 < \alpha \leq 2$. SαS is the Markovian stochastic process which starts in 0, has stationary independent increments and is H-sssi with $H = 1/\alpha$, i.e. $L_\alpha(ct) = c^{1/\alpha} L_\alpha(t),\ t \geq 0$. The probability density function for SαS is

$$w_\alpha(x,\ t) = \frac{1}{2\pi} \int_{-\infty}^{\infty} \mathrm{d}k\ \mathrm{e}^{\mathrm{i}kx} \exp(-\sigma|k|^\alpha t) \tag{2.58}$$

where $\sigma > 0$ is a scaling parameter.

For SαS it is known that 'a law 1/α' can be defined for the fractional structural function $S_v(\tau, \alpha) = M[L_\alpha(t+\tau) - L_\alpha(t)]^v$ in the following manner for $0 < \alpha < 2$:

$$S_v(\tau, \alpha) = \begin{cases} \tau^{v/\alpha} V(v; \alpha), & v < \alpha \leq 2 \\ \infty, & v \geq \alpha \end{cases} \tag{2.59}$$

where $V(v;\ \alpha)$ is defined as

$$V(v; \alpha) = \frac{\sigma^{v/\alpha}}{2\pi} \int_{-\infty}^{\infty} \mathrm{d}\xi |\xi|^v \int_{-\infty}^{\infty} \mathrm{d}\varsigma \exp(\mathrm{i}\xi\varsigma - |\varsigma|^\alpha) \tag{2.60}$$

Note that $V(v; \alpha)$ can be easily estimated [46] and results in

$$V(v, \alpha) = \frac{2\sigma^{v/\alpha}}{\pi v} \sin\left(\frac{\pi v}{2}\right) \Gamma(1+v) \Gamma\left(1 - \frac{v}{\alpha}\right), \quad v < \alpha \leq 2 \tag{2.61}$$

Following the generalization of ordinary Brownian motion to fractional Brownian motion (FBM) fulfilled by Mandelbrot in Reference [45], the fractional Levy motion (FLM) process can be defined by the following fractional Riemann–Louville integral:

$$L_{\alpha,H}(t) = \frac{1}{\Gamma(H + \frac{1}{2})} \int_0^t \mathrm{d}L_\alpha(\tau)(t-\tau)^{H-1/2} \tag{2.62}$$

where $L_\alpha(t)$ is the ordinary symmetric α-stable Levy motion (SαS). (The definition of the fractional integral can be found for instance in Reference [47].)

It should be noted that FLM is the generalization of the well-known fractional Brownian motion, which can be obtained from Equation (2.62) for $\alpha = 2$. Therefore the role played by FLM among the stable processes is similar to the role played by FBM among the Gaussian processes. The FLM process can be defined as $\Delta L_{\alpha,H}(\tau) = \{L_{\alpha,H}(t+\tau) - L_{\alpha,H}(t),\ \tau \geq 0\}$, which is the time-continuous stationary process. Some important properties of the FLM process and its increments can be proved by the following theorem.

Theorem 2.1[53]
FLM is the H-sssi process with the Hurst exponent $H - \frac{1}{2} + 1/\alpha$. Therefore in accordance with definition (2.62), FLM is the $H - \frac{1}{2} + 1/\alpha$- sssi process. This gives the following consequence.

Consequence 2.1
The increments process $\{L_{\alpha,H}(t_2) - L_{\alpha,H}(t_1)\}$ is self-similar with the Hurst exponent $H - \frac{1}{2} + 1/\alpha$. In fact, it is easy to show that for $t_2 \geq t_1$ and $c > 0$,

$$L_{\alpha,H}(ct_2) - L_{\alpha,H}(ct_1) \stackrel{d}{=} c^{H-1/2+1/\alpha}[L_{\alpha,H}(t_2) - L_{\alpha,H}(t_1)]$$

i.e. the increments process is self-similar with the same Hurst exponent $H - \frac{1}{2} + 1/\alpha$.

2.5.1.2 Probability Density Function for FLM

Theorem 2.2
The probability density function $w_{\alpha,H}(x,t)$ of the FLM process is defined as

$$w_{\alpha,H}(x,t) = \frac{1}{2\pi} \int_{-\infty}^{\infty} \mathrm{d}k\ \mathrm{e}^{ikx} \exp(-\overline{\sigma}|k|^\alpha t^{\alpha(H-1/2)+1}) \tag{2.63}$$

Applying the Taylor series expansion to Equation (2.63), gives

$$w_{\alpha,H}(x,t) = \delta(x) - \frac{1}{\pi} \sum_{n=1}^{\infty} \frac{(-\overline{\sigma})^\pi}{n!} \frac{t^{\alpha\pi(H-1/2)+n}}{|x|^{\alpha\pi+1}} \sin\left(\frac{\pi\alpha n}{2}\right) \Gamma(\alpha n + 1) \tag{2.64}$$

This series can be used to investigate the asymptotic behaviour $\omega_{\alpha,H}(x,t)$ as $|x| \to \infty$.

The PDF for FBM can be found from Equation (2.63) as a specific case for $\alpha = 2$:

$$w_{2,H}(x,t) = \frac{1}{2\pi} \int_{-\infty}^{\infty} \mathrm{d}k\ \mathrm{e}^{ikx} \exp(-\overline{\sigma} k_B^2 t^{2H}) = \sqrt{\frac{1}{4\pi\overline{\sigma}_B t^{2H}}} \exp\left(-\frac{x^2}{4\pi\overline{\sigma}_B t^{2H}}\right)$$

where $\overline{\sigma}_B$ for $\alpha = 2$ is obtained as $\overline{\sigma}_B = \sigma/[2H\Gamma^2(H + \frac{1}{2})]$.

2.5.2 Algorithm of Fractional Levy Motion Modelling

Let ε be the i.i.d. SαS random variables with $\alpha \in (0,2]$. Then the vector $\boldsymbol{a}$ is obtained for which $a_1 = 1$ and $a_n = n^{\beta} - (n-1)^{\beta}$, where $\beta = H - 1/\alpha$; α and H are in turn the parameters of the formed FLM. The process $MA(\infty)$ is defined in the following manner: $X_n = \sum_{i \in Z_+} a_i \varepsilon_{n-i}$. Then under the condition that $\sum_{i=1}^{\infty} |a_i|^{\alpha} < \infty$, or that $H = 1/\alpha + \beta < 1$, a similar process will exist [50].

Let $S_n(t) = \sum_{j=1}^{\lceil nt \rceil} X_j = \sum_{i=0}^{\infty} X_i (\sum_{j=1-i}^{\lceil nt \rceil - i} a_j)$ be the partial sum of the X_n process. Under the appropriate limitations imposed on a_i coefficients the normalized $S_n(t)$ realization converges in the sense of the finite-dimensional distribution to the self-similar process $L_{H,\alpha}(t)$, i.e. to FLM. In the case of finite variance for the innovation ε process, as a result, at the limit the process converges to fractional Brownian motion. For given n, m,

$$\begin{pmatrix} Y_{m,1} \\ Y_{m,2} \\ \dots \\ Y_{m,m} \end{pmatrix} = \mathbf{A} \begin{pmatrix} \varepsilon_0 \\ \varepsilon_{-1} \\ \dots \\ \varepsilon_{1-m} \end{pmatrix} \tag{2.65}$$

where

$$\mathbf{A} = \begin{pmatrix} a_1 & a_2 & \dots & a_{m-1} & a_m \\ a_2 & a_3 & \dots & a_m & a_1 \\ \dots & \dots & \dots & \dots & \dots \\ a_m & a_1 & \dots & a_{m-2} & a_{m-1} \end{pmatrix}$$

is a circulant matrix of size $m \times m$. Let $S_n^Y(t)$ be a step function such as $S_n^Y(k/n) = \sum_{i=1}^{k} Y_{m,i}, 0 \le k \le n$. Then as $n, m \to 0$, $S_n^Y(t)/n^H \sigma \overset{\text{fdd}}{\rightarrow} L_{H,\alpha}(t)$, where $\overset{\text{fdd}}{\rightarrow}$ designates the convergence in finite-dimensional distributions. The proof of this statement can be found in References [49] and [50].

On the basis of this discussion the algorithm will be formulated for FLM realizations generation. The $\boldsymbol{Y}$ vector in Equation (2.65) with the $m \times l$ dimension can be easily calculated on the basis of the fast Fourier transform (FFT). Let $a = (a_1, \dots, a_m)$ and $e = (\varepsilon_0, \varepsilon_{1-m}, \varepsilon_{2-m}, \dots, \varepsilon_{-1})$. Moreover, let $\bar{\boldsymbol{a}} = \text{FFT}(\boldsymbol{a})$ and $\bar{\boldsymbol{e}} = \text{FFT}(\boldsymbol{e})$ be the FFT for appropriate vectors. Due to the $\mathbf{A}$ matrix of $m \times m$ size being cumulant, $Y = \text{IFFT}$ (inverse fast fourier transform) $(\boldsymbol{v})$, where $\boldsymbol{v} = (v_1, \dots, v_m)$ and $v_i = \bar{\boldsymbol{a}}_i \bar{\boldsymbol{e}}_i$. The known generation algorithm can be formed in the following manner.

Algorithm 2.1

Step 1. Use the FFT for a DFT (discrete Fourier transform) calculation of a. Let $b_m = \text{DFT}(a) = (b_1, \dots, b_m)$.

Step 2. Generate the m-dimensional random vector $\boldsymbol{e}$, the elements of which are the i.i.d. SαS random variables found with the help of the following relations [51,52]:

If $\alpha = 2$, then $\varepsilon_i = \sqrt{2} G_i$.
if $\alpha = 1$, then $\varepsilon_i = \text{tg}[\pi(X_i - 0.5)]$.
if $\alpha > 0$, then $\varepsilon_i = \left\{ \dfrac{\cos[(1-\alpha)X_i]}{E_i} \right\}^{1/\alpha - 1} \dfrac{\sin(\alpha X_i)}{[\cos(X_i)]^{1/\alpha}}$.

Here G_i are random variables distributed according to the Gaussian law with zero mean and unit variance, E_i are exponential random variables with a unit mean, X_i are random variables distributed according to uniform law over the interval [0,1], ε_i are symmetrical α-stable random variables and $i = 1, \ldots, m$.

Step 3. Use the FFT for calculation of the DFT from e_m. Let $f_m = \text{DFT}\,(e_m) = (f_1, \ldots, f_m)$.

Step 4. Let $\mathbf{v} = (b_1 f_1, \ldots, b_m f_m)$.

Step 5. Use the FFT I for calculation of the IDFT (inverse DFT) of v_m. Let $(y_1, \ldots, y_m) = \text{IDFT}\,(v) = Y_{m,j}, j = 1, \ldots, m$.

Step 6. Calculate the cumulative sum for y_i and carry out the normalization with the $n^H \sigma$ coefficient to obtain the required realization.

The operating speed of this algorithm is actually defined on the operating speed of the FFT algorithm and has the temporal complexity $O(m \log m)$. Another feature of this algorithm is the fact that it allows $L = \lfloor m/n \rfloor$ FLM to be obtained simultaneously, which essentially reduced the time costs during the network traffic imitation.

As an example, in Figure 2.32 FLM realizations are shown that have been obtained with the help of this algorithm. As a comparison the realizations were obtained for different α and H parameters, which are shown in the figure. The properties being determined by the Hurst exponent depend on LRD characteristics of the examined process, as in the cases of FBM and FGN, but the α parameter (Levy parameter) is responsible for the tail distribution heaviness and as it decreases more noticeable bursts are demonstrated in the plots.

2.5.3 Fractal Traffic Formation Based on FLM

Norros [51,52] used fractional Brownian motion for the formation of the Gaussian self-similar model of teletraffic. The Norros model definition can be formulated in the following form.

Definition 2.1

The continuous integral process of arrivals $\hat{A}(t)\; t \in (0, \infty)$ is defined as

$$\hat{A}(t) = mt + \sqrt{am} B_H(t) \tag{2.66}$$

where $m > 0$ and $a > 0$ are constants and $B_H(t)$ is the process of fractional continuous Brownian motion with the Hurst exponent H.

2.5.3.1 Fractional Levy Motion with Symmetrical Stable Innovations

Sometimes the process of fractional Brownian motion used in the Norros model has been substituted for the symmetrical self-similar stable process. This traffic model [54] is constructed based on the fractional Levy motion defined by Equation (2.62).

With the help of FLM the traffic volume $\hat{A}(t)\; t \in (0, \infty)$, which arrives during the period $[0, t)$ into the traffic channel, can be described by the analogy with the Norros model as

$$A(t) = mt + (\bar{\sigma} m)^{1/\alpha} L_{\alpha,H}(t) \tag{2.67}$$

Similar wording in the case of FBM is referred to as the Norros model. Considering fractional Levy motion instead of FBM, this form will be the natural generalization of the well-known

Norros model, i.e. FBM itself is the specific case of FLM at $\alpha = 2$. This traffic model is specified by four parameters: the intensity $m > 0$; the Levy parameter $\alpha \in (0;\ 2]$, which determines the distribution tail heaviness and, as seen in Figure 2.32, is responsible for sharp hits in data; the scaling parameter $\bar{\sigma} > 0$, which determines the traffic value dispersion around the intensity mean; and the Hurst exponent $H \in [0,\ 1)$.

Use of the generalized non-Gaussian self-similar stochastic FLM process essentially extends the family of traditional fractal models of network traffic. The condition $H > 1/\alpha$ is the positive LRD, $H < 1/\alpha$ is the negative LRD and $H = 1/\alpha$ is an independent process. However, since FLM is the process with infinite variance (and infinite mean value if $\alpha < 1$), it is necessary to be very careful with the interpretation of m and a. Besides, as FLM is more bursty than FBM, the probability is that $A[i] < 0$ is more for any given m and a.

2.5.3.2 Fractional Levy Motion with Nonsymmetrical Innovations [54]

The examined model flexibility can be additionally increased by substituting symmetrical random variables for nonsymmetrical ones. However, this increases the number of parameters that should be estimated for this model. As a result, the arrivals process is described by the following equation:

$$A[i] = m + \sqrt{am}(L_{\alpha,\beta,H}[i])$$

Here $L_{\alpha,\beta,H}(t)$ is the fractional stable Levy noise (FSLN), which is defined in integral form as

$$L_{\alpha,\beta,H}(t) = \int_{-\infty}^{\infty} [(t - x - 1)^{H - 1/\alpha} - (t - x)^{H - 1/\alpha}] L_{\alpha,\beta}(\mathrm{d}x) \tag{2.68}$$

where $L_{\alpha,\beta}(\mathrm{d}x)$ is the independent and identically distributed stable process and β is the parameter characterizing the distribution asymmetry. As above, m is the average arrivals intensity and a is the scaling coefficient.

2.5.3.3 Norros Model for the Case of Arbitrary Distribution

The definition assumes that $\hat{A}(t_1) \geq \hat{A}(t_2)$, if $t_1 > t_2$ for any $t_1 > 0,\ t_2 > 0$, m is the average intensity of the process arrivals and a is the scaling coefficient. The advantages of the model (2.66) are in the economy (three parameters) and in the fact that the assumption concerning the Gaussian distribution allows analytical expressions to be obtained for lower limits of the buffer filling probability. The model's shortcoming is that it can generate traffic having the Gaussian distribution and also that there is no direct estimation method for the α parameter. At the same time, in most cases the assumption of the Gaussian traffic distribution character is not fulfilled, and the Norros model adaptability should be extended to the case of arbitrary non-Gaussian distribution. That is why it is expedient to consider the example of the transformation of the classical Norros model with FBM for the case of the arbitrary probability distribution, which can be found from the measured data.

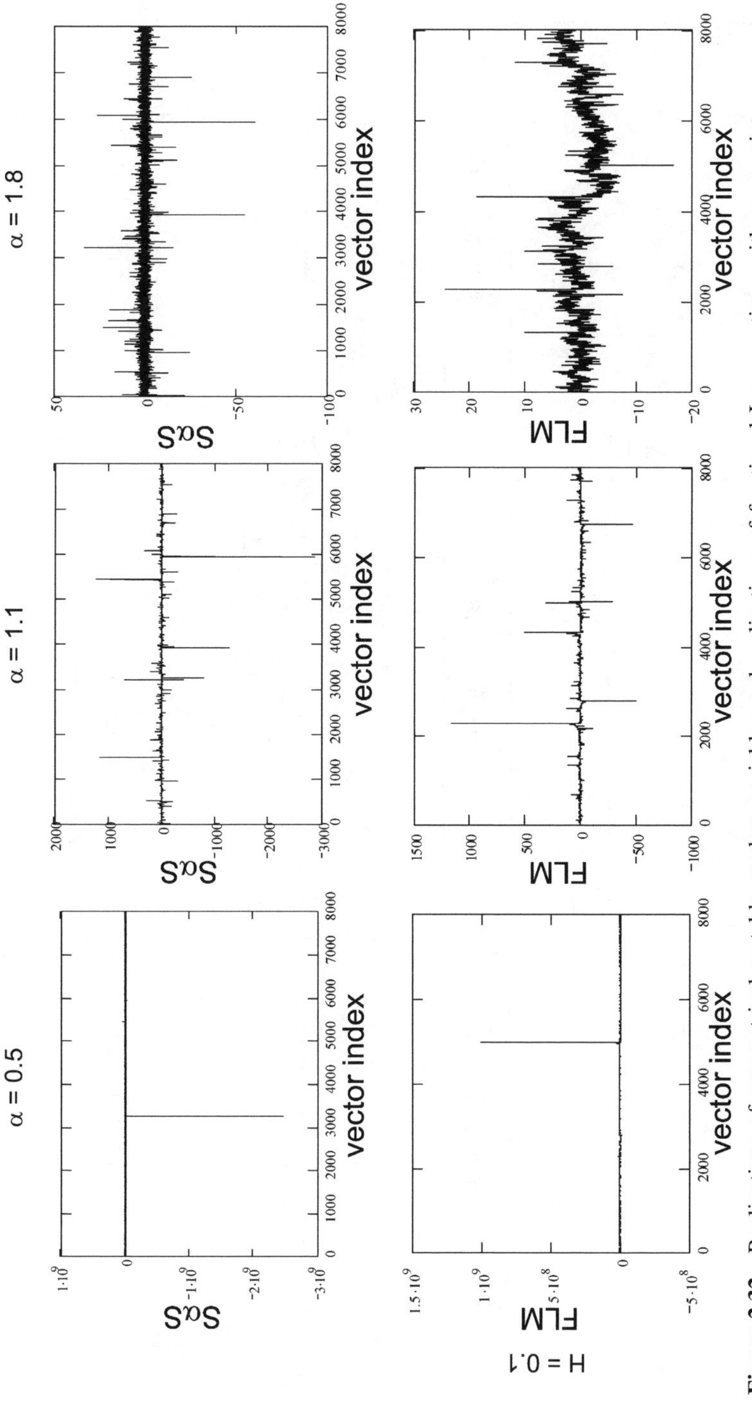

Figure 2.32 Realizations of symmetrical α-stable random variables and realizations of fractional Levy motions with appropriate parameters determined on this basis

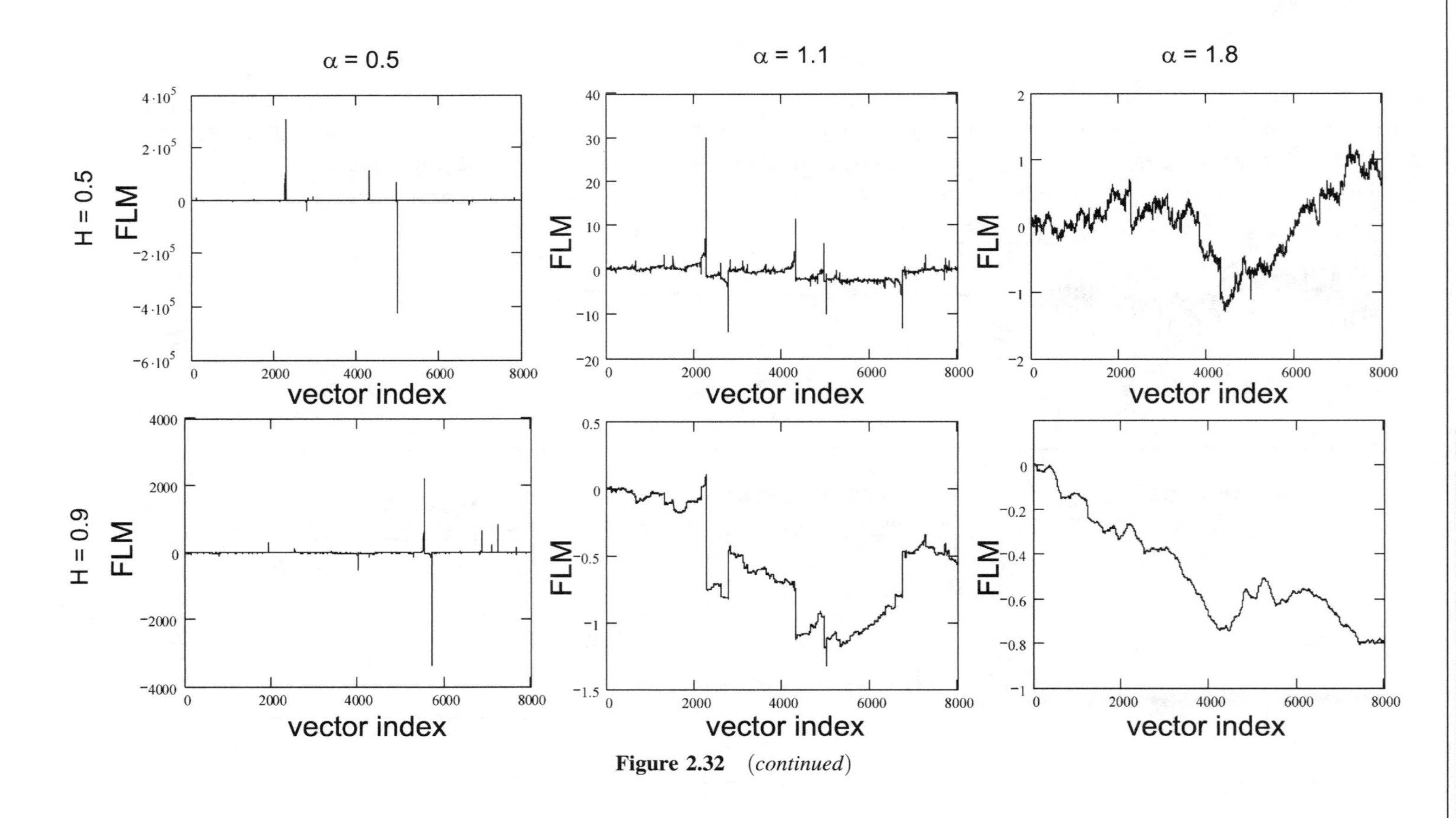

Figure 2.32 *(continued)*

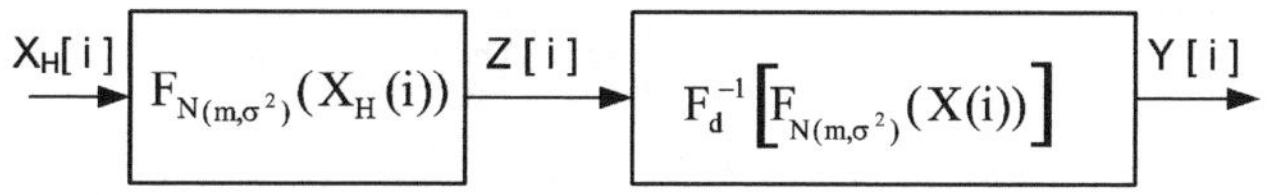

$X_H[i]$ → $F_{N(m,\sigma^2)}(X_H(i))$ → $Z[i]$ → $F_d^{-1}\left[F_{N(m,\sigma^2)}(X(i))\right]$ → $Y[i]$

Figure 2.33 Algorithm structure for the non-Gaussian fractal process formation

Assuming that $F_d(\cdot)$ is the cumulative distribution function (DF) of the required output process, the fractal process $Y[i]$ can be found with the help of mapping:

$$Y[i] = F_d^{-1}(F_{N(m,\sigma^2)}(X_H[i])) \tag{2.69}$$

Here $X_H[i]$ is the fractional Gaussian noise with the appropriate Hurst exponent H and is the DF of the Gaussian distribution $N(m, \sigma^2)$.

$$F_{N(m,\sigma^2)} = F(X) = \int_{-\infty}^{X} \frac{1}{\sqrt{2\pi}\sigma[x]} \mathrm{e}^{-(X - m[x])^2/2\sigma^2[x]} \mathrm{d}x$$

The modelling strategy for the non-Gaussian fractal process can be illustrated by Figure 2.33. The algorithm itself can be formulated in the following manner.

Step 1. The FGN $X_H[i]$ is transformed into the $Z[i] = (F_{N(m,\sigma^2)}(X_H[i]))$ process having a uniform distribution with the help of the nonlinear transform $F_{N(m,\sigma^2)}$.
Step 2. Using the experimental estimation found from available data, the DF of the $F_d(\cdot)$ process with the required non-Gaussian distribution is obtained.
Step 3. The inverse function $F_d^{-1}(\cdot)$ for obtaining the output values $Y[i] = F_d^{-1}$ $(F_{N(m,\sigma^2)}$ $(X_H[i]))$ with the required distribution law is determined.
Step 4. The modelling efficiency is estimated with the data statistical processing approach.

As and example, consider the non-Gaussian fractal process formation with the following distribution:

$$w(Y) = \begin{cases} Aw_1(Y) & \text{for } Y_1 \leq Y \leq \infty \\ Bw_2(Y) & \text{for } 0 < Y < Y_1 \end{cases} \tag{2.70}$$

where $w_1(Y) = N(m, \sigma^2)$ is the Gaussian distribution and $w_2(Y) = \lambda\,\mathrm{e}^{-\lambda Y}$ is the exponential distribution; $A, B =$ constant are the normalizing coefficients, ensuring the normalization $\int_0^\infty w(Y)\,\mathrm{d}Y = 1$. The PDF form $w(Y)$ is shown in Figure 2.34(a).

It the DF F(Y) corresponding to $\omega\,(Y)$ is assessed analytically, then

$$F(z) = \int_0^z w(Y)\ \mathrm{d}Y = \begin{array}{ll} -\frac{1}{\lambda}\mathrm{e}^{-\lambda\, z} & \text{for } z \leq z_1 \\ \int_{z_1}^{z} \frac{1}{\sqrt{2\pi}\sigma} \exp\left[-\frac{(Y-a)^2}{2\sigma^2}\right] \mathrm{d}Y & \text{for } z > z_1 \end{array} \tag{2.71}$$

The DF form corresponding to Equation (2.71) is shown in Figure 2.34(b). Since the expression for the inverse function $F^{-1}(z)$ has the complex analytical form $F(z)$ to obtain the inverse

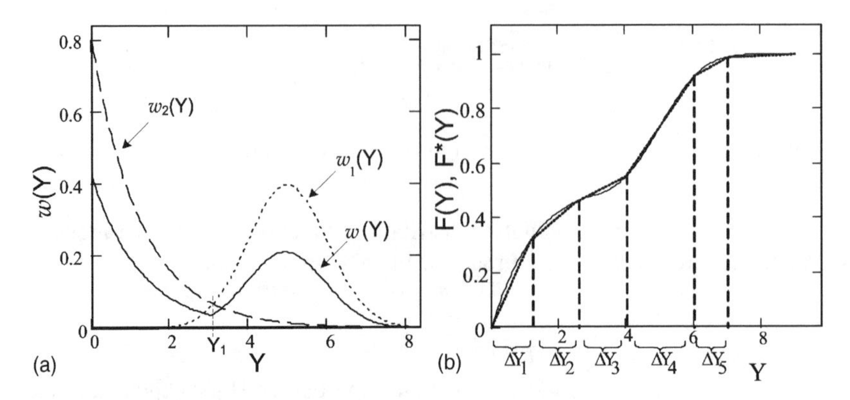

Figure 2.34 (a) PDF and (b) CDF of the process $Y(t)$ with parameters $\lambda = 0.8$; $m = 5$; $\sigma = 1$

function, it is suitable to apply the piecewise-linear approximation

$$F^*(z) = \begin{cases} a_1 z + b_1 & \text{for } z \in \Delta z_1 \\ a_i z + b_i & \text{for } z \in \Delta z_i \\ \vdots & \\ a_N z + b_N & \text{for } z \in \Delta z_N \end{cases}$$

where a_i and b_i are coefficients in the ith piece of approximation $Z = Y_i$; $\Delta Z_i = \Delta Y_i$. The inverse function will be respectively equal to

$$F^{-1}(z) = \begin{cases} \dfrac{z - b_1}{a_1} & \text{for } z \in \Delta z_1 \\ \dfrac{z - b_i}{a_i} & \text{for } z \in \Delta z_i \\ \dfrac{z - b_N}{a_N} & \text{for } z \in \Delta z_N \end{cases}$$

The statistical modelling results are examined for fractional noise with the PDF in the form (2.70) in accordance with the algorithm presented in Figure 2.33. The FGN realization at the generator input is shown in Figure 2.35.

Figure 2.36 shows the histogram of instantaneous amplitudes and Figure 2.37 shows the plot of the covariance function of the formed random sequence $\{y_i, i = 1...N\}$ on the log–log scale. Assuming that the plot slope exceeds -0.5, it can be confirmed that the process at the generator output shown in Figure 2.33 really has fractal properties. This show that it is possible to form fractal processes with the given distribution law.

2.6 Models of Multifractal Network Traffic

There are different approaches suitable for multifractal process modelling. For the first time multiplicative cascades were used as traffic multifractal models [59,60]. This class of model is the very well known among the multifractal processes. The simplest example of the multifractal process is a binomial cascade, which is defined with the aid of a binary tree-like structure [60,61]. Having combined this process with FBM, a new class of fractional Brownian motions

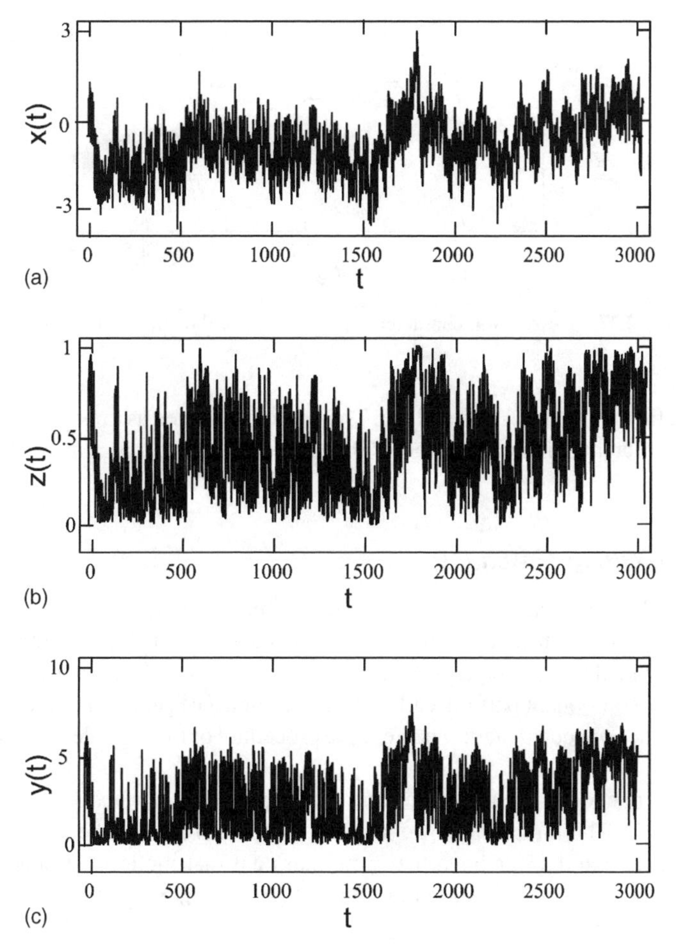

Figure 2.35 (a) FGN realization at the generator input. (b) Realization of the random process $z(t)$ at the converter output for the characteristic $F_{N(m,\sigma^2)}(X_H[i])$. (c) Realization of the random process $y(t)$ at the converter output for the characteristic $F_d^{-1}(F_{N(m,\sigma^2)}(X_H[i]))$

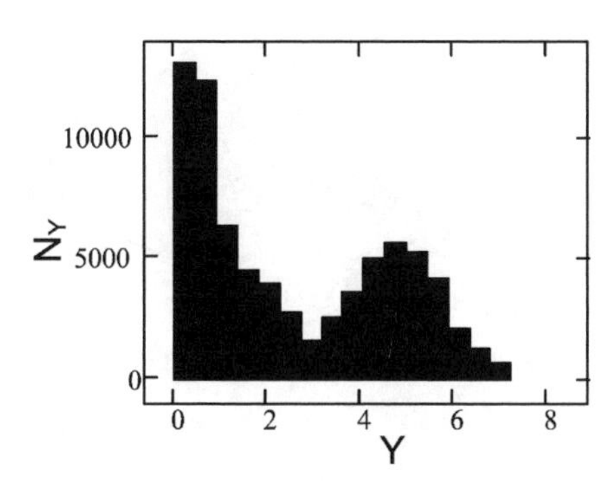

Figure 2.36 Histogram of the simulated random process

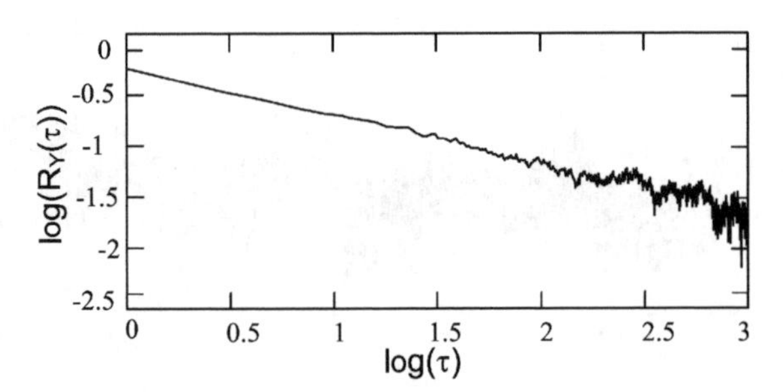

Figure 2.37 Correlation characteristic (log–log) of the output random process

in the multifractal time space can be defined. The process obtained as a result has several unique properties. In particular, it is able to cover LRD and the multifractal range independently of each other.

2.6.1 Multiplicative Cascades

The simplest multifractals are usually constructed by means of an iterative procedure, called a multiplicative cascade. Consider the unit interval related to the unit mass. At the stage $k = 1$ the unit interval is divided into two equal subintervals and masses r and $1 - r$ are connected to them respectively. The component part r is called the multiplier. The same principle is used for each subinterval and the connected mass. The iterative procedure of the cascade structure is shown in Figure 2.38.

Multipliers r are chosen as independent random variables R, located in [0;1] with the PDF $F_R(x)$, $M[R] = 1/2$. The multiplier r is chosen so that it will lead to the symmetrical density function, therefore r and $1 - r$ have the same marginal distribution. At stage k the dyadic interval with length $\Delta t_k = 2^{-k}$ beginning at $t = 0, \eta_1, \ldots, \eta_k = \sum \eta_i 2^{-i}$ has the mass (the measure)

$$\mu(\Delta t_k) = R(\eta_1)R(\eta_1, \eta_2) \cdots R(\eta_1, \ldots, \eta_k)$$

where $R(\eta_1, \ldots, \eta_i)$ corresponds to the multiplier at the ith stage. An example of the cascade construction at different stages is depicted in Figure 2.39.

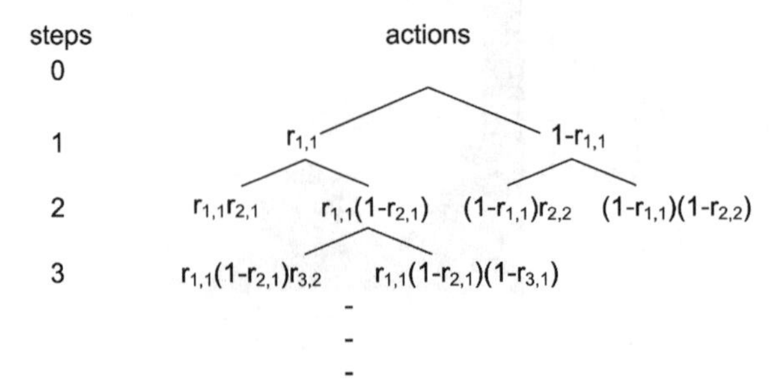

Figure 2.38 Iterative procedure for cascade construction

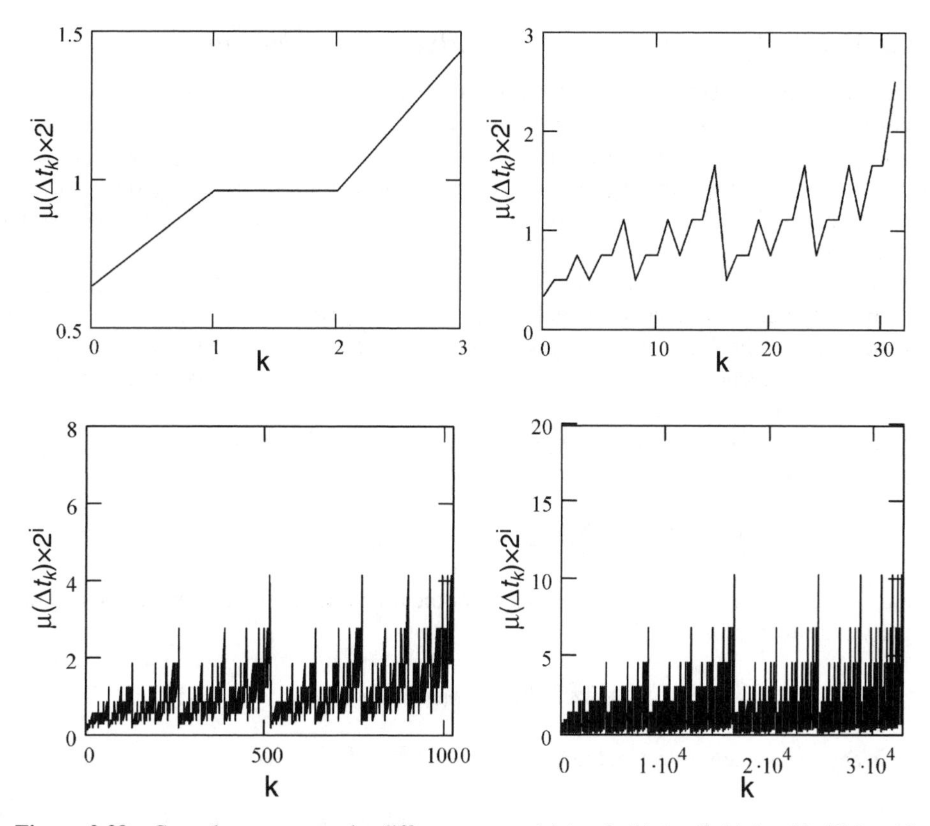

Figure 2.39 Cascade structure under different stages: (a) $i = 2$; (b) $i = 5$; (c) $i = 10$; (d) $i = 15$

Since multipliers are independent and homogeneously distributed, it is easy to show that the measure μ satisfies the following scaling relation:

$$M[\mu(\Delta t_k)^q] = (M[R^q])^k = (\Delta t_k)^{-\log_2 M[R^q]}$$

which defines the multifractal process with the scaling function $\tau_0(q) = -\log_2 M[R^q]$.

It should be noted that the above-examined multifractal process is also called the conservative cascade. The important property of the random cascade consists in its independent structure resulting from its construction. If the multipliers used in the construction have the same fixed value $r_0 (0 < r_0 < 1)$, the obtained multiplicative measure is referred to as *binomial*.

The binomial measure is the deterministic cascade. Its scaling function is defined as $\tau_0(q) = -\log_2[r_0^q + (1 - r_0)^q] + 1$. Moreover, if the iteration keeps the mass at the average value only, i.e. the multipliers at each mass division are also independent and identically distributed but have half the mean value, the appropriate measure is called *canonical* [61].

From the point of view of network modelling, the conservative cascades are of great interest. The binomial cascades are not the worst consideration due to their determined structure. The canonical cascade could not be used because it is a random independent process while the network traffic flows are long-range dependent. In this investigation the conservative cascade was used as 'a laying brick' for the traffic model construction.

2.6.2 Modified Estimation Method of Multifractal Functions

The multifractal model of network traffic offered in Reference [58] will now be examined. This model is a combination of the multiplicative cascade and the random process having the one-dimensional log-normal distribution and possesses all the important properties demonstrated in the traffic including LRD, multifractality and log-normality. The model is flexible enough to cover all multifractal features including the scaling function $\tau_0(q)$ and the moment coefficient $c(q)$.

A full description of the multifractal model consists of the scaling function $\tau_0(q)$ and the moment coefficient $c(q)$. The estimation method based on absolute moments provides a simple approach for the checking scaling property and also for estimation of the scaling function. However, the moment coefficient is also necessary for the analysis and with regard to this the following alteration was introduced.

The definition of the multifractal processes [62] imposes the stationarity condition on the increments. Therefore, it is easy to check the following relation for increment moments:

$$M[|Z^{(\Delta t)}|^q] = c(q)(\Delta t)^{\tau(q)+1} = c(q)(\Delta t)^{\tau_0(q)}, \quad q > 0 \tag{2.72}$$

where $Z^{(\Delta t)}$ designates the increment process for the time sample Δt. Therefore, the following equality is correct for $m = 1, 2, \ldots$:

$$M[|Z^{(m\Delta t)}|^q] = c(q)(m\Delta t)^{\tau_0(q)} \tag{2.73}$$

It Δt is chosen as a unit interval, then

$$\log M[|Z^{(m)}|^q] = \tau_0(q) \log m + \log c(q), \quad q > 0 \tag{2.74}$$

Based on this property, the method is reduced to the following: the increment process $Z_1, Z_2, \ldots, Z_n$ is introduced and its appropriate real aggregated sequence $\{Z^{(m)}\}$ is defined at the aggregation level m:

$$Z_k^{(m)} = Z_{(k-1)m+1} + Z_{(k-1)m+2} + \cdots + Z_{km}, \quad k, m = 1, 2, \ldots \tag{2.75}$$

If the sequence $\{Z_k\}$ has a scaling property, the absolute moments function $M[Z^{(m)}]^q$ with respect to m on the log–log plot in accordance with Equation (2.74) should be shown with the straight line. This line slope gives the estimation for $\tau_0(q)$ and the segment truncated on the coordinate axis is the log $c(q)$ value. This method is illustrated in Figure 2.40.

Note that it is not necessary to estimate $c(q)$ and $\tau_0(q)$ for all positive values of q, which is impossible to do in principle. In fact, the maximum value for q should be considered in accordance with the finite given queue length related to queue length probability.

2.6.3 Generation of the Traffic Multifractal Model

It is assumed that the multifractal analysis of real data, found as a result of network traffic measurement, demonstrates its multifractal properties described by the scaling function $\tau_0(q)$ and the moment coefficient $c(q)$. The obvious problem of cascade modelling is to find the appropriate probability distribution for R multipliers, so that$-\log_2(M[R^q]) = \tau_0(q)$. However,

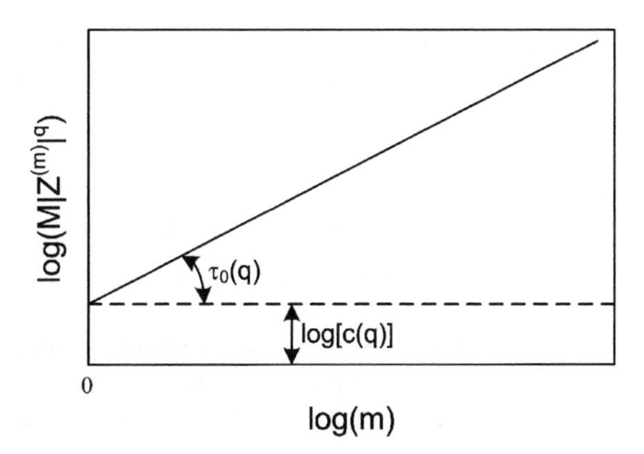

Figure 2.40 Method for estimation of the scaling function $\tau(q)$ and the moment coefficient $c(q)$

the cascade model covers multifractal properties only, given with the aid of the scaling function $\tau_0(q)$, and cannot render any information concerning the moment coefficient $c(q)$.

The main idea of the suggested traffic model consists of the following.

1. At the first stage 2^N artificial data $\mu(\Delta t_N)$ are generated with the aid of the multiplicative cascade with multipliers obtained on the basis of random distributed R.
2. At the second state the data 2^N in length series of the model are multiplied in couples by the cascade series and by independent and identically distributed random samples of the positive random variable Y with the same length.

Variable Y is chosen independent of the cascade measure $\mu(\Delta t_N)$. Therefore the series marked as $X(\Delta t_N)$ satisfies the equality [58]

$$M[X(\Delta t_0)^q] = M[Y^q]2^{N(q+\log_2 M[R^q])}\Delta t_0^{-\log_2 M[R^q]} \tag{2.76}$$

where Δt_0 designates the unit interval of the modelled traffic. As a result, the problem of model parametrization consists in finding random variables R and Y:

$$\begin{aligned} -\log_2(M[R^q]) &= \tau_0(q) \\ M[Y^q] &= c(q) \end{aligned} \tag{2.77}$$

The model is related to multifractal network traffic due to the following reasons:

1. The model is based on the multiplicative cascade structure, which on closer examination looks like the mechanisms of TCP/IP protocol functioning. The similar mechanism (as described in various studies of network traffic [1, 59, 60]) is the main reason for traffic multifractality over small scales.
2. The modelled traffic can be presented as the product of the random peak flow rate Y and the bursty structure measure $\mu(\Delta t_N)$ on the modelled timescale Δt_N.

For multifractal traffic the scaling function $\tau_0(q)$ and the logarithm of the moment coefficient $c(q)$ can be estimated, for example, with the aid of the simple absolute moment method [58]. The estimated functions can be notated as $\tilde{\tau}_0(q)$ and $\log \tilde{c}(q)$ respectively. Then, with regard to Equations (2.72) and (2.76) for $\Delta t_0 = 1$, relations (2.77) can be transformed to the form

$$-\log_2(M[R^q]) = \tilde{\tau}_0(q) \tag{2.78}$$

$$\begin{aligned} \log M[Y^q] &= \log \tilde{c}(q) - (q + \log_2 M[R^q])N \log 2 \\ &= \log \tilde{c}(q) - [q - \tilde{\tau}_0(q)]N \log 2 \end{aligned} \tag{2.79}$$

Analysis of different traces of the measured traffic with multifractal properties shows that the choice of R in the form of the symmetrical random variable with the beta distribution $\beta(\alpha, \alpha)$ over the interval [0;1] characterized by the only $\alpha > 0$ parameter is enough to model the estimated scaling function. In this case,

$$\tau_0(q) = \log_2 \frac{\Gamma(\alpha)\Gamma(2\alpha + q)}{\Gamma(\alpha + q)\Gamma(2\alpha)} \tag{2.80}$$

where $\Gamma(z) = \int_0^{+\infty} x^{z-1} e^{-x} dx$, $z > 0$, is the gamma function.

The random variable Y will be chosen in the form of the random variable with the log-normal distribution for m and σ parameters:

$$w(Y) := \frac{1}{Y\sigma\sqrt{2\pi}} \exp\left[-\frac{(\ln Y - m)^2}{2\sigma^2}\right], \quad Y > 0$$

Since the qth moment of the log-normal distribution has $M[Y^q] = e^{mq + \sigma^2 q^2/2}$ form, in accordance with Equation (2.79) the m and σ parameters should satisfy the equation

$$mq + \frac{\sigma^2 q^2}{2} = \log \tilde{c}(q) - \left[q - \log_2 \frac{\Gamma(\alpha)\Gamma(2\alpha + q)}{\Gamma(\alpha + q)\Gamma(2\alpha)}\right] N \log 2 \tag{2.81}$$

It should be noted that the random variable Y distribution can be arbitrarily chosen, but it will not change the model properties. The log-normal distribution was chosen as it has the simplest log-moment function.

Finally, the presented multifractal model has three parameters (α, m, σ) that fully define the scaling function $\tau_0(q)$ and moment coefficient $c(q)$ in the form of the following functions:

$$\tau_0(q) = \log_2 \frac{\Gamma(\alpha)\Gamma(2\alpha + q)}{\Gamma(\alpha + q)\Gamma(2\alpha)} \tag{2.82}$$

$$c(q) = e^{mq + \sigma^2 q^2/2} 2^{N\{q - \log_2[\Gamma(\alpha)\Gamma(2\alpha + q)/\Gamma(\alpha + q)\Gamma(2\alpha)]\}} \tag{2.83}$$

The scaling function $\tau_0(q)$ plots for $\alpha =$ constant are shown in Figure 2.41. The function

$$\ln c(q) = \left(mq + \frac{\sigma^2 q^2}{2}\right) + \left\{q - \log_2 \left[\frac{\Gamma(\alpha)\Gamma(2\alpha + q)}{\Gamma(\alpha + q)\Gamma(2\alpha)}\right]\right\} N \ln(2), \quad \text{for } m = 0.6; \sigma = 0.2; N = 20$$

is shown in Figure 2.42.

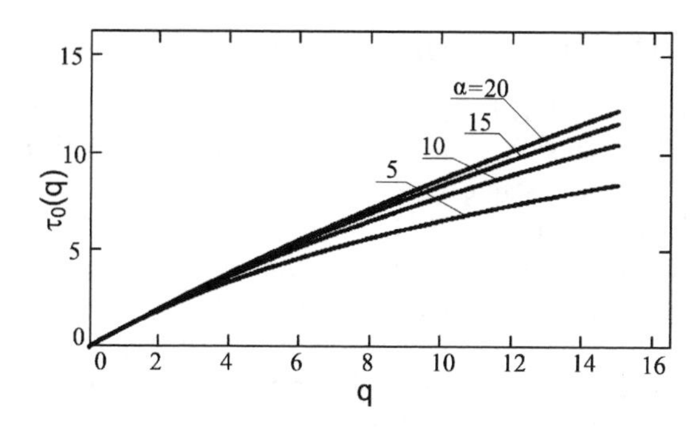

Figure 2.41 Scaling function $\tau_0(q)$ for α = constant

Thus, the modelling process represents a combination of the multiplicative cascade and the independent and identically distributed log-normal process. The resulting traffic model is capable of covering all multifractality characteristics determined by its scaling function and the moment coefficient.

The Gaussian process with the scaling property is multifractal and its parameters have the following form:

$$\tau(q) = \frac{q}{2}[\tau(2) + 1] - 1$$

$$c(q) = \frac{[2c(2)]^{q/2}}{\sqrt{\pi}} \Gamma\left(\frac{q+1}{2}\right)$$

For FBM at $q = 2$, $c(2) = 1$ and $\tau(2) = 2H - 1$ and then

$$\tau(q) = qH - 1$$

$$c(q) = \frac{2^{q/2}}{\sqrt{\pi}} \Gamma\left(\frac{q+1}{2}\right) \tag{2.84}$$

From this in the Gaussian case $\tau_0(q) = \tau(q) + 1 = qH$. Thus, the accurate and approximated mathematical models of the multifractal traffic are offered on the basis of the combination of multiplicative cascades and measured statistical characteristics of telecommunication traffic.

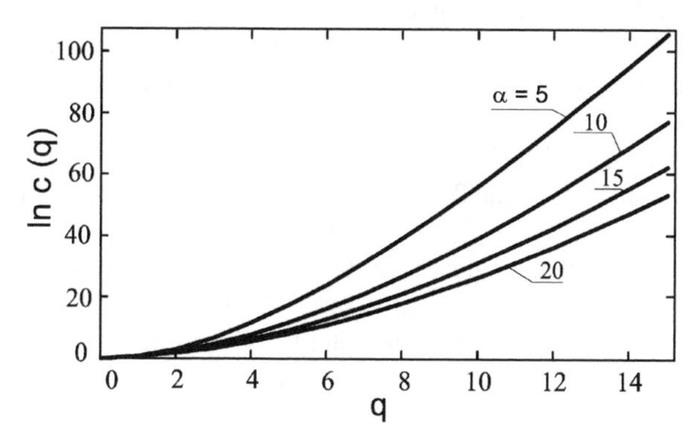

Figure 2.42 Function ln $c(q)$ for different α values

2.7 LRD Traffic Modelling with the Help of Wavelets

The algorithm for synthesis of self-similar traffic flow on the basis of the inverse discrete wavelet transform (IDWT) lies at the heart of modelling LRD traffic with the help of wavelets. It consists of the formation of the discrete time series $V(t) : v(t_0), v(t_1), \ldots, v(t_{N-1})$, consisting of N samples over the given time interval $(0, T]$ with the resolution Δt such as $\Delta t = T/N$, with the help of scaling and wavelet coefficients. Let, for example, the informational transmission rate of the data, which arrive during the time interval $(t_n, t_{n+1}] - v(t_n)$ (bits/s), correspond to each n-th sample $(n = 0, 1, 2, \ldots, N-1)$ of the series $V(t)$. The interval size $\Delta t = t_{n+1} - t_n$ is defined in accordance with the parameters of the original traffic flow.

The time series V(t) is represented in the form

$$V(t) = V_J(t) + \sum_{j=1}^{J} D_j(t) \tag{2.85}$$

where

$$V_J(t) = \sum_{k=0}^{n_0/2^J-1} s_{J,k}\varphi_{J,k}(t)$$

is the initial approximation function corresponding to the scale $J(J \leq J_{\max})$; $s_{J,k} = \langle V(t), \varphi_{J,k}\rangle$ is the scaling coefficient equal to the scalar product of the initial series $V(t)$ and the scaling function of the coarsest scale J, displaced by k scale units to the right from the coordinate origin;

$$D_j(t) = \sum_{k=0}^{n_0/2^j-1} d_{j,k}\psi_{j,k}(t)$$

is the detailed function of the jth scale; and $d_{j,k} = \langle V(t), \psi_{j,k}\rangle$ is the wavelet coefficient of scale j, equal to the scalar product of the initial series $V(t)$ and the wavelet of scale j, displaced by k scale units to the right from the coordinate origin.

The time series $V(t)$, synthesized with the help of the IDWT, is presented as the superposition of the detailed functions $D_j(t)$, determined for each jth scale, $j = 1, 2, \ldots, J$, where $J \leq J_{\max}$, and the initial approximation function $V_j(t)$ corresponding to the selected J scale. In accordance with Equation (2.86), the synthesis problem for the time series $V(t)$ divides into two stages:

First stage: the synthesis of the initial approximation $V_j(t)$ on the J scale.
Second stage: the synthesis of the detailed functions $D_j(t)$ on scale $j = 1, 2, \ldots, J$.

To carry out the first stage it is necessary to know the distribution function of the scale coefficients over the J scale of the set time interval $(0, T]$. In the general case, the given distribution function may have any form. The conditions, presented to the distribution function of the scale coefficients, are:

(a) its positiveness;

(b) the equality of its expectation to the product of the average transmission rate in the maximal loading period V and the normalized multiplier $2^{j/2}$, corresponding to the J scale:

$$M[s_{J,k}] = V2^{J/2}, \quad k = 0, 1, \ldots, K_j - 1. \tag{2.86}$$

To carry out the second stage it is necessary to know the distribution functions of wavelet coefficients for each jth scale, $j = 1, 2, \ldots, J$. The following requirements are presented to the wavelet coefficient values:

(a) The PDF corresponding to them should have the symmetric form with respect to the expectation.
(b) The expectation of the wavelet coefficients is equal to zero. These requirements should be fulfilled for the distribution function of the wavelet coefficients over each scale $j = 1, 2, \ldots, J$.

As a result of the possible theoretical distribution analysis for the normalized wavelet coefficients two distributions are allocated that correspond most closely by to the appropriate results of the empirical observations: Gaussian and Cauchy. Knowledge of the wavelet coefficient distribution law for each scale of its wavelet transform is necessary for the solution of the synthesis problem of the self-similar time series. The wavelet coefficient variance over the jth scale is determined from the expression

$$\mu_j = M[d_x(j,k)^2] = \frac{1}{K_j}\sum_{k=0}^{K_j-1} |d_j|^2 = 2^{(2H-1)+c} \tag{2.87}$$

where $c = \log_2 C_W = \log_2[c_f C(\alpha, \psi)]$ is the parameter that is independent of the j scale (see Section 1.4.1.4 expressions (1.69) to (1.71)). It is possible to show that for the specific case, when detailed functions of all J scales are independent and formed with the help of wavelet coefficients, distributed in accordance with Gaussian law, the following equation is valid:

$$c = \log_2\left(\sigma^2_{V(t)} \frac{1 - 2^{(2H-2)}}{2^{2H-2)} - 2^{(2H-2)(J+1)}}\right) \tag{2.88}$$

where $\sigma^2_{V(t)}$ is a variance of the time series $V(t)$.

In the case of synthesis of the refining functions on the base of wavelet coefficients having another distribution, the given c value is recommended to use as the initial approximation to the true value, which allows synthesis of the series $V(t)$ with the required parameters. Thus, the Hurst exponent, the mean value, the transmission rate variance and the peak rate of the traffic are used as initial data for synthesis of self-similar traffic flow on the basis of the inverse discrete wavelet transform method.

2.8 *M/G/*∞ Model

The concept of the $M/G/\infty$ model was introduced in Reference [61] and is capable of creating approximately self-similar traffic. Let $\{X_t\}_{t=0,1,2,\ldots}$ be the denumerable process denoting the user number in the $M/G/\infty$ system at time moment t. If F is the distribution function of the user service, the correlation function for X_t can be found from the relation

$$R(k) = \lambda \int_k^{\infty} [1 - F(x)]\mathrm{d}x \tag{2.89}$$

where λ is the Poisson process intensity for the users arriving in the system.

The $M/G/\infty$ process represents an adequate alternative to the existing traffic models due to the following reasons:

1. Formula (2.89) shows that the traffic with time correlation in the wide range of timescales can be simply modelled with the help of tail behaviour control of the arbitrary distribution $F(x)$. For example, if the $F(x)$ distribution has heavy tails (or is exponential), the $M/G/\infty$ process is long-range dependent (or short-range dependent) [61].
2. The $M/G/\infty$ process models some prevailing applications very well, such as Telnet and FTP [61]. The $M/G/\infty$ process is the model that yields the mathematical interpretation.

2.8.1 $M/G/\infty$ Model and Pareto Distribution

Consider users with an independent time service found from the Pareto distribution with the location parameter k and the shape parameter $1 < \alpha < 2$. The correlation function can be written from Equation (2.90) as

$$R(\tau) = \lambda \int_{\tau}^{\infty} \left(\frac{k}{x}\right)^{\alpha} \mathrm{d}x = \frac{\lambda k^{\alpha}}{\alpha - 1} \tau^{(1-\alpha)}$$

It is known that the $\{X_t\}_{t=0,1,2,...}$ process is asymptotically self-similar if $R(\tau) \sim \tau^{-D} L(\tau)$, $\tau \to \infty$ for $0 < D < 1$, and L is the slowly varying function at infinity. Therefore, for $k \geq 0$ and $1 < \alpha < 2$ the denumerable process for the $M/G/\infty$ model with service times corresponding to the Pareto distribution is asymptotically self-similar and, hence, long-range dependent.

The process is exactly self-similar if $R(\tau) = \frac{1}{2}[(\tau + 1)^{2H} - 2\tau^{2H} + (\tau - 1)^{2H}]$ for $\frac{1}{2} < H < 1$. In this case the $\{X_t\}$ process and the aggregated (integrated) process $\{X_t^{(m)}\}$ have similar correlation functions. As as result, for Pareto service times and for arbitrary arrival intensity λ the denumerable process for the $M/G/\infty$ model is not exactly self-similar.

From Reference [62] it is known that $\{X_t\}$ has a Poisson marginal distribution with a mean value $\lambda\mu$ where μ is the expected service time. For the $M/G/\infty$ model with Pareto service times, for $\alpha > 1$ the expected service time is $\alpha k/(\alpha - 1)$. However, when $\{X_t\}$ has a Poisson marginal distribution the expected service time is $\lambda \alpha k/(\alpha - 1)$.

2.8.2 $M/G/\infty$ Model and Log-Normal Distribution

Consider the $M/G/\infty$ model for service times with the F distribution function. It is shown above that if F has the Pareto distribution, the denumerable process found from the $M/G/\infty$ model is asymptotically self-similar and, hence, long-range dependent. It will now be demonstrated that if the service time has the log-normal distribution, the denumerable process from the $M/G/\infty$ model is not long-range dependent.

The distribution function for the log-normal distribution (with the scale and shape parameters equal to 1) can be written in the form

$$P[X \geq x] \sim \frac{1}{\sqrt{2\pi}\log x} \mathrm{e}^{-\log^2 x/2} \tag{2.90}$$

From Equations (2.90) and (2.91),

$$R(\tau) \sim \lambda \int_{\tau}^{\infty} \log^{-1} x \frac{1}{(2\pi)^{1/2}} \mathrm{e}^{-\log^2 x/2} \mathrm{d}x \sim \frac{\lambda}{(2\pi)^{1/2}} \int_{\tau}^{\infty} \frac{1}{\log x \, x^{(\log x)/2}} \mathrm{d}x$$

The denumerable process from the $M/G/\infty$ model with log-normal service times is long-range dependent only if $\sum_{\tau=T}^{\infty} R(\tau) = \infty$. For large T,

$$\sum_{\tau=T}^{\infty} R(\tau) \sim \sum_{\tau=T}^{\infty} \frac{\lambda}{(2\pi)^{1/2}} \int_{\tau}^{\infty} \frac{1}{\log x \, x^{(\log x)/2}} \mathrm{d}x$$
$$\sim \frac{\lambda}{(2\pi)^{1/2}} \sum_{\tau=T}^{\infty} \sum_{x=\tau}^{\infty} \frac{1}{\log x \, x^{(\log x)/2}} \sim \frac{\lambda}{(2\pi)^{1/2}} \sum_{x=\tau}^{\infty} \frac{x - T + 1}{\log x \, x^{(\log x)/2}}$$

Since the sum $\sum_{x=1}^{\infty} 1/x^2$ is finite and

$$\frac{x - T + 1}{\log x \, x^{(\log x)/2}} \leq \frac{x}{x^{(\log x)/2}} \leq \frac{1}{x^2}$$

for large enough x, then the sum $\sum_{\tau=T}^{\infty} R(\tau)$ is also finite, and the denumerable process from the $M/G/\infty$ model with log-normal service times is not long-range dependent.

The fulfilled analysis shows that the nature of the $M/G/\infty$ queue completely changes if the service time PDF is log-normal rather than Pareto.

References

[1] V. Paxson, 'Fast, approximate synthesis of fractional Gaussian noise for generating self-similar network traffic', *Computer Communication Review*, **27**, October 1997, 5–18.

[2] J. Beran, 'A test of location for data with slowly decaying serial correlations', *Biometrika*, **76**, 1989, 261–269.

[3] D.P. Heyman, A. Tabatabai and T.V. Lakshman, 'Statistical analysis and simulation study of video teleconference traffic in ATM networks', *IEEE Transactions on Circuits and Systems for Video Technology*, **2**, 1992, 49–59.

[4] G. Ramamurthy and B. Sengupta, 'Modeling and analysis of a variable bit rate video multiplexor', in Proceedings of INFOCOM '92, Florence, Italy, 1992, pp. 817–827.

[5] M. Hayes, '*Statistical Digital Signal Processing and Modeling*', John Wiley & Sons, Ltd, 1995.

[6] R. Grunerifelder, J. Cosrnas, S, Manthorpe and A. Odinma-Okafor, '*Characterization of video codecs as autoregressive moving average processes and related queueing system performance*', *IEEE Journal on Selected Areas in Communications*, **9**, April 1991, 284–293.

[7] D.R. Brillinger, '*Time Series: Data Analysis and Theory*', Holden Day, San Francisco, CA, 1981.

[8] D.R. Cox, 'Long-range dependence: a review', in editors, '*Statistics: An Appraisal*' (eds H.A. David and H.T. David), Iowa State University Press, 1984, pp. 55–74.

[9] J. Beran, '*Statistics for Long-Memory Processes*', Chapman & Hall, New York, 1994.

[10] J. Geweke and S. Porter-Hudak, 'The estimation and application of long memory time series models', *Journal of Time Series Analysis*, **4**, 1983, 221–238.

[11] J. Beran, 'Statistical methods for data with long-range dependence', *Statistical Science*, **7**(4), 1992, 404–416. With discussions and rejoinder, pp. 404–427.

[12] P.M. Robinson, 'Log-periodogram regression of time series with long range dependence', *The Annals of Statistics*, **23**, 1995, 1048–1072.

[13] P.J. Brockwell and R. A. Davis, '*Time Series: Theory and Methods*', 2nd edn, Springer-Verlag, New York, 1991.

[14] J.R.M. Hosking, 'Fractional differencing', *Biometrika*, **68**(1), 1981, 165–176.

[15] B.K. Ryu and S.B. Lowen, 'Point process models for self-similar network traffic, with applications', *Stochastic Models*, 1998.

[16] S.B. Lowen, 'Fractal renewal processes as a model of charge transport in amorphous semiconductors'. *Phys. Rev. B*, **46**, 1992, 1816–1819.

[17] S.B. Lowen, 'Fractal stochastic processes', PhD thesis, Columbia University, 1992.

[18] S.B. Lowen and M.C. Teich, 'Doubly stochastic Poisson point process driven by fractal shot noise', *Physical Review A*, **43**, 1991, 4192–4215.

[19] S.B. Lowen and M.C. Teich, 'Fractal renewal processes generate $1/f$ noise', *Physical Review E*, **47**, 1993, 992–1001.

[20] S.B. Lowen and M.C. Teich, 'Estimation and simulation of fractal stochastic point processes', *Fractals*, **3**, 1995, 183–210.

[21] M. Bartlett, 'The spectral analysis of point process', *J. Roy. Stat. Soc. B*, **25**(2), 1963, 264–296.

[22] B.K. Ryu, 'Fractal network traffic: from understanding to implications', PhD thesis, Columbia University, 1996.

[23] A. Papoulis, '*Probability, Random Variables, and Stochastic Processes*', 3rd edn, McGraw Hill, 1991.

[24] J.M. Berger and B.B. Mandelbrot, 'A new model for the clustering of errors on telephone circuits', *IBM Journal of Research Development*, **7**, 1963, 224–236.

[25] B.B. Mandelbrot, 'Self-similar error clusters in communications systems and the concept of conditional systems and the concept of conditional stationarity', *IEEE Transactions on Communications Technology*, **COM-13**, 1965, 71–90.

[26] D. Veitch, 'Novel methods of broadband traffic', in Proceedings of Globecom '93, Houston, TX, 1993, pp. 1057–1061.

[27] B.K. Ryu and S.B. Lowen, 'Point process approaches to the modeling and analysis of self-similar traffic: Part I–Model construction', in Proceedings of IEEE INFOCOM '96, San Francisco, CA, 1996.

[28] R. Jain and S.A. Routhier, 'Packet trains: measurements and a new model for computer network traffic', *IEEE Journal on Selected Areas in Communications*, **4**, 1986, 986–995.

[29] S.I. Resnick, 'Heavy tail modeling and teletraffic data', Preprint, School of ORIE, Cornell University, Ithaca, NY, 1995.

[30] A. Karasaridis and D. Hatzinakos, 'On the modeling of network traffic and fast simulation of rare events using α-stable self-similar processes', in Proceedings of the IEEE SP Workshop on '*Higher-Order Statistics*', Banf, Canada, 1997, pp. 268–272.

[31] X. Yang, A.P. Petropulu and V. Adams, 'Ethernet traffic modeling based on the power-law Poisson model', in 33rd Annual Conference on '*Information Sciences and Systems*', Baltimore, MD, March 1999.

[32] G. Samorodnitsky and M.S. Taqqu, '*Stable Non-Gaussian Processes: Stochastic Models with Infinite Variance*', Chapman & Hall, New York, London, 1994.

[33] W. Willinger, V. Paxson and M.S. Taqqu, 'Self-similarity and heavy tails: structural modeling of network traffic', in '*A Practical Guide to Heavy Tails: Statistical Techniques and Applications*' (eds, J. Adler, R. E. Feldman, and M. S. Taqqu), Birkhauser, Boston, MA, 1998.

[34] M.S. Taqqu, W. Willinger and R. Sherman, 'Proof of a fundamental result in self-similar traffic modeling', *Computer Communication Review*, **27**, 1997, 5–23.

[35] B.B. Mandelbrot, 'Long-run linearity, locally Gaussian processes, H-spectra and infinite variances', *International Economic Review*, **10**, 1969, 82–113.

[36] M.S. Taqqu and J. Levy, 'Using renewal processes to generate long-range dependence and high variability', in '*Dependence in Probability and Statistics*' (eds, E. Eberlein and M. S. Taqqu), Birkhauser, Boston, MA, 1986, pp. 73–89.

[37] C.W.J. Granger, 'Long memory relationships and aggregation of dynamic models', *Journal of Econometrics*, **14**, 1980, 227–238.

[38] S.B. Lowen and M.C. Teich, 'Fractal shot noise', *Phys. Rev. Lett.*, **63**, 1989, 1755–1759.

[39] S.B. Lowen and M.C. Teich, 'Power-law shot noise', *IEEE Transactions on Information Theory*, **IT-36**(6), 1990, 1302–1318.
[40] B.E.A. Saleh and M.C. Teich, 'Multiplied-Poisson noise in pulse, particle, and photon detection', *Proceedings of IEEE*, **70**, 1982, 229–245.
[41] B.K. Ryu and S.B. Lowen, 'Modeling, analysis, and generation of self-similar traffic with the fractal-shot-noise-driven Poisson process', in Proceedings of IASTED on '*Modeling and Simulation*', Pittsburgh, PA, 1995.
[42] B.V. Gnedenko, 'To theory of attraction domain for stable laws' (in Russian), *Uchenie zapiski MGU*, **30**, 1939, 61–82.
[43] V.M. Zolotarev, '*Stable Laws and Its Applications*' (in Russian) Znanie, Moscow, 1984, p. 64; '*New in a Life, Science, Engineering*' (in Russian), Series in Mathematica, Cybernetics, No. 11.
[44] P. Levy, 'Random functions: general theory with special reference to Laplacian random functions', *University of California Published Statistics*, **1**, 1953, 331–390.
[45] B.B. Mandelbrot and J.W. Van Ness, 'Fractional Brownian motions, fractional noises and applications', *SIAM Rev.*, **10**, 1968, 422–437.
[46] A. Karasaridis and D. Hatzinakos, 'On the modeling of network traffic and fast simulation of rare events using stable self-similar processes', in Proceeding Signal Process Workshop on '*HOS*', Banff, Alberta, Canada, 1997.
[47] A. Karasaridis and D. Hatzinakos, 'Broadband heavy-traffic modeling using stable self-similar processes', in Proceedings of 2nd Canadian Conference on '*Broadband Research*' (CCBR), Ottawa, Canada, 1998, pp. 157–168.
[48] F. Avram and M.S. Taqqu, 'Weak convergence of moving averages with infinite variance', in '*Dependence in Probability and Statistics: A Survey of Recent Results*' (eds E. Eberlein and M. S. Taqqu), Birkhauser, Boston, MA, 1986, pp. 399–416.
[49] J.M. Chambers, C.L. Mallows and B.W. Stuck, 'A method for simulating stable random variables', *Journal of the American Statistical Association*, **71**, 1976, 340–344.
[50] J.M. Chambers, C.L. Mallows and B.W. Stuck, 'Correction to: a method for simulating stable random variables', *Journal of the American Statistical Association*, **82**, 1987, 704.
[51] I. Norros, 'On the use of fractional Brownian motion in the theory of connectionless networks', *Journal of Selected Areas in Communications*, **13**(6), 1995, 953–962.
[52] I. Norros, 'A Storage model with self-similar input', *Queuing Systems*, **16**, 1994, 387–396.
[53] N. Laskin, I. Lambadaris, F.C. Harmantzis and M. Devetsikiotis. 'Fractional Levy motion and its application to network traffic modeling', *Computer Networks*, **40**, 2002, 363–375.
[54] A. Karasaridis and D. Hatzinakos, 'Network heavy traffic modeling using α-stable self-similar process', *IEEE Transactions on Communications*, **49**(7), 2001, 1203–1214.
[55] A.C. Gilbert, W. Willinger, and A. Feldmann, 'scaling analysis of conservative cascades, with applications to network traffic', *IEEE Transactions on Information Theory*, **45**(3), 1999, 971–991.
[56] R.H. Riedi, 'Multifractal processes', in *Theory and Applications of Long Range Dependence* (eds P. Doukhan, G. Oppenheim and M. S. Taqqu), Birkhauser, Boston, MA, 2002.
[57] A. Fisher, L. Calvet and B.B. Mandelbrot, 'Multifractality of Deutschmark/US dollar exchange rates', Yale University, 1997, Working Paper.
[58] Trang Dinh Dang, 'New results in multifractal traffic analysis and modeling', PhD Dissertation, Budapest, Hungary, 2002.
[59] A. Feldmann, A.C. Gilbert and W. Willinger, 'data networks as cascades: investigating the multifractal nature of Internet WAN traffic', *ACM Computer Communication Review*, **28**, 1998, 42–55.
[60] R.H. Riedi and W. Willinger, 'Toward an improved understanding of network traffic dynamics', in '*Self-Similar Network Traffic Analysis and Performance Evaluation*', (eds K. Park and W. Willinger), Wiley–Interscience, New York, June 1999.
[61] V. Paxon and S. Floyd, 'Wide-area traffic: the failure of poisson modelling', in Proceedings of the ACM Sigcomm '94, London, 1994, pp. 257–268.
[62] D.R. Cox and V. Isham, '*Point Process*', Chapman & Hall, London, 1980.

3

Self-Similarity of Real Time Traffic

3.1 Self-Similarity of Real Time Traffic Preliminaries

The real time traffic self-similarity will be considered by examples of two of the most important components of modern telecommunication networks: voice and video traffic.

The voice service is one of the most important services in modern communication networks. The traffic characteristics generated by separate voice sources strongly depend on the used voice encoder (codec). Two classes of voice codecs will be singled out as well as the flows generated by them. The traffic flows with a constant bit rate (e.g. G.711 waveform codecs) are related to the first class. Voice traffic flows produced at codec outputs using pause suppression and generating the active (ON) and passive (OFF) periods following each other will be referred to as the second class. The main purpose of the voice codec is to produce the analogue-to-digital (A/D) conversion of the signal as well as its digital compression. The hybrid codecs (e.g. GSM 6.10, G.723.1 and G.729A) are used in Internet telephony more often than others [1,2]. The main point is that similar codecs generate the audio frames with the constant bit rate. In the case when the pause suppression circuit is used, the codecs can operate in two modes: the pause mode with zero bit intensity (or reduced bit rate for several codec types) and the active mode with the compressed digital flow rate. Independently of the mode the frame duration and the frame size remain constant. From the point of view of modelling, the second class is important. Therefore the traffic models of the VoIP ON/OFF-type codecs will be discussed.

There are several advantages in the case of voice transmission using VoIP technology: the reduced communication cost, the use of joined IP infrastructure, the use in multimedia applications, etc. Moreover, VoIP applications use the advantages of packet switching in the network. For example, highly effective network utilization can be achieved for these applications by keeping this quality of the networks using channel switching.

For VoIP applications the network control system should support the quality of service (QoS). The design and analysis of QoS method performance for VoIP require the correct traffic models.

The problem of packet voice traffic modelling is not new and has been considered in many of the earlier publications. In Reference [1] the method of the classical algorithm for voice activity detection (VAD) when the algorithm parameters are fixed in time is discussed. The codec traffic based on the classical VAD algorithms consists of the voice periods and the intervals between

Self-Similar Processes in Telecommunications O. I. Sheluhin, S. M. Smolskiy and A. V. Osin
© 2007 John Wiley & Sons, Ltd

them. The length of these periods can be successfully modelled using the exponential distribution, which means using the usual Markovian model with two states for the separate sources. To analyse the performance of the systems with VoIP traffic it is necessary to attract the analysis of the appropriate queuing models. The total traffic arrives in a queue in the queueing system and this total traffic represents the multiplexed traffic of input channels. Thus the arriving packets will be served in accordance with the used service discipline. In the case where only the voice flows arrive in the queue, the FIFO (first in first out) discipline can be used. However, if the channels are divided between data and voice packets, as a rule the priority service is used. To obtain the analytical results for the queueing systems (e.g. the distribution of the packet waiting time in the queue) the simple, accessible to analytical interpretation and accurate approximation of the additive process of the packet arrivals is needed. The Poisson model is one of the simplest models, which is widely used in the classical theory of traffic modelling. However, traffic in the data networks shows other characteristics, leading to the Poisson approximation being used in specific conditions only.

The aggregated voice traffic can frequently be considered as the superposition of many separate independent ON/OFF sources, which transmit with the same intensity but with durations that are distributed according to the heavy tail distribution. At the limit, for the infinite number of sources, the ON/OFF model converges to FGN. Therefore, the investigation of aggregated VoIP traffic is carried out below on the basis of this viewpoint.

Nevertheless, these models are difficult to parametrize using the network parameters. In particular, they do not take into account the reason for burst rises in the network traffic. The information on the connection layer allows a careful analysis of traffic bursts to be carried out.

Traffic bursts arise in aggregated traffic models (including ON/OFF models) as a result of the large number of connections starting to transmit the bytes or the packets simultaneously. In this chapter an estimation will be made of the statistical characteristics of voice traffic on the basis of real measurements in the telecommunication networks at the connection layer, as well as at the packet layer. Simulation results of similar processes using the NS2 software package will also be considered.

The results show that in the case of a high workload the Poisson model does not correspond to the modelling results and the aggregated process is strongly correlated and demonstrates the properties of long-range dependence. The estimate on self-similarity carried out with the help of various tests shows that the self-similarity property of total traffic should be taken into account during modelling of a high intensity of call arrivals. It is shown that the fractional Gaussian noise can be used as well as more complicated multicomponent models to simulate the aggregated VoIP traffic.

3.2 Statistical Characteristics of Telecommunication Real Time Traffic

3.2.1 Measurement Organization

The organization and measurement results for voice and video traffic are considered for the example of a distributed telecommunication network, the simplified configuration of which is shown in Figure 3.1. The telecommunication network (TN) covers a considerable terrain and combines a large number of various protocols of the link layer. The figure shows only the generalized network structure, demonstrating the possible tracks of voice calls that are passing. There are terrestrial and satellite communication channels inside the network.

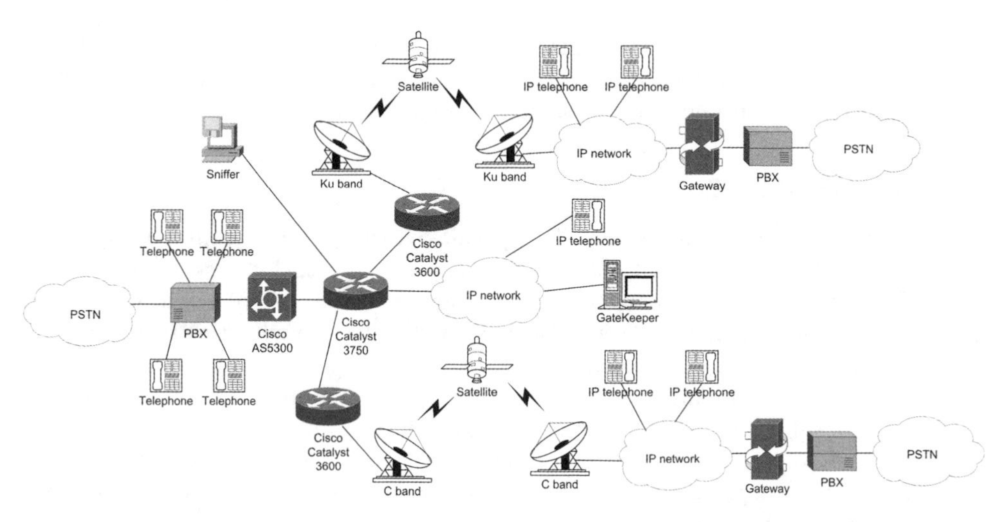

Figure 3.1 Telecommunication network configuration at measurement realization

An experiment on research of the statistical characteristics of TN traffic was carried out in the following way [3]. In the network configuration shown in Figure 3.1 the device (Cisco Catalyst 3750) is the network 'kernel', the free port of which was connected through the RJ-45 interface to the personal computer (PC) with the measurement software and the scheduler, which initiated the chosen traffic measurement program every three hours. At the link layer the connection was realized over the fast Ethernet protocol. The chosen device port was configured as the SPAN port. As measurement software, the sniffer software 'windump' was used, which catches all packets falling into the interface. After 48 million packets were 'sniffed' the sniffer stopped. The total measurement time was equal to 96 hours. As a result 35 files were recorded. Since the sniffer software caught total traffic passing through Cisco Catalyst 3750, the recorded (by the IP addresses known in advance) log-files were subjected to filtering in order to allocate the voice packets.

For this purpose a special filter was created to which the parameters were added for analysis of the packets having the sizes of 74 bytes, 78 bytes, 84 bytes and 88 bytes only. These packet sizes are defined by the VoIP codec used in the system, with a frame size equal to 10 bytes, since encapsulation in the VoIP system was realized as two or three frames in one IP packet. The 4 byte difference in the sizes of voice frames can be explained by the presence or the absence of four MPLS (multiprotocol label switching) bytes in the heading of the link layer.

When measuring the traffic transmitted in IP packets by the transport layer protocols, such as TCP, UDP, ICMP (Internet control message protocol) and OSPF (open shortest path first), as well as by the application layer protocols, such as HTTP (hypertext transfer protocol), FTP (file transfer protocol) and SMTP (simple mail transfer protocol), time series were obtained, which demonstrate the measured traffic used in the analysis in bytes per second.

The analysis shows that TCP data of the transport layer dominate in the traffic and the character of these data determines the characteristics of IP traffic. Besides TCP traffic there is a large fraction of UDP traffic, but this traffic type is smoother than TCP and has no strong impact on IP traffic dynamics.

The measured data represented the Ethernet frames and the UDP, and the RTP (real time protocol) over it was used at the transport layer. During the research the frames for connection which made the TCP operate at the transport layer were not taken into consideration.

3.2.2 Pattern of TN Traffic

To substantiate the analysed network model a quantitative analysis of the traffic obtained at the single measurement was carried out. For this purpose the media access control (MAC) addresses as well as corresponding devices were registered, the activity of which was fixed in the corresponding traffic log-file. The following traffic characteristics were registered:

1. *Total traffic*. This traffic, measured in bytes and packets, demonstrates the packet number (or the total packet size in bytes) for which the MAC address of the appropriate device or the MAC address of the destination was present in the field of the source's MAC address as well as similar values in percentage of the total packet number, fixed when measurements were taken or drawn from their total number in bytes respectively.
2. *Input traffic*. This traffic, classified during the analysis as incoming into the switch and used for the measurement, was fixed in the packet form as well as in the total volume in bytes.
3. *Output traffic*. This is the traffic that is classified at the analysis as outgoing from the switch and used for measurement. This parameter is also fixed in packets and in the total volume in bytes.

On the basis of the obtained data it is possible to draw the conclusion that the traffic fixed at the measurement is asymmetrical in terms of input–output. Therefore, it can be stated that not all the channels connected to the switch, the port of which was used as the SPAN port, were involved in the measurement. A similar asymmetry could occur for one of the two following reasons:

1. Several ports were not involved in the measurement, but some traffic came through these ports and this traffic fraction was switched to the ports, traffic which was overdirected to the SPAN port. As a result, these packets are fixed once and are falsely classified as incoming to the switch.
2. Several ports were not involved in the measurement but the traffic through them was switching and passed to the ports, taking part in the measurements. As a result, these packets are fixed once during the measurement and are correctly classified as incoming to the switch.

To illustrate the above-mentioned situation, Figure 3.2 explains the reasons for the asymmetry occurring in terms of input–output. The simplified measurement structure shown in Figure 3.2 will be analysed. Several devices, which will be conditionally subdivided into data sources and destinations, are connected to the ports of the switch, shown in the figure centre. The SPAN session is open on port 4 of the switch. The terminal with the loaded sniffer software is connected to this port. Ports 11 and 8 are involved in the SPAN session, but ports 1 and 3 are not involved. Therefore, during the measurement only the traffic of the ports involved in the SPAN session (i.e. 11 and 8) will be fixed. In the case where the device connected to port 11 is the traffic source and when the device connected to Port 8 is the destination, both traffic ingoing to port 8 and outgoing from port 11 will be fixed in the log-file with the measured traffic. In the case where the device connected to port 3 (not involved in the SPAN session) is the traffic

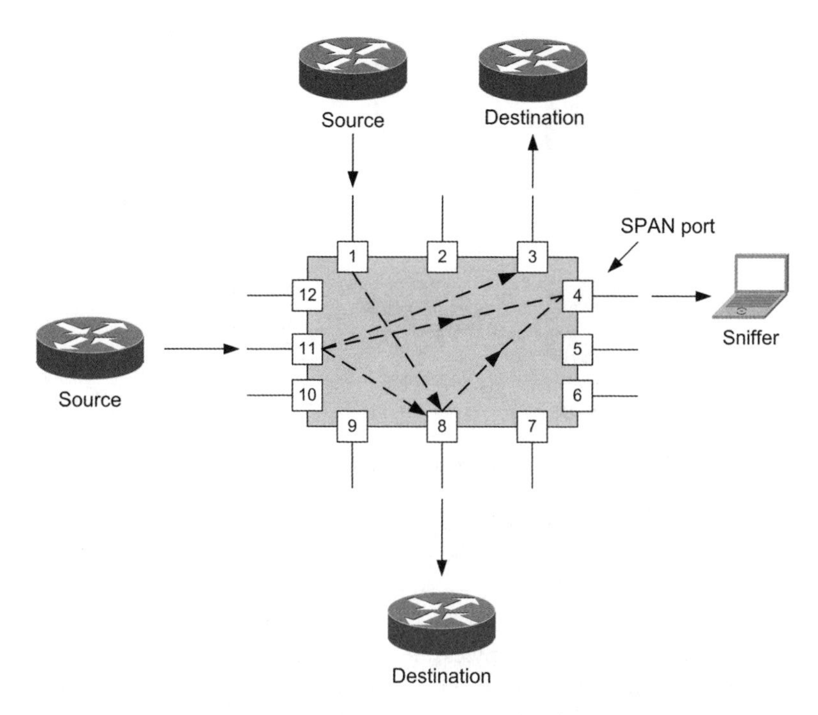

Figure 3.2 Diagram explaining the asymmetry

destination in the log-file, which is generated by sniffer software during the measurements, the traffic coming to port 11 will be fixed only. In the case where the device connected to port 1 (not involved in the SPAN session) is the traffic source and the device connected to port 8 (involved in the SPAN session) is the destination, the situation will be the inverse: in the obtained log-file of the measured traffic only the traffic from port 8 to the source is fixed, but the traffic on the input of the switch is not fixed. Each of the considered cases of passing the traffic through the switch is shown by dotted lines in Figure 3.2.

For the following analysis only the ports whose traffic was overdirected to the measuring SPAN port will be considered. Therefore, the traffic was analysed in the channels with the equipment that has the MAS address defined unambiguously.

As a second example, the generalized structural diagram of the VoIP network is shown in Figure 3.3, with the help of which the measurements of voice traffic were carried out. In this figure the clouds represent the telephone network for general use with the traffic sources in the form of voice users and IP networks, which are also the sources of telephone calls (IP telephones). The traditional telephony networks were connected to the IP network by the gateway (Cisco Catalyst 3500) through which all telephone calls in the system passed. Therefore, the measuring equipment was connected to one port of the switch for which the SPAN session was created to derive the traffic passing through the switch to the measuring port.

The order of the experiment realizations and the stages used to obtain statistical data are shown in Figure 3.4:

Step 1. Two users communicate being the PSTN (public switched telephone network) clients. The traffic created by them is transmitted from one PSTN segment to another with the

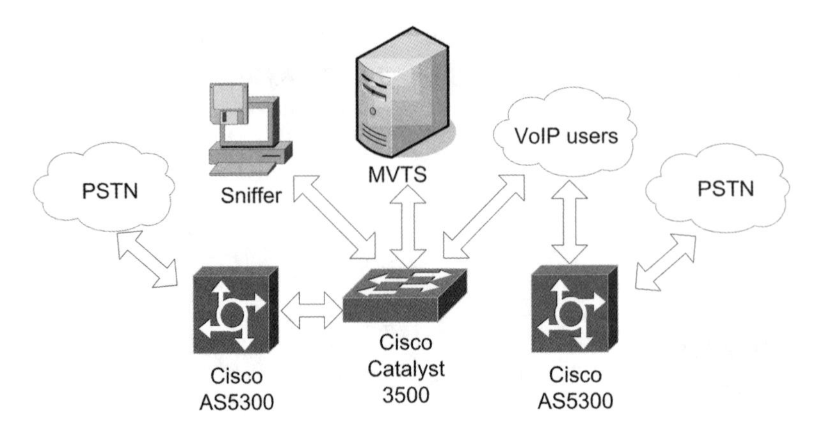

Figure 3.3 Generalized structure of the VoIP network

help of the IP network. When the traffic arrives at the central switch of the transit IP segment, it is diverted to the SPAN port and is locked by the sniffer software, being saved on the hard disk of the measuring PC.

Step 2. The primary analysis of all measurement data results is carried out (the time marks, the transmitted bytes, the received bytes, the statistics on protocols, etc.). The recorded log-file is processed and presented in a form suitable to export to the database server with the help of specific software.

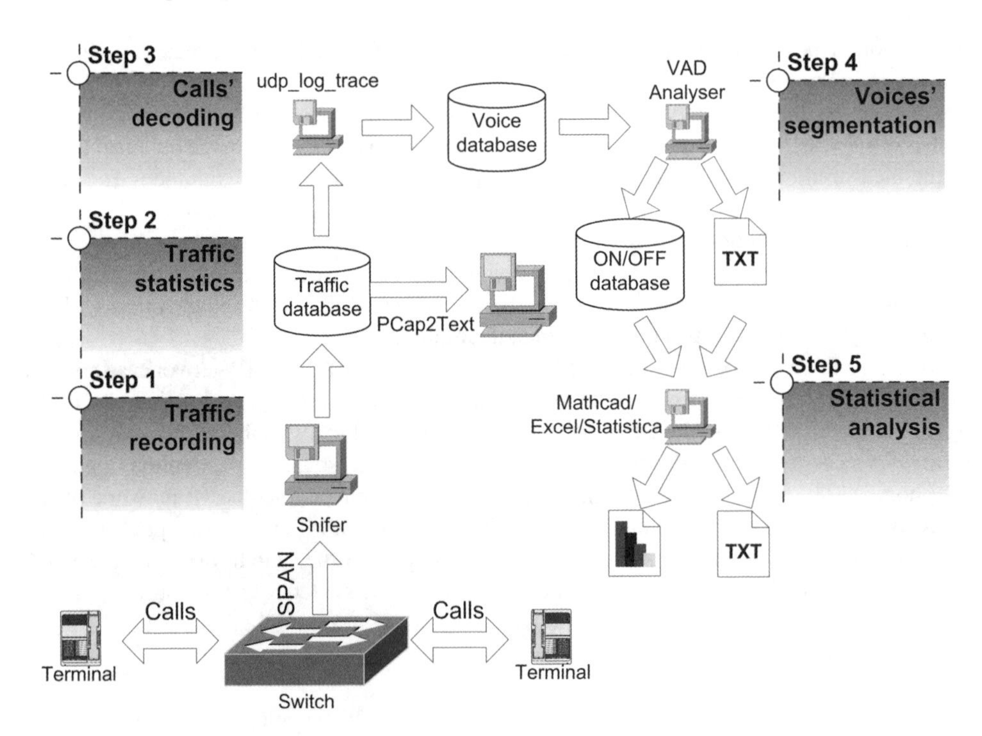

Figure 3.4 Structure of the hardware–software complex

Step 3. With the help of the software, the allocation of the separate telephone calls from the total traffic (obtained in step 1) is carried out and are saved in the form of audio files on the PC hard disk for future segmenting and obtaining of the multiplexed information flows.

Step 4. The combined algorithm of voice information segmenting is applied to the created database for the purpose of allocating ON/OFF periods in the voice.

Step 5. The statistical analysis of the obtained information is conducted.

To fulfil the multiplexing experiment 100 sources of voice information with a duration of more than 5 minutes were chosen. Each source contains the information passing to the channel from the users participating in the telephone dialogues, which are active at the considered time moment. The voice information coding was fulfilled with the help of the G.711 codec, at the output of which the packets were packed in IP packets, with each four G.711 packets for one IP packet. As a result, the multiplexing of IP packets with a size of 214 bytes was considered.

The combined algorithm for voice division into ON/OFF periods was chosen as the algorithm of segmentation of the voice information sources. This algorithm makes decisions on the basis of the energy level in the analysed frame and the zero level transition number for the given frame. Each voice source was divided into 10 ms duration frames and the decision was made concerning each frame. Because four G.711 frames of 5 ms each were used to form one IP packet, two G.711 were present in each frame and it was necessary to use two blocks of segmented information to form the IP packet. Thus, if at least one of the two subsequent segmented information blocks is active, both blocks formed the IP packet and were considered to carry significant information. If no blocks following each other were defined as active, the IP packet for this information was not created. The schedule of formation of the IP packets is sketched in Figure 3.5.

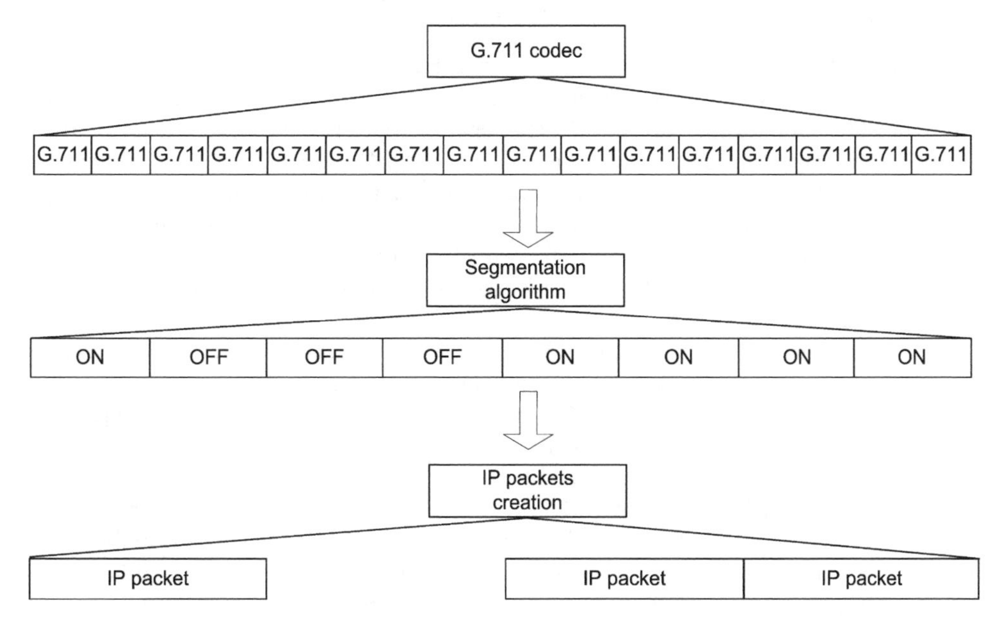

Figure 3.5 Schedule of IP packet formation on the basis of G.711 codec information

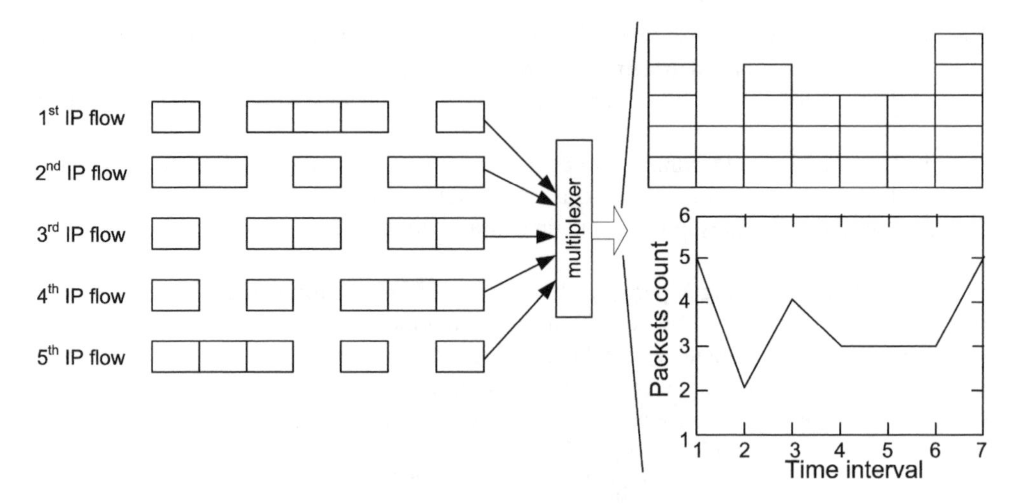

Figure 3.6 Multiplexing process at the IP packet layer

After carrying out the schedule of formation of the IP packets for each source the multiplexing operation was conducted for five IP flows, which is illustrated in Figure 3.6. After forming the multiplexed IP packets flow from different voice sources, the statistical characteristics of the obtained arrival processes were estimated.

3.3 Voice Traffic Characteristics

3.3.1 Voice Traffic Characteristics at the Call Layer

Voice traffic at the call layer consists of the call arrival process and the call duration process. The data at the call layer were obtained on the basis of the call detail record(cdr)-logs analysis, the record of which was made at the network gatekeeper [3]. Each log line presented the information about the separate VoIP telephone conversations. Analysed cdr-logs contained the time interval beginning from 02.02.2004 to 21.06.2004 covering the total 455 874 calls.

As a result of processing the cdr statistics file, the periodic data structure at the call layer was discovered. The analysis showed that all data studied could be divided into two categories: calls effected during working days and calls over weekends and holidays. Figures 3.7(a) to (c) show the call traffic realizations.

These figures show the call intensity versus the call sequence index. As the sequence resolution is equal to 1 s, the intensity has the dimension of call numbers per second. The call traffic histogram shown in Figure 3.8 during a typical working day has a complicated structure, which can be explained by the call affiliation with different groups, each of which can be characterized by its own law of call distribution.

The call arrival process can be divided into two random processes: the process of the interarrival time and the process of call duration. The realisation of the first process over the whole period of measurement is shown in Figure 3.7. The periodic bursts correspond to long time intervals when there were no calls at all. This behaviour can be explained by the reduced user's activity at night. Therefore, there are time intervals of approximately 1–1.5 hours when the

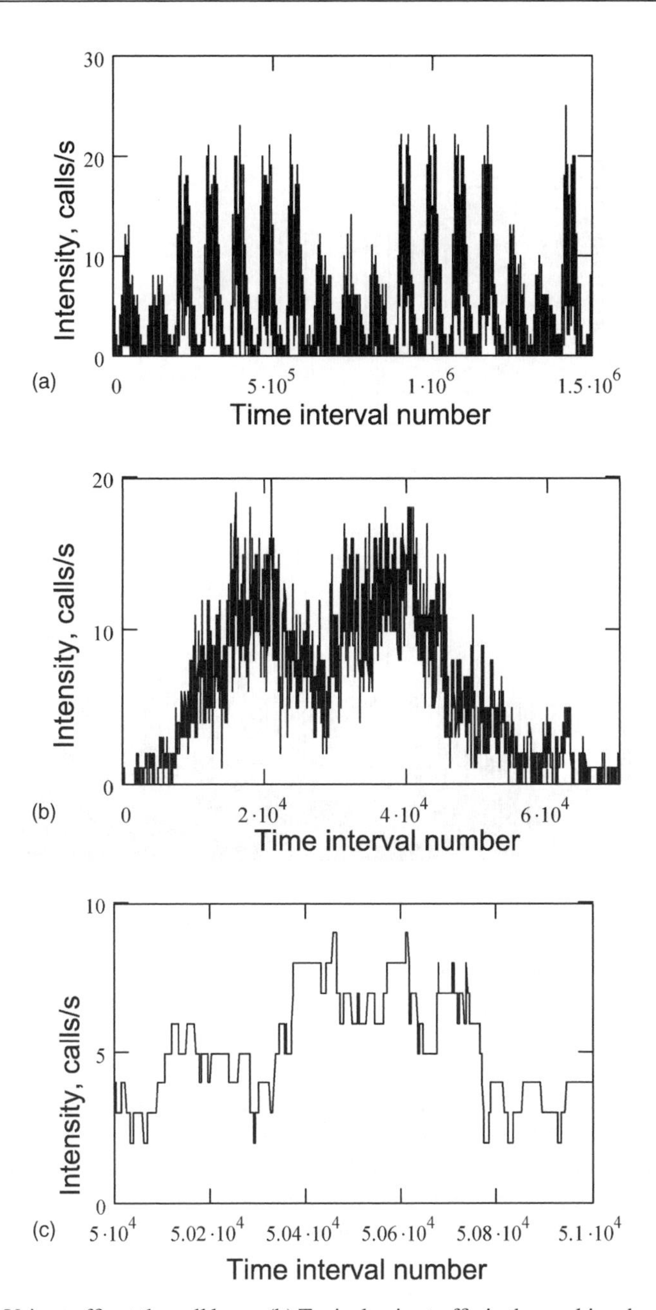

Figure 3.7 (a) Voice traffic at the call layer. (b) Typical voice traffic in the working day. (c) Voice traffic fragment at the call layer at the smaller time interval

system remains idle and no calls are observed. The process of the inter-arrival time is shown in Figure 3.9 for 3000 calls. The sample statistics of the process shown in Figure 3.9 are illustrated in Table 3.1. Since the gatekeeper registered the statistics even for those calls that were not finished for various reasons (the user changed his or her mind concerning the call but the number

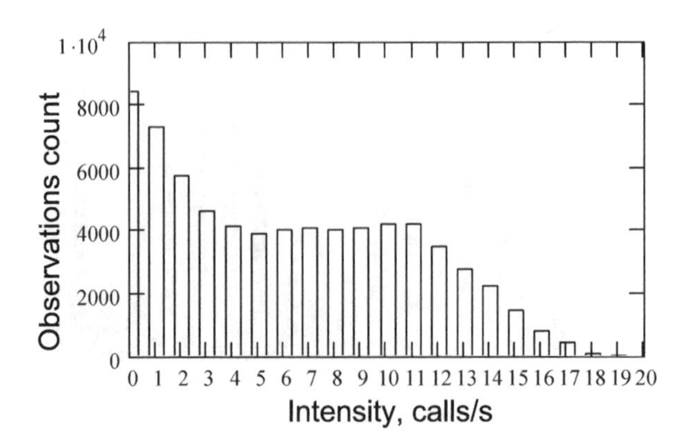

Figure 3.8 Histogram of calls in the working day

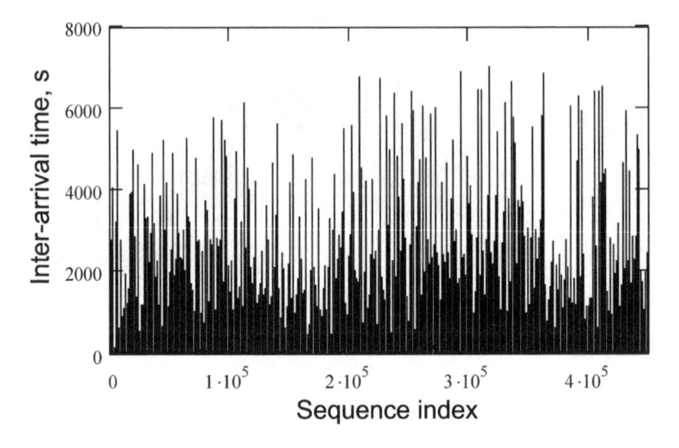

Figure 3.9 Values sequence of the inter-arrival time

was already sent or called, the receiver of the call did not answer the call, etc.), calls shorter then 10 s were excluded from the consideration.

The statistical analysis of experimental data shows that the inter-arrival time are concentrated basically within the range of 0–500 s covering 99.5 % of all the calls. The inter-arrival time are mainly concentrated within the range of 0–100 s covering 96.2 % of all calls.

Table 3.1 Sample statistics of realization shown in Figure 3.9

All observations	Observations number	Sample mean (s)	Minimal value (s)	Maximal value (s)	Standard deviation (s)
Duration >10 s	455 878	26.5	0	7033	120.8

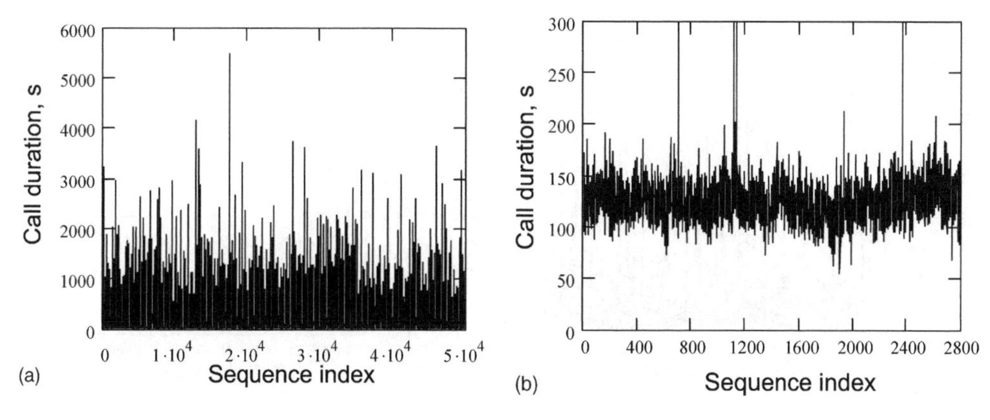

Figure 3.10 (a) Sample path of the call duration process for 50 000 calls. (b) Call duration (the aggregation level 160)

A preliminary statistical analysis of the call duration process will be carried out. Figure 3.10 shows the realization of the call duration over the whole measurement period. The main sample statistics of the call duration process is shown in Table 3.2.

The data analysis at the connections layer (Table 3.2) allows an intensive study of the traffic bursts to be conducted. In the aggregated traffic models (including ON/OFF models) the traffic bursts occur due to a large number of users who begin to transmit information simultaneously (in bytes or in packets), i.e. the bursts appear as a result of the 'constructive interference' of a large quantity of connections.

The initial calls sequence plotted in Figure 3.7(b) can be divided into two processes: the call duration process (Figure 3.11) and the process of the time inter-arrival time (Figure 3.12). The analysis shows that the distributions of the call duration process and that of the time inter-arrival time differ essentially from exponential and are well described by the distributions having 'heavy tails'. The Pareto distribution particularly gives a good approximation.

As a rule, the information at the connection layer contained in the widely available traffic traces was ignored in the course of the classical analysis of the aggregated traffic (LRD, fractal, multifractal). Nevertheless, an account of this information allows a more correct statistical study of voice traffic to be conducted, thus increasing the accuracy of the bursty network traffic modelling.

3.3.2 Voice Traffic Characteristics at the Packet Layer

The analysis of the total voice traffic at the packet layer in the distributed telecommunication network (Figure 3.1) demonstrates the composition of several processes in it (nonstationary as a rule) and shows that it can generally be presented in the form of two nonstationary components

Table 3.2 Sample statistics realization shown in Figure 3.10(a)

All observations	Observations number	Sample mean (s)	Minimal value (s)	Maximal value (s)	Standard deviation (s)
Duration $>$ 10 s	455 874	123.2	0.00	9942.0	200.03

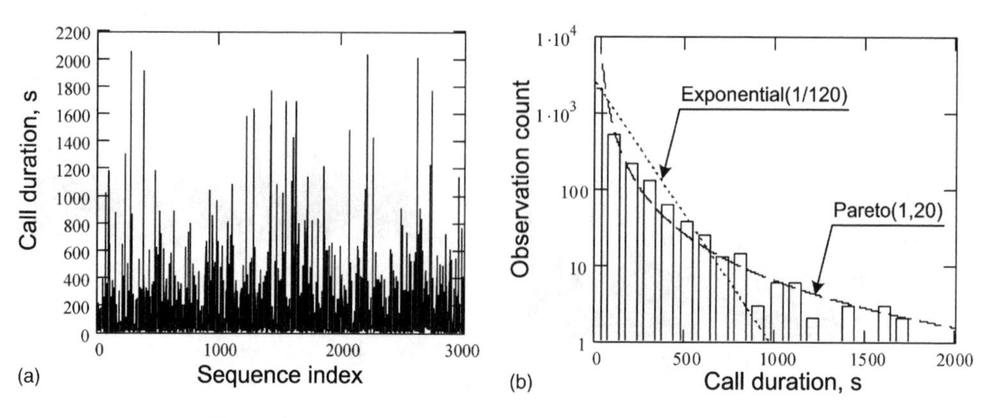

Figure 3.11 (a) Call duration process and (b) its distribution

[3]. An illustration of this type of traffic is shown in Figure 3.13 (data1, data2, etc., differ from each other by the traffic aggregation level).

The first component is conditioned by the traffic at the call layer over the large (minutes and hours) time intervals and describes the periodic structures of daily loading that are usually observed in the backbone channel. As a rule, this component has a strongly pulsating non-Gaussian (in the general case) structure, the correlation properties of which are restricted by daily traffic fluctuations.

The second component is present over small (seconds and minutes) time scales only, has a long-range character and disappears in the case of increasing the time scale for daily loading usually observed in the backbone channel.

Figure 3.14 shows the realization for VoIP network traffic when 100, 50, 25 and 10 IP flows are multiplexed. The stationarity estimation of the multiplexed voice flows by the inversion method discussed in Chapter 1 is demonstrated in Figure 3.15.

It can be seen that in the case of the single voice source the hypothesis on stationarity is rejected. This agrees with the known results concerning random sequence nonstationarity on the output of the voice encoder. In the case of 10 sources the estimation curve lies at the

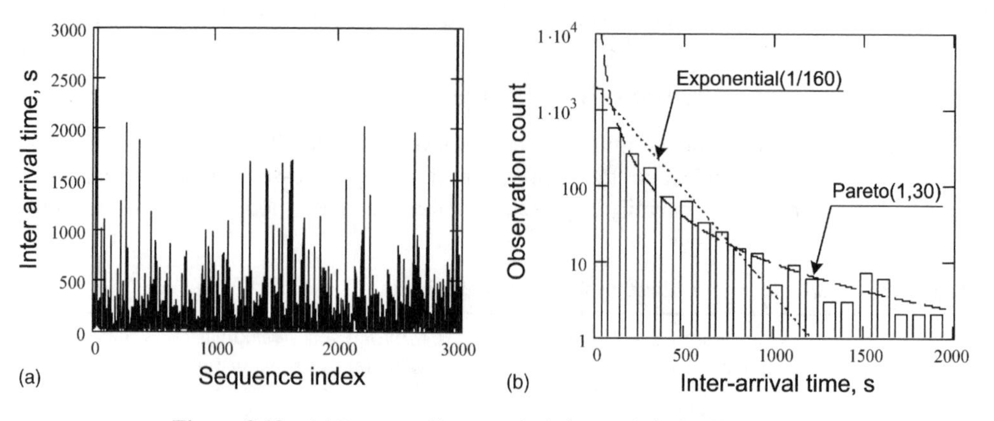

Figure 3.12 (a) Process of inter–arrival time and (b) its distribution

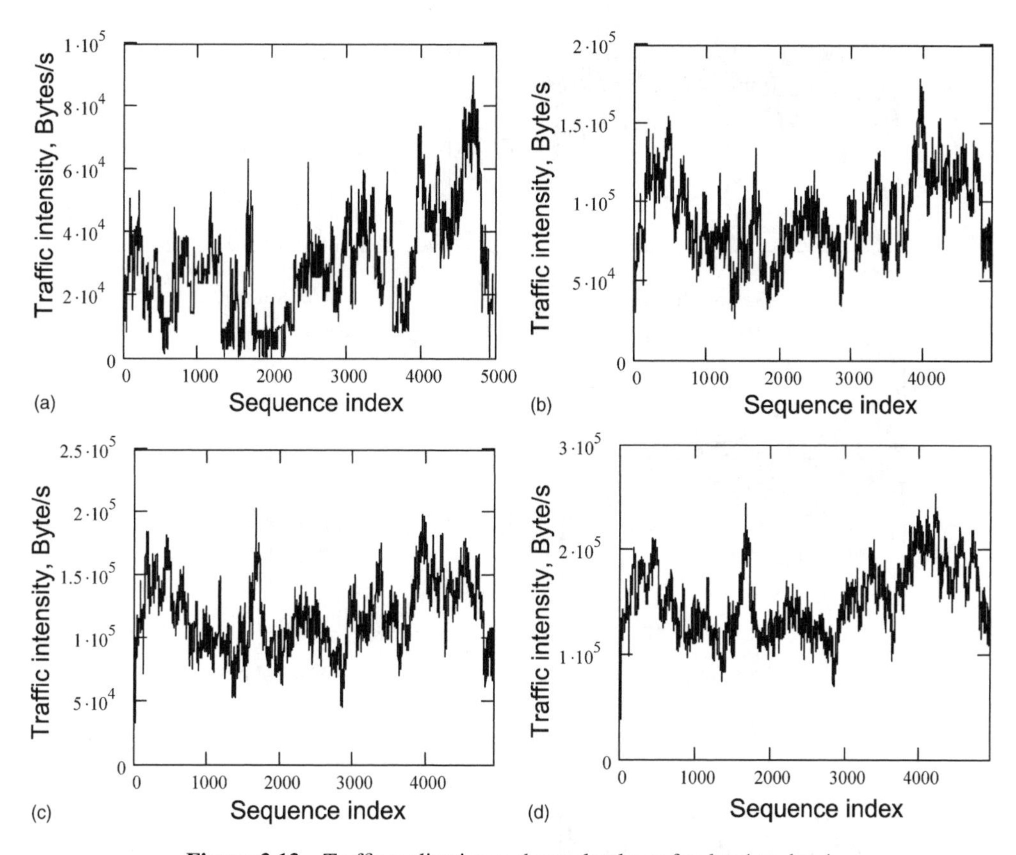

Figure 3.13 Traffic realization at the packet layer for data1 to data4

boundary of the confidence band for the significance layer $\alpha = 0.05$. However, in the case of 10 and more multiplexed sources the hypothesis on voice traffic stationarity can already be accepted [4].

Figure 3.16 shows the results of a Hurst exponent estimation for various numbers of sources being aggregated.

3.4 Multifractal Analysis of Voice Traffic

3.4.1 Basics

There is no doubt that the network traffic is created by a large number of the independent separate sources. The simple ON/OFF model assumes that these sources switch between the two states: the ON state in which the sources create the traffic with the constant rate and the OFF state in which they are silent. The aggregation of similar traffic gives the total traffic loading observed, for example, for a gateway. For this model, ON periods described by the heavy tails distributions lead to LRD similar to that observed for real traffic. This assumption is well-founded due to convincing results of modelling. However, ON/OFF models are correct at the

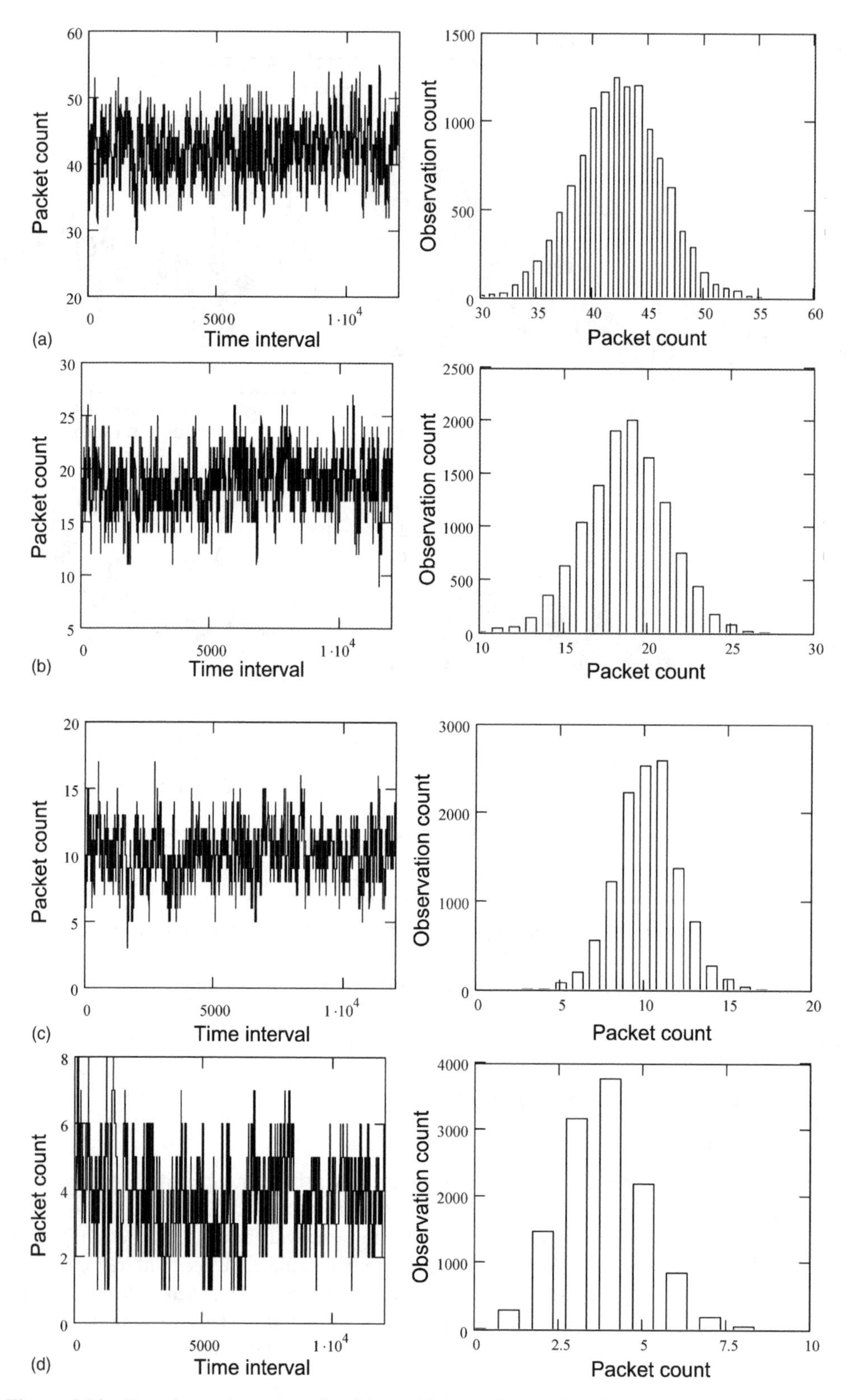

Figure 3.14 Experimental sample paths of the multiplexed flows and its histograms for: (a) 100 sources; (b) 50 sources; (c) 25 sources; (d) 10 sources

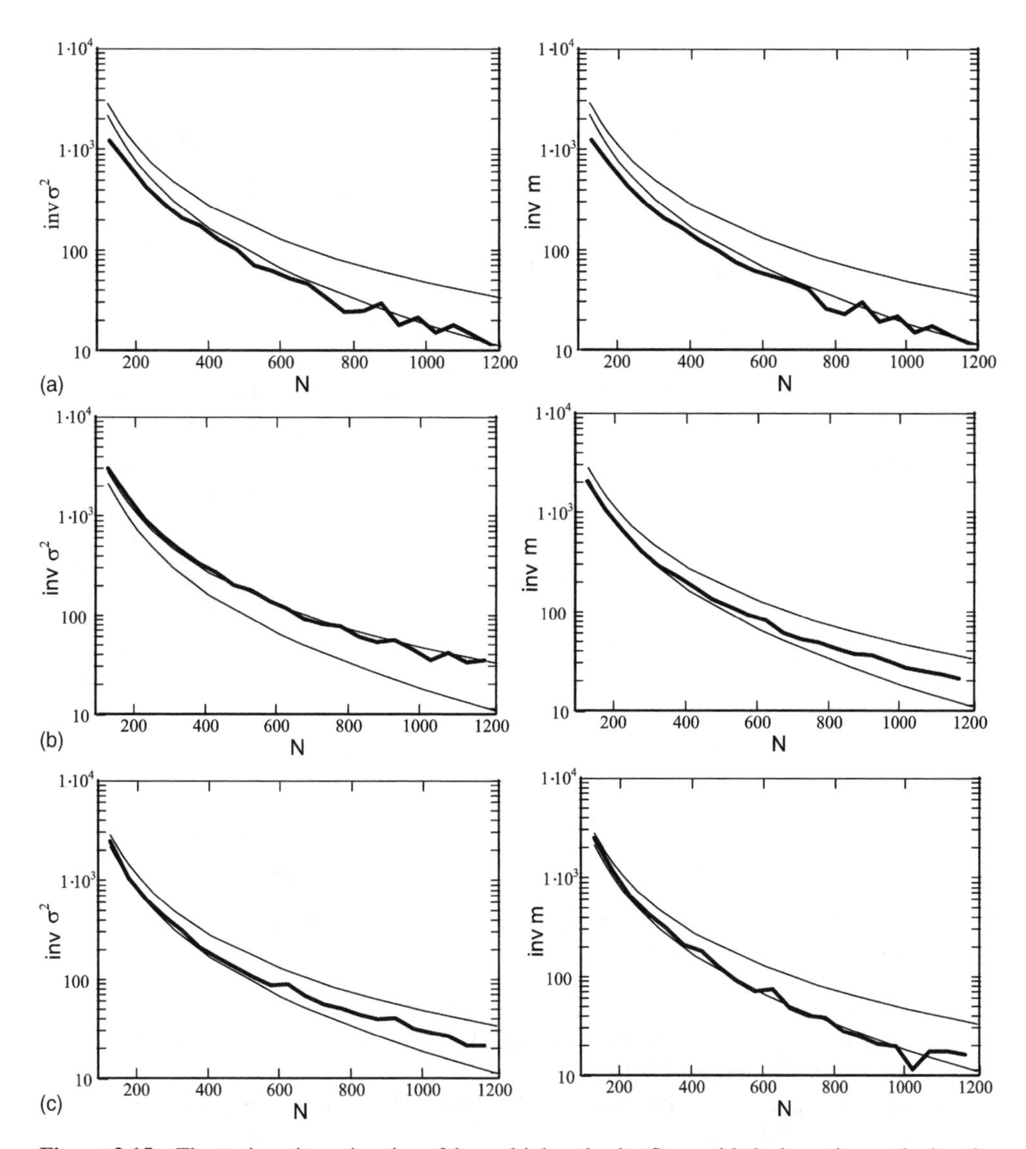

Figure 3.15 The stationarity estimation of the multiplexed voice flows with the inversion method on the variance (σ^2) and the expectation value (m) of data blocks: (a) 1 source; (b) 10 sources; (c) 100 sources

limit of large time scales only (seconds and more). They do not take into account the real queueing and the multiplexing that are happening in the network.

A more complete description of network traffic data requires an understanding of its dynamics, not only for large but also for small time scales (hundreds of milliseconds and even less). Under these conditions the multifractal analysis has an evident advantage compared to the standard statistical approaches since it gives information about local as well as global properties of the observed data. This is explained by the fact that the packet flow over the detailed time scales is mainly formed by the protocols and the transparent mechanisms of

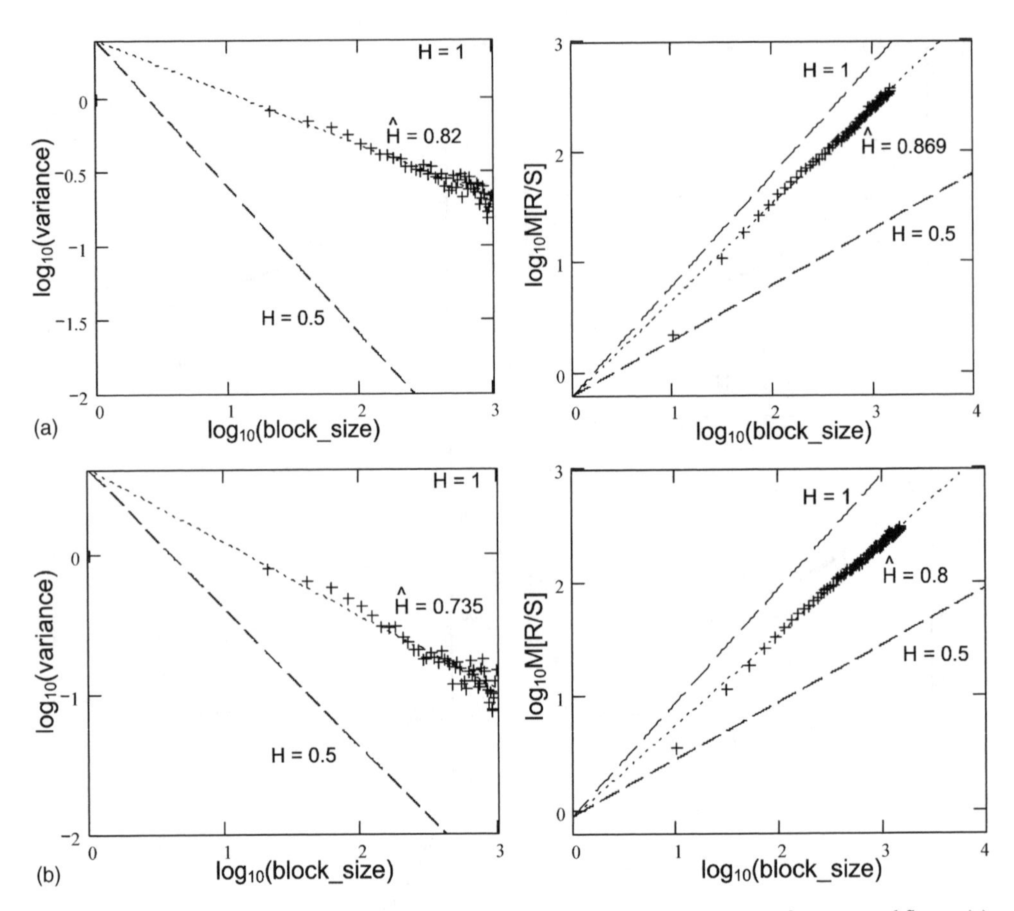

Figure 3.16 The results of a Hurst exponent estimation for a different number of aggregated flows: (a) 10 sources; (b) 25 sources; (c) 50 sources; (d) 100 sources

congestion avoidance (e.g. the TCP), which regulate the complicated interactions between various connections in the network [5].

Consider the data $(Z_i)_{i=1}^{N}$ as the μ measure discretization over [0;1] with resolution $N = 2^n$ and define the partition sum

$$S_m^Z(q) = \sum_{k=1}^{N/m} \left(\overline{Z}_k^{(m)}\right)^q$$

where Z is the vector of data for which the multifractal spectrum is created; $\overline{Z}_k^{(m)} = \sum_{l=1}^{m} Z_{(k-1)n+l}$ is the μ measure discretization over the scale $\delta_m = m/N$; and $m = 1, 2, 2^2, \ldots, 2^n$ is the size of the partitioning block. As a result of partition sum plotting, in the graphical view the curve family representing the partition sums for various q values is obtained.

If $\log S_m^Z(q)$ linearly depends on $\log(m)$ at approximation it can be said that the data demonstrate multifractal scaling, i.e. Z_i is a multifractal. The slope of the approximating

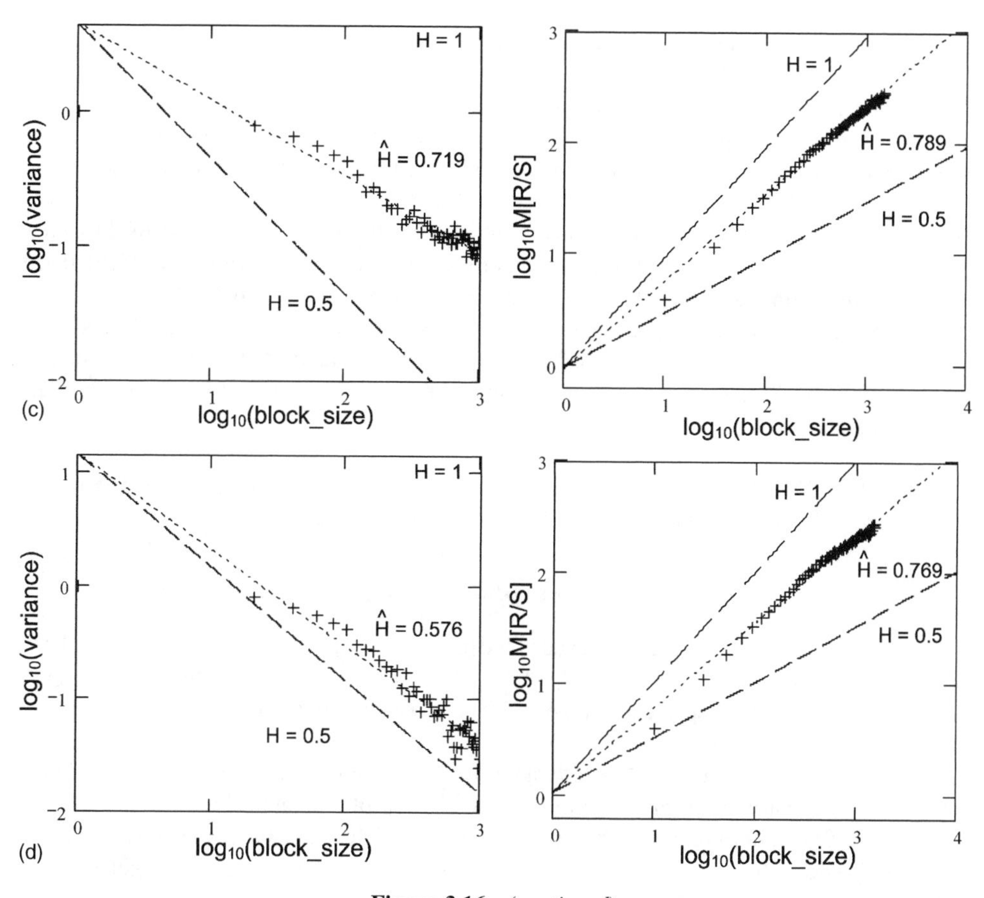

Figure 3.16 (*continued*)

straight line can be obtained using linear regression, which is denoted $\tau(q)$:

$$\log S_m^Z(q) \approx \tau(q) \log m + c(q)$$

Since $\tau(q)$ has a small slope which changes very insignificantly, its plot in the range $[\frac{1}{2};2]$ can seem almost linear. Therefore, the Legandre transform with respect to the partition function $\tau(q)$, which is designated as f_L, will be more informative. As a result, the multifractal spectrum $f_L(\alpha)$ can be found by Legandre transform with respect to the partition function $\tau(q)$: $f_L(\alpha) = \inf_{q \in R}[\alpha q - \tau(q)]$. Thus, the multifractal spectrum $f_L(\alpha)$represents the frequency measure of the singularity exponent $\alpha(t)$ till t time moment and shows the probability of the singularity exponent definite meaning. Accordingly $\tau(q) \leq \inf_{\alpha}[q\alpha - f_L(\alpha)]$. This approach is referred to as 'the multifractal analysis based on the increments'.

3.4.2 Algorithm for the Partition Function $S_m(q)$ Calculation

Step 1. The examined data array (data), the range of scale changing (m_begin; m_end) and the step of variation (m_step) are introduced into the program for the partition function calculation. If the partition function is plotted on m, a value is set for the moment (q) for which it is calculated.

Step 2. The initial realization is divided into blocks, the size of which at the first cycle iteration is chosen equal to the left boundary of the variation interval m (m_begin).

Step 3. For each of the obtained blocks the sum of the realization values (sum) included in it is calculated.

Step 4. The found sums are raised to power q and the result is consecutively summed for each block. As a result of summing over all blocks in which the realization was divided, the partition function ordinate is found for the given m value of step 2.

Step 5. Increasing m by the m_step value, the scaling resolution of the examined realization is changed and step 2 is repeated. The iterations are repeated until m reaches the m_end value. As a result, the partition function ordinates will be found for the given scale range.

The multifractal spectra construction should now be executed for the data obtained as a result of the voice traffic measurements in various telecommunication networks.

3.4.3 Multifractal Properties of Multiplexed Voice Traffic

The multifractal scaling of the multiplexed voice traffic for varying source numbers considered in Figure 3.14 can be studied with the help of the partition function. The functions $S_m(q)$ shown in Figure 3.16 illustrate the presence of the multifractal scaling for any q in the case when m was chosen equal to 10, 20, 30, ...,1000.

The scaling function has the linear character in all four plots in Figure 3.17, and the slight deviation from linearity is observed for the very detailed resolution of the log–log plot only. For each case presented in Figure 3.17, the functions $\tau(q)$, which at a visual inspection practically coincide and look linear (Figure 3.18(c)), are obtained with the help of weighted linear regression matching. From Figure 3.18(b) it can be seen that the function $c(q)$ for the small aggregation degree of the sources (<25) has a pronounced nonlinear character, but with the increase in the multiplexing degree the function $c(q)$ becomes more and more linear. To show the properties of the multifractal scaling more clearly the Legandre spectrum $f_L(\alpha)$ was plotted for the examined data (see Figure 3.18(a)). The curves shown in the figure demonstrate the narrowing of the multifractal Legandre spectrum with the increase in the multiplexed sources. This proves that the multifractal scaling region is reduced with an increase in the multiplexed voice sources [5].[1]

It should be mentioned that for a monofractal process such as fractional Gaussian noise (or process of increments of fractional Brownian motion) the singularity exponent $\alpha(t)$ is the constant value H for any t. Therefore, this situation can be considered as the degenerated case of multifractality. The appropriate partition function $\tau(q) = qH - 1$ for FGN is the linear function of q. Since $\alpha(t)$ is equal to H for any t in the FGN case, its multifractal spectrum should look like a single point in the $(H, 1)$ plane.

[1]Similar research concerning the Legendre spectrum of multiplexed traffic on the example of LAN was described in the recent book by P. Murali Krishna, Vikram M. Gadre and Uday B. Desai, *Multifractal Based Network Traffic Modeling*, Kluwer Academic Publishers,2003 pp. 119–120. However, the results obtained there, as well as their explanation, give cause for doubt.

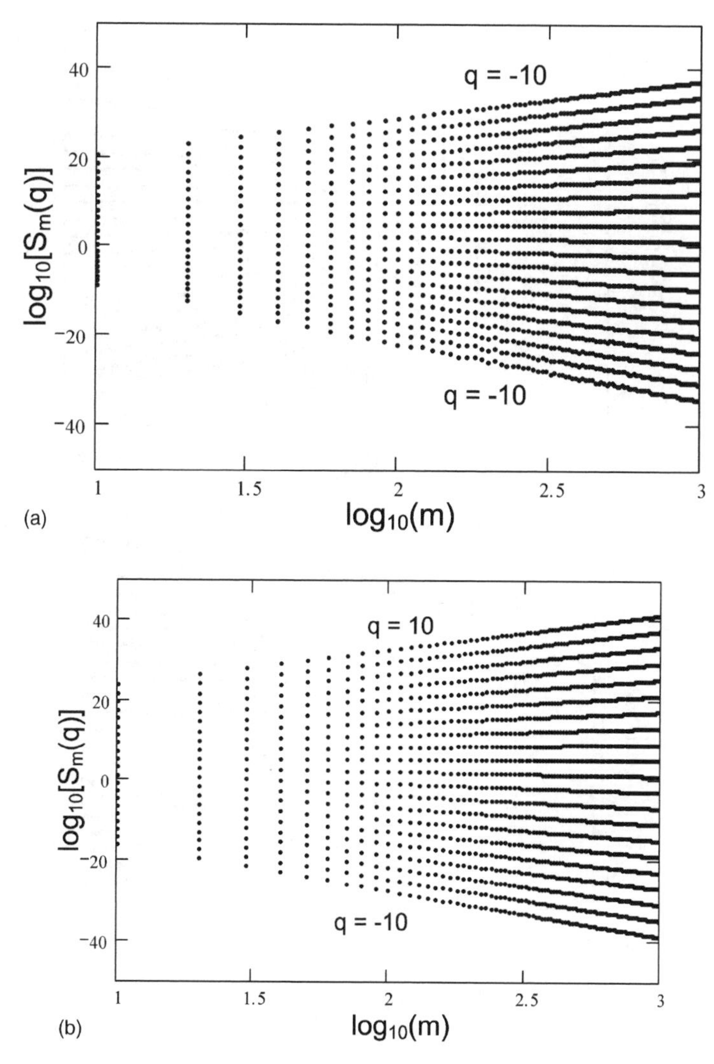

Figure 3.17 The study of the multifractal scaling with the help of the partition function $S_m(q)$ on m in the log–log scale. Looking top-down, it is found that q changes from 10 to -10 with the unitary step. The functions are shown for varying numbers of multiplexed sources: (a) 10; (b) 25; (c) 50; (d) 100

FGN will be used as the 'testing' process and the multifractal spectrum will be compared for real and artificial traffic. The partition function is the convex function of q for the multifractal process and the singularity exponent $\alpha(t)$ can accept the wide value range. The convex curves of the partition function show that real and artificial traffic are multifractal processes and the partition function for FGN is a linear function due to its monofractal behaviour. The FGN spectrum demonstrates the value of probability: $P[\alpha(t) = H] \approx 1$. For real traffic the spectra would quite probably demonstrate the presence of wide-range singularity exponents. Increasing the number of multiplexed sources shows that the resulting process is better described by FGN.

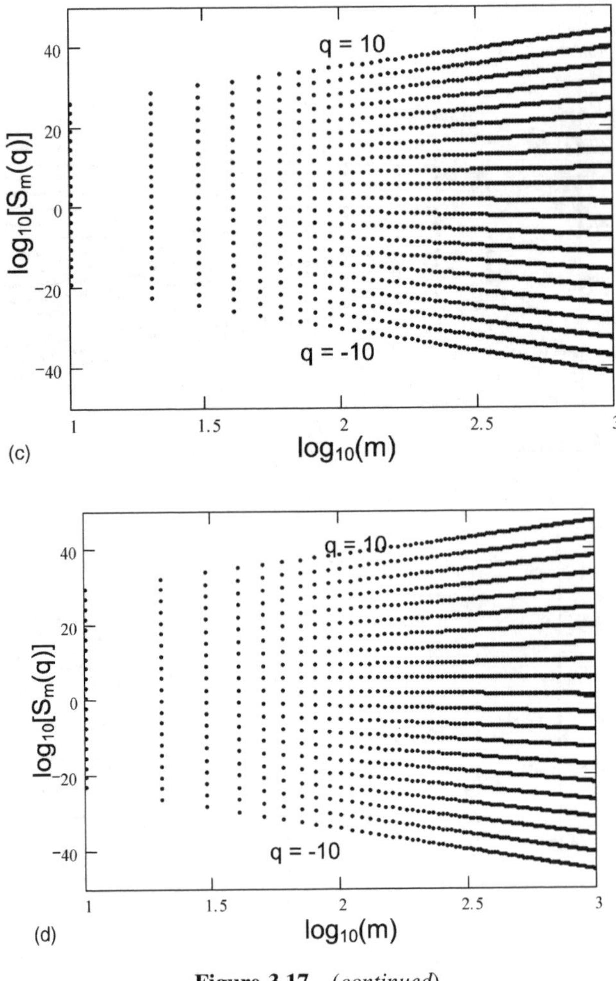

Figure 3.17 *(continued)*

3.4.4 Multifractal Properties of Two-Component Voice Traffic

Figure 3.19 shows the results of the multifractality estimation for two-component voice traffic presented in Figure 3.13. An analogous tendency can be observed for two-component traffic similar to the case of multiplexed voice sources: when the aggregation flow degree increases, the spectrum becomes narrower and, consequently, the multifractal scaling region decreases.

3.5 Mathematical Models of VoIP Traffic

3.5.1 Problem Statement

The problem of packet voice traffic modelling is not novel and has been studied in various publications. In Reference [1] the approach to the classical algorithm for voice activity detection (VAD) is described when the algorithm parameters are fixed in time. The codec

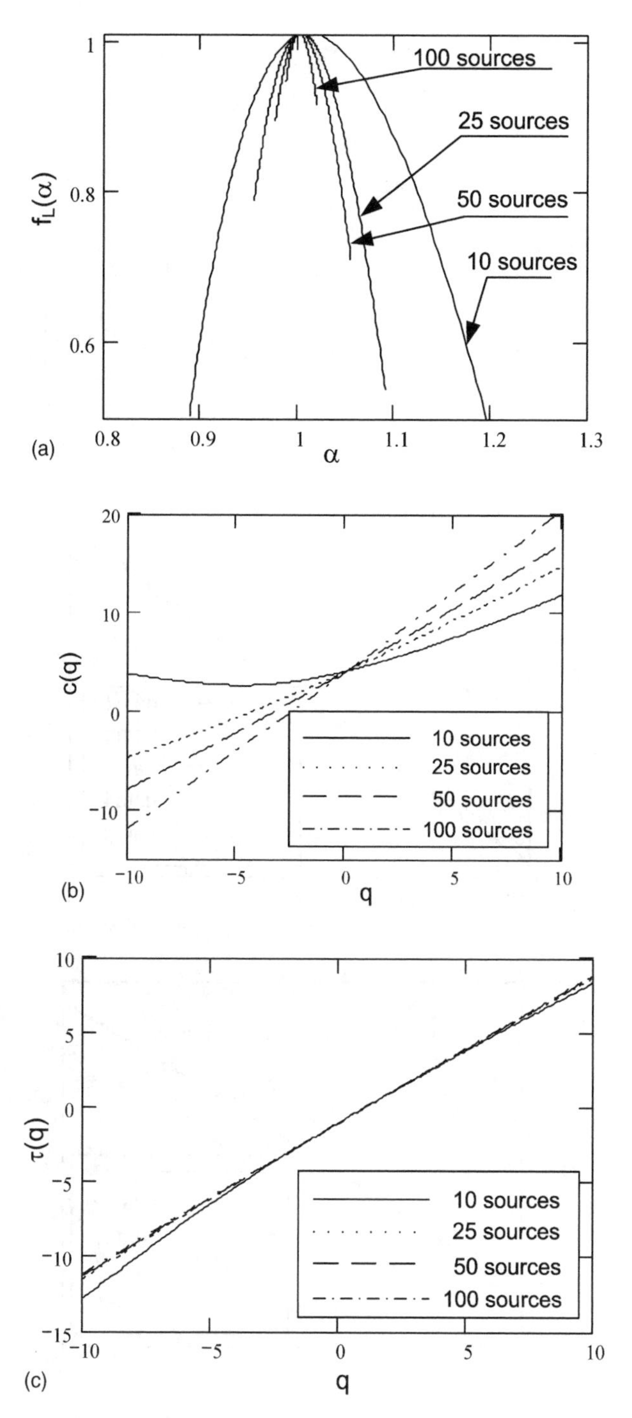

Figure 3.18 Studying the multifractal scaling: (a) multifractal Legendre spectrum; (b) $c(q)$ function; (c) $\tau(q)$ function

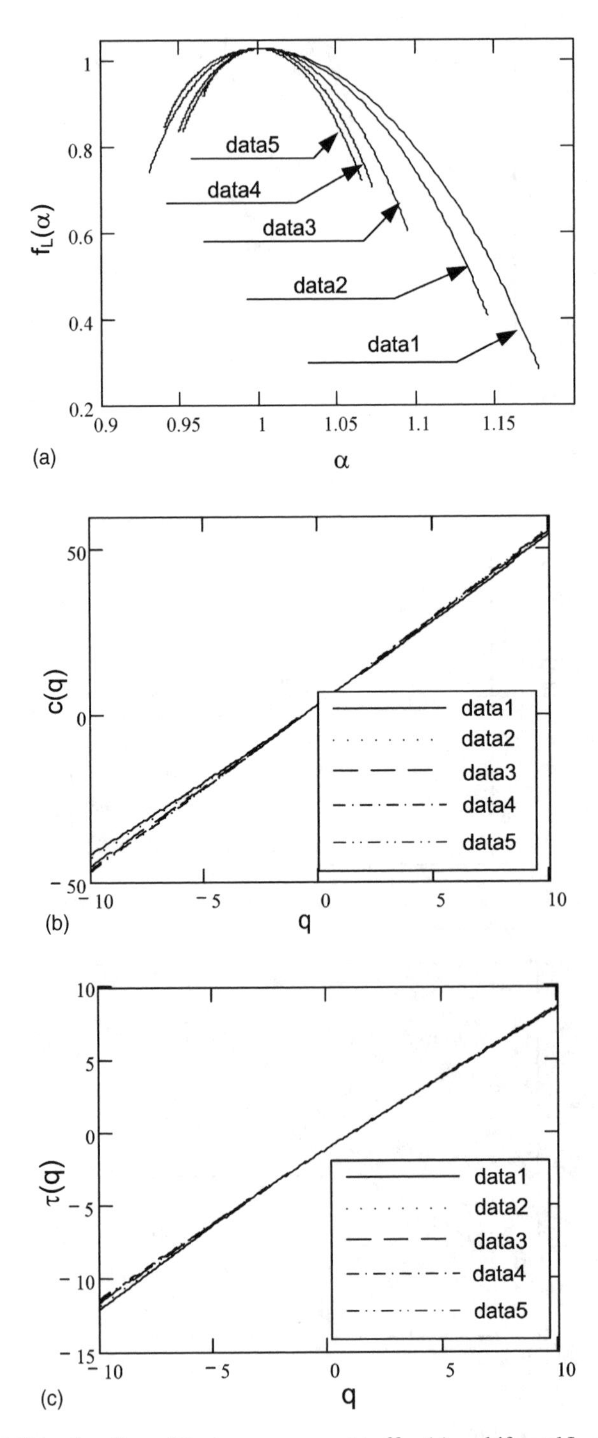

Figure 3.19 Multifractal scaling of the two-component traffic: (a) multifractal Legendre spectrum; (b) $c(q)$ function; (c) $\tau(q)$ function

traffic based on classical VAD algorithms consists of the voice periods and the intervals between them. The length of these periods can be successfully modelled using the exponential distribution, which implies the usage of the ordinary Markovian model with two states for each separate source.

Nevertheless, the ON/OFF sources analysis demonstrated for various voice codecs shows that those codecs that use the voice activity detectors and implement the dynamic and adapting coding principle behave in a different manner from the waveform codecs. The obtained results imply the use of heavy tails distributions. To model these ON/OFF periods it was suggested in some publications [6] that the generalized Pareto distribution should be used.

Since traffic aggregation for a large number of the 'heavy tailed' ON/OFF sources is self-similar, the flows of the aggregated VoIP traffic were examined. To form adequate models, it was required to execute the parametrization on the basis of real data. The models, which correctly described the salient traffic characteristics, are useful for analysis and modelling and allow a deep understanding of the network dynamics, thereby helping during design and control.

Most of the contemporary research dealing with analysis and modelling of network traffic aimed at describing the aggregated traffic behaviour in which all simultaneously active connections are united in one flow. A typical aggregated series consists of the packet number in the time unit over some intervals. Many publications showed that the aggregated traffic demonstrates fractal or self-similar scaling, i.e. the traffic looks as statistically similar over any time scale. The discovery of self-similar traffic behaviour resulted in new fractal models of the aggregated traffic. Fractional Gaussian noise (the most widely used fractal model) is the Gaussian process with a strictly scaled structure. Due to its Gaussian properties, it can be used in analytical investigations of queuing. The additional argument showing the presence of FGN in the networks is that the aggregated traffic can often be considered as the superposition of a large number of separate independent ON/OFF sources, which transmit with similar intensity but with the duration distributed in accordance with heavy tails distributions [7]. At the limit of the infinite source number the ON/OFF model converges to FGN.

The obtained information concerning the traffic characteristics at the connection layer, contained in widely available traffic traces, was, as a rule, ignored in classical aggregated analyses (LRD, fractal, multifractal). Taking into account this information when creating traffic models leads to a new approach to the analysis and modelling of pulsating network traffic [8].

3.5.2 Voice Traffic Models at the Call Layer

Consider the continuous Markovian chain (MC) $\xi(t)$ with continuous time $T \in \mathbb{R}_+[0,\infty)$ and the finite state set $X = \{0,\ldots,N\}$ (see Figure 3.20). The random vectors sequence will be given as $\{(\xi_n, T_n), n = 0, 1, \ldots\}$, where ξ_n possesses the values from the $X(\xi_n \in X)$ set and $T_n \in \mathbb{R}_+$ is the semi-Markovian sequence if

$$\begin{aligned} &P\{\xi_n = j, T_n < t | \xi_0 = i_0; T_0 < t_0, \xi_1 = i_1, \ldots, \xi_{n-1} = i, T_{n-1} < t_{n-1}\} \\ &\quad = P\{\xi_n = j, T_n < t | \xi_{n-1} = i\} = Q_{ij}(t) \end{aligned} \tag{3.1}$$

for arbitrary natural n, any $t_0, t_1, \ldots, t_{n-1} \in \mathbb{R}_+$ and any $i, j, i_0, \ldots i_{n-1} \in X$. The ξ_n component of the $\{(\xi_n, T_n)\}$ sequence will be called the leading component and the T_n component the

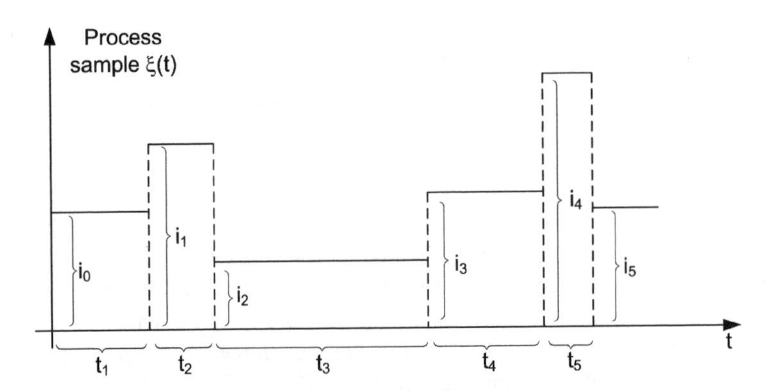

Figure 3.20 Plot of semi-Markovian process sample path

accompanying component (nested component of this sequence). The function $Q_{ij}(t)$ will be called the transition function of the $\{(\xi_n,\ T_n)\}$ sequence.

The modelling schedule of the random process $\{\xi(t), t_n \in \mathbb{R}_+\}$ over the semi-Markovian sequence $[(\xi_n, T_n)]$ can be fulfilled in the following way. It will be assumed that $T_0 = 0$ with the probability $P\{T_0 = 0\} = 1$. It will be considered that $\xi(t) = \xi_0$ if $0 \leq t < T_1$, $\xi(t) = \xi_1$ if $T_1 \leq t < T_1 + T_2$, $\xi(t) = \xi_2$ if $T_1 + T_2 \leq t < T_1 + T_2 + T_3$, etc. It will also be assumed that over the arbitrary time segment $[0, t]$ with the probability $P = 1$ the finite number of the transitions for process $\xi(t)$ occurs. The process $\{\xi(t), t \geq 0\}$ constructed in such a manner is referred to as the semi-Markovian process, constructed according to the semi-Markovian sequence.

The trajectory of the constructed process can be formed as follows, where the initial distribution is $\{p_i(0) = P\{\xi_0 = i\}, i \in X\}$:

Step 1. Draw the variables (realizations) (ξ_1, T_1) corresponding to $Q_{i_0j}(t)$ distribution. It will be assumed that the pair (i_1, t_1) is the result of such a draw. In this case suppose that the process $\xi(t)$ realization in the time interval $[0, t_1)$ possesses the value i_0.

Step 2. Draw the vector (ξ_2, T_2) values according to the distribution Q_{i_1j}. It is assume that they are equal to (i_2, t_2). As a result, the process $\xi(t)$ realization in the time interval $(t_1, t_1 + t_2)$ is set equal to i_1, etc.

The typical trajectory of the $\xi(t)$ process is shown in Figure 3.20, where the semi-Markovian process trajectory demonstrates that the stepping functions are continuous from the right side. The matrix $Q_{ij}, i, j = \overline{1, N}$, is the most important characteristic of the semi-Markovian processes. The methods of its description and the appropriate definitions are now given.

Let it be $P\{\xi_n = j|\xi_{n-1} = i\} = p_{ij} \neq 0$. Then

$$\begin{aligned} Q_{ij}(t) &= P\{\xi_n = j, T_n < t|\xi_{n-1} = i\} \\ &= P\{T_n < t|\xi_n = j, \xi_{n-1} = i\}P\{\xi_n = j|\xi_{n-1} = i\} = p_{ij}F_{ij}(t) \end{aligned} \tag{3.2}$$

where

$$F_{ij}(t) = P\{T_n < t|\xi_n = j, \xi_{n-1} = i\} \tag{3.3}$$

is the distribution function (DF) of the $\xi(t)$ process staying time in the state i, if it is known that its next state will be j. Thus, the process $\{\xi_n\}$ is the MC with discrete time and the transition probabilities (p_{ij}) matrix. If the Markovian chain does not contain the absorbing states, for arbitrary $i \in X$ then $p_{ij} = P\{\xi_n = j|\xi_{n-1} = i\} = \lambda_{ij}/\lambda_i, i \neq j; p_{ii} = 0$. Here p_{ij} is the probability that the MC, being in state i, will transit to state j at the next state changing moment. It can be seen from Equation (3.3) that the transition function $Q_{ij}(t)$ of the arbitrary semi-Markovian sequence can be written in the form

$$Q_{ij}(t) = p_{ij}F_{ij}(t) \tag{3.4}$$

where the distribution function of the arbitrary (not necessary Markovian) random variable plays the part of $F_{ij}(t)$.

Thus, the voice and video conference traffic at the call layer can be described and modelled by the semi-Markovian process, which is fully characterized by its elements: the transition probabilities matrix (p_{ij}), the DF matrix $F_{ij}(t)$ and the initial distribution $\{P_i(0), i \in X\}$. Note that the semi-Markovian process is the stochastically continuous uniform MC with continuous time, if and only if the transition function $Q_{ij}(t)$ for $\forall i, j \in X$ can be presented in the form $Q_{ij}(t) = p_{ij}(1 - \mathrm{e}^{-\lambda_i t})$; i.e. the semi-Markovian process is the MC with continuous time, if and only if the staying time in each of its states does not depend on the type of the next state and it is distributed in accordance with the exponential law.

In the Markovian case $F_{ij}(t) = P\{T_n < t|\xi_n = j, \xi_{n-1} = i\} = 1 - \mathrm{e}^{-\lambda_i t}$; i.e. for the analysed process $\xi(t)$ the staying time in the arbitrary state $i \in \mathrm{X}$ does not depend on which process state will be next. Moreover, the staying time in state i follows the exponential distribution λ_i parameter, which depends inclusively on the $i \in \mathrm{X}$ state. Thus, in the case of Markovian processes $\xi(t)$, the DF $F_{ij}(t)$ does not depend on j and has the form $F_{ij}(t) = 1 - \exp(-\lambda_i t)$, while in a more general (non-Markovian) case the DF may be arbitrary.

It is possible to estimate several statistical characteristics of semi-Markovian processes of great importance for the traffic description and modelling at the call layer. Thus, the voice traffic at the call layer can be described and modelled by the semi-Markovian process with the transition function of the following kind: $Q_{ij}(t) = p_{ij}F_{ij}(t)$, $i, j = 1, 2, \ldots, N$, where $F_{ij}(t) = P\{T_n < t|\xi_n = j, \xi_{n-1} = i\}$ is the function of the staying time distribution for the $\xi(t)$ process in state i if it is known that state j will be its next state j: $p_{ij} = P\{\xi_n = j|\xi_{n-1} = i\} = \lambda_{ij}/\lambda_i, i \neq j$; $p_{ii} = 0$ is the probability that the MC, being in state i, will transit to state j in the regular moment of state changing, where $\{P_i(0), i \in X\}$; N is the state number of the Markovian chain.

When the number of multiplexed channels is small ($N < 20$), the voice and video conference traffic at the call layer can be described and modelled by the semi-Markovian process, which is fully characterized by its elements: the transition probabilities matrix (p_{ij}), the DF matrix $F_{ij}(t)$ and the initial distribution $\{P_i(0), i \in X\}$.

3.5.3 Estimation of Semi-Markovian Model Parameters and the Modelling Results of the Voice Traffic at the Call Layer

The analysis of the experimentally obtained plots for the call traffic realizations shows that in the analysed voice traffic process at the call layer it is possible to allocate a finite number of

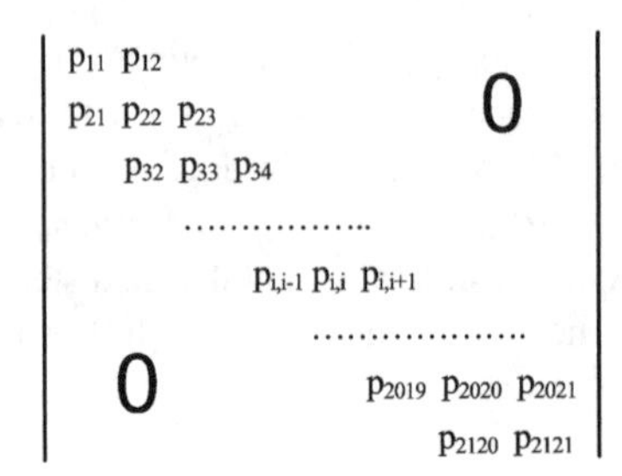

Figure 3.21 Matrix of the transition probabilities for traffic modelling at the call layer for semi-Markovian distribution of call arrival intervals for realization

states N (in the examined realizations $N = 21$), each of which corresponds to the number of active calls for the given time moment [3].

Conducting the call traffic approximation with the help of the Markovian matrix $N \times N$ (of size 21×21 in the experiment), it is possible to estimate the parameters of the transition probabilities matrix (Figure 3.21) on the basis of the experimental data examinations using the approach offered in Reference [8]. After obtaining the transition probability matrix, modelling of the process described by the state graph (Figure 3.22) will be executed with the help of the matrix and the obtained data will be compared with the experimental data.

Figure 3.23 shows the realization obtained using the transition probability matrix estimated on the basis of experimental data. The histogram of the amplitude distribution (layer number) for this case is shown in Figure 3.24.

To check the distribution adequacy of the experimental and modelled processes the 'quantile–quantile' plot (or briefly the QQ plot) will be found, which is usually used to find the most convenient distribution from the chosen distribution family. Applying the quantile analysis, it is possible to show good conformity between the modelled process and the experimental one. At the same time, the researches show that the correlation properties of the process, obtained as a result of modelling the Markovian and the real processes, essentially differ and require mathematical model refinement for voice traffic at the packet layer.

3.5.4 Mathematical Models of Voice Traffic at the Packet Layer

Knowing VoIP traffic properties, various types of self-similar processes can be used for similar traffic modelling [9]. The accomplished researches show that when increasing the series

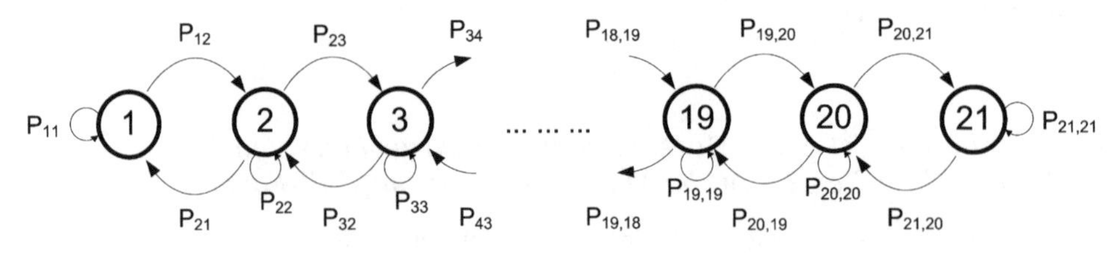

Figure 3.22 State graph of the Markovian chain

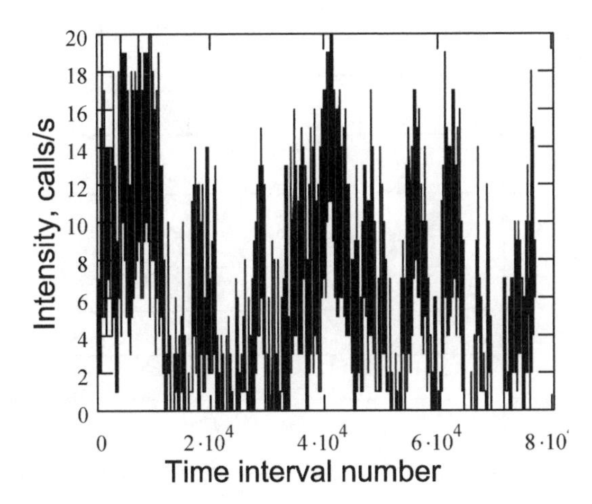

Figure 3.23 Averaged call traffic in the working day obtained with the help of the Markovian matrix of the transition probabilities

number, the aggregated process for the small time scale approaches FGN. In this formulation, with the help of the byte series, long-range dependence of the aggregated process occurs when ON and OFF periods have heavy tails distributions.

A similar model was used in many publications to approximate the byte series with ON/OFF periods having heavy tails distributions. In particular, Figure 3.25 presents a large number of active byte series, in each of which the active ON periods are followed by the nonactive OFF periods. ON and OFF periods in the series are independent of each other and have similar statistical properties.

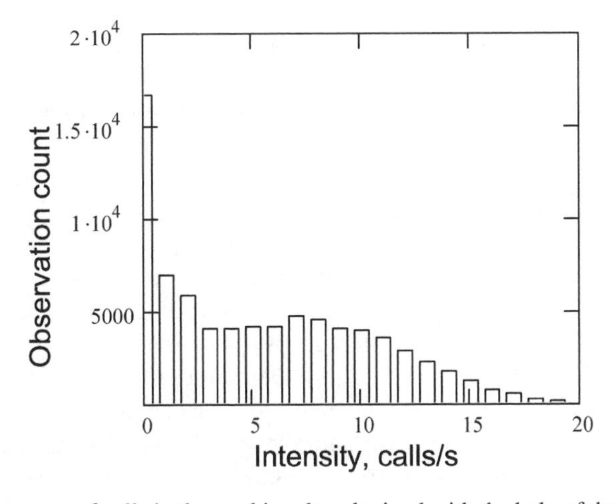

Figure 3.24 Histogram of calls in the working day obtained with the help of the Markovian model

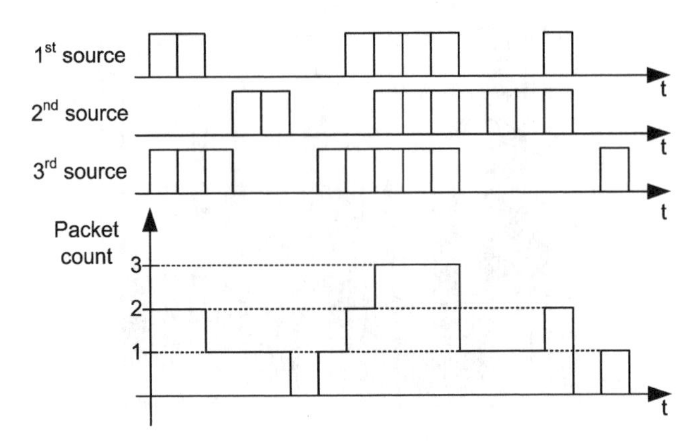

Figure 3.25 Schematic demonstration of the process of the input traffic generation (observed three packet sources)

Consider the simplest way to model self-similar traffic using fractional Gaussian noise (FGN) as an example. This model is defined as

$$X(t) = m + \sigma G_H(t) \tag{3.5}$$

where $G_H(t)$ designates centralized FGN with the Hurst exponent H, the mean value m, the r.m.s. (root mean square) deviation σ and the correlation function $R(k) = (\sigma^2/2)[(k+1)^{2H} - 2k^{2H} + (k-1)^{2H}]$. Thus, model (3.5) has three parameters for the complete description: H, m and σ. FGN will be parameterized on the basis of real measurements in order to achieve identical input flow parameters and the model. The obtained parameters will be used in the algorithm of FGN modelling.

In the real networks, instead of a fixed number of byte series over the full time interval the series arrive and move away from the system at random time moments. The series arrival time moments are obtained from the nonstationary Poisson process and their life times are distributed exponentially with a large mean value. As a result, over small time scales the traffic demonstrates self-similarity and long-range dependence properties due to separate ON/OFF processes. However, this model is nonstationary if the series arrival intensity changes over time. Moreover, since the partial series have a restricted duration (due to daily fluctuations), the correlation events over large time scales are discarded by the model.

To simplify the nonstationary model estimation the FGN sequence is considered as the result of superposition of a large number of independent and identically distributed FGN processes, each of which corresponds to the active traffic series. Although there is a difference between the byte series and the traffic series, i.e. the byte series are the process of 0–1 alteration but the traffic series are processes of the FGN type, in both cases the aggregated traffic corresponds to the traditional FGN model. Therefore the difference is not so important in the case of aggregated versions modelling. Using the FGN traffic series simplifies the theoretical description of the nonstationary model. As a result the model represents the superposition of the large number of active segments in the FGN traffic series, the duration of which is distributed exponentially. The self-similarity and correlation properties over the small time scale are the same as for FGN, but over large time scales these properties disappear.

To develop the nonstationary model the aggregated traffic $X(St)$ is considered over the time interval $(St, S(t+1))$ in the form of samples of the large number $(N(t) = N)$ of independent and identically distributed traffic series. Each series by itself is the FGN process with general parameters α, σ^2 of the Gaussian distribution and with the Hurst exponent H. As a result, the aggregated nonstationary FGN model for network traffic samples with the time scale S for various time moments $t = 0, \pm 1, \pm 2, \ldots$ can be written as

$$X(St) = m\,\hat{N}\,S + \sigma S \sum_{i=1}^{\hat{N}} G_{H,i}(t), \quad t \in T \tag{3.6}$$

Here $\hat{N}$ is the estimation of the summarized FGN number and is determined using the semi-Markovian call model; T is the duration of the interval corresponding to state i of the Markov Chain (MC), defined by the DF $F_{ij}(t) = P\{T_n < t | \xi_n = j, \xi_{n-1} = i\}$; and $G_H(t)$ is FGN with the Hurst exponent H, the mean value m, the r.m.s. deviation σ and the correlation function $R(k) = (\sigma^2/2)[(k+1)^{2H} - 2k^{2H} + (k-1)^{2H}]$.

Consider the process of the input traffic modelling formed as the aggregated flow, where separate elements can be presented in the form of the packet sequence, the principle of which is shown in Figure 3.25. In the case under examination the nonstationarity was introduced by means of the packet series arrival intensity changing in time and by means of changing the summarized partial flows number.

The positive features of the self-similar models similar to Equation (3.6) consist mainly in their ability to cover LRD saving at the possibility of theoretical analysis. However, the self-similar models such as FBM/FGN have some shortcomings. In particular, Gaussian distribution signifies the possibility of the presence of negative values in the data. Besides, it is necessary to pay attention to the degenerate multifractal properties of similar models. At the same time, the use of similar models for aggregated processes allows the long-range dependence properties to be described. Various test applications for various configurations of the aggregated voice traffic allows the conclusion to be made that the self-similarity property of the summarized traffic should be taken into account when modelling the high intensity of call arrivals.

3.6 Simulation of Voice Traffic

3.6.1 Simulation Structure

Consider the TN of the equivalent structural diagram for imitation in the Network Simulation 2 (ns2) medium, which is shown in Figure 3.26. The experimental studies of the voice traffic for two codec types, G.723 and G.729B, using the voice activity detector were based on the simulation [10]. The voice sources were subjected to processing (reduced to an ON/OFF presentation) and the processes of the ON and OFF periods were obtained from the packet flows. The mean values of the ON/OFF periods depend on the VAD mechanism adjustments. The standard equipment adjustments manufactured by Cisco Systems were used during the measurements. The distribution functions for the duration of the ON/OFF periods obtained from the experimental data have practically a similar form for different codecs and almost coincide with each other. This means that the used voice codec type affects the main characteristics of the ON/OFF periods in the packet sources insignificantly.

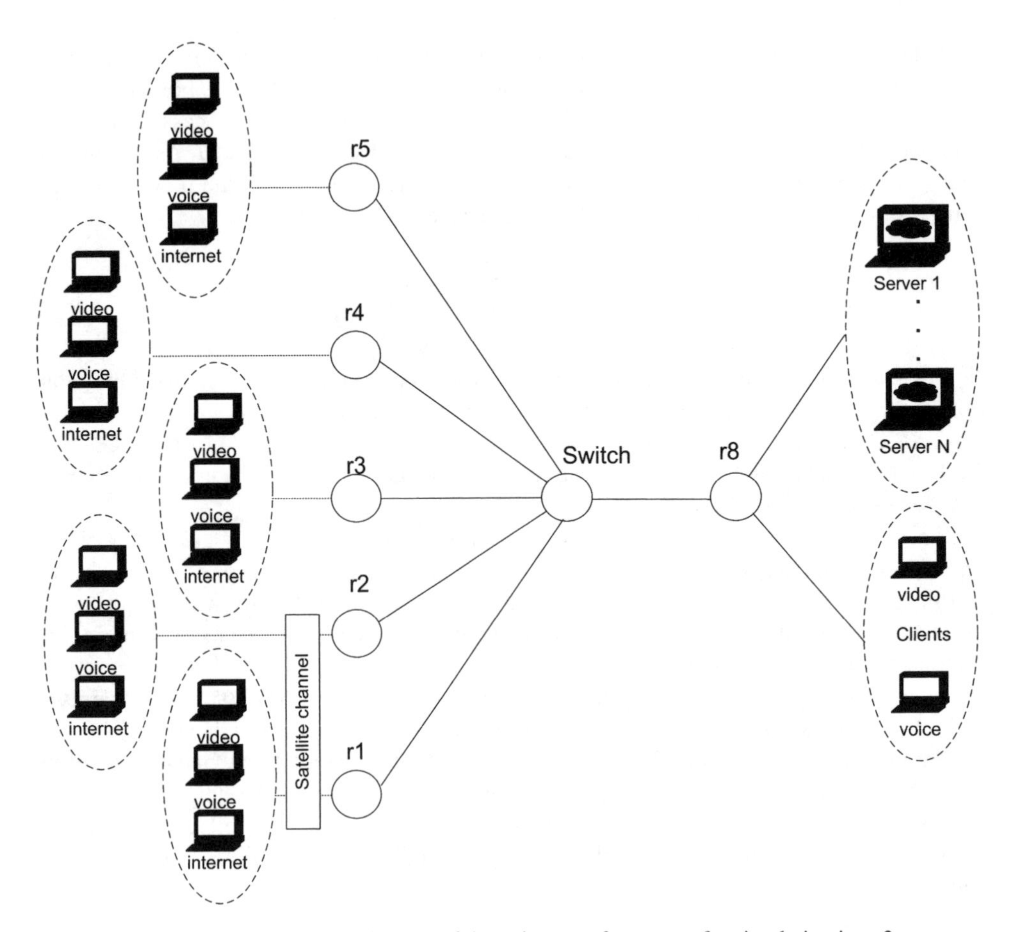

Figure 3.26 Structural diagram of the voice transfer system for simulation in ns2

It was found that the duration of the ON/OFF periods does not correspond to the exponential distribution as it is assumed by the classical models of voice traffic. The distributions of the ON/OFF period duration for the summarized VoIP traffic obtained on the basis of the experimental data are not exponential and the heavy tails distributions should be used as a model. Pareto distribution has been widely adopted:

$$w(x) = \frac{ab^{\alpha}}{x^{\alpha+1}}, \quad \text{for } x \geq b \tag{3.7}$$

where α is the shape parameter of Pareto distribution and b is the scaling parameter.

The traffic generator using Pareto distribution to simulate the ON/OFF period duration, realized in the ns2 system, was chosen for the description of operation of the voice sources. The generator parametrization was fulfilled on the basis of the experimental data analysis: the

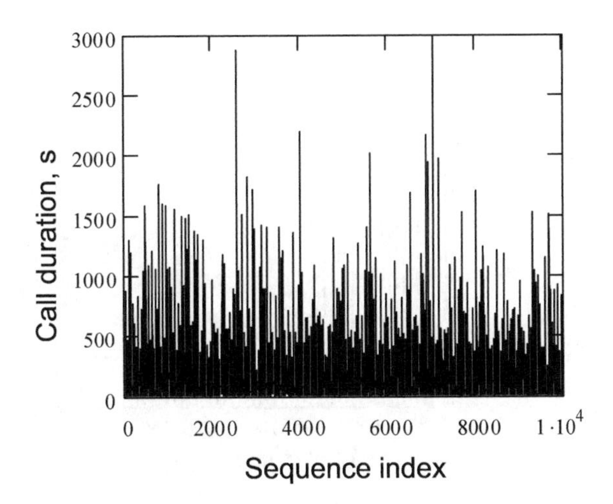

Figure 3.27 Sample path of the call duration random process

averaged activity interval was equal to 500 ms and the averaged pause interval to 1500 ms. The Pareto distribution parameter α was changed for various experiments depending on the necessary self-similarity degree of the multiplexed flow.

The statistical characteristic estimation of existing sources was carried out to set the traffic source parameters correctly. The real sources analysis was conducted on the basis of the cdr-logs analysis, which contained the information of more than a half-year period. The information about the telephone calls effected from the single IP address was allocated for the problem set. It was assumed that the call statistics for all the addresses that were present at cdr-logs are identical. In accordance with the model offered earlier, an analysis of the main characteristics of the call arrival process was carried out: the process of the call duration (duration) and the process of inter-arrival time (inter_time).

Figure 3.27 shows an example of the call duration random process realization (the duration is measured in seconds and the serial number of the arriving call is indicated on the x axis). Figure 3.28 shows the histogram of the call duration random process, the realization of which is shown in Figure 3.27 (the frequency of the burst in the appropriate histogram interval is indicated in the log scale). It is seen that the histogram envelope deviates from a straight line, which proves the difference between the obtained distribution and the exponential one. Table 3.3 presents the main selective statistics for the call duration random process.

Figure 3.29 shows the random process demonstrating the inter-arrival time (the inter-arrival time is shown in seconds and the serial number of the appropriate inter–arrival time is indicated on the x axis). It is seen that the abnormally large observations occur in the data while the main part of the data has values that are less by approximately three orders. The abnormal observations can be explained by the cdr-file registeration in the continuous mode (in the daytime and at night). As the call intensity at night is essentially less than in the daytime, the intervals between the calls at night increase compared to the daytime measurements.

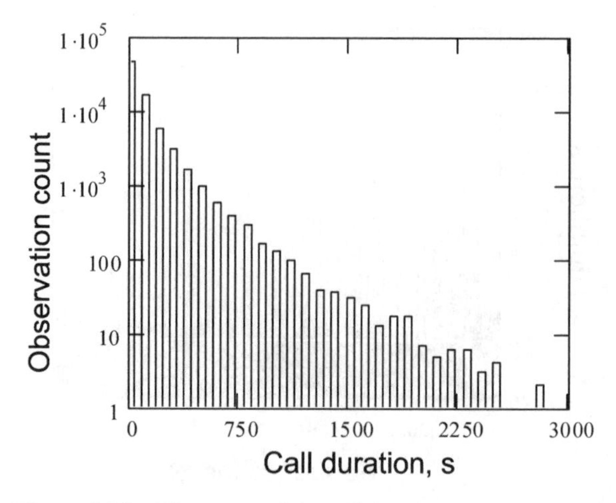

Figure 3.28 Histogram of the call duration random process

Table 3.3 Sample statistics for the call duration process

Parameter	Points number	Mean value (s)	Minimal value (s)	Maximal value (s)	Standard deviation (s)
Duration (34_1)	74 143	129.9	0.00	4511.0	182.1

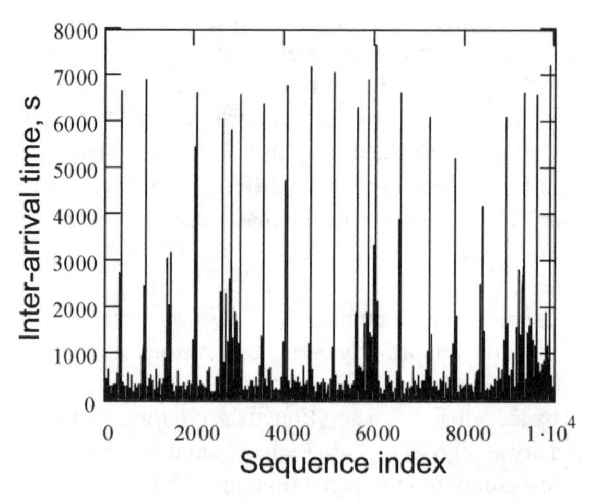

Figure 3.29 Random process sample path for the inter–arrival time duration

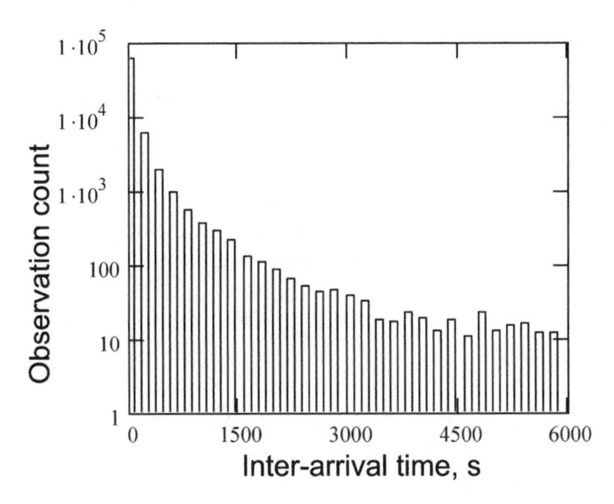

Figure 3.30 Random process histogram for the inter-arrival time duration

Figure 3.30 shows the random process histogram for inter-arrival time duration(the frequency of the hit in the appropriate histogram interval is indicated in the logarithmic scale). It can be seen from the figure that the histogram envelope essentially deviates from the straight line; therefore the assumption concerning the exponential distribution character should be rejected. The histogram clearly demonstrates the heavy tail presence in the distribution. Research shows that the Pareto distribution can be chosen when modelling such a distribution. Table 3.4 gives the main sample statistics for the process shown in Figure 3.29. It can be seen that the maximal inter-arrival time arrivals was equal to approximately three hours, which is quite possible for night time only.

3.6.2 Parameter Choice of Pareto Distributions for the Voice Traffic Source in ns2

The Pareto generator of ON/OFF traffic (POO_Traffic) is the traffic generator (at the application layer) which is built into the Otcl class of the Application/Traffic/Pareto. POO_Traffic generates the traffic in accordance with ON and OFF periods distributed according to Pareto law. The packets are transmitted at a constant rate during ON periods, but during OFF periods they are not transmitted. Both ON and OFF periods ensue from Pareto distribution using constant size packets. Such sources were used to generate the aggregated traffic demonstrating LRD.

Table 3.4 Sample statistics for the process of inter–arrival time

Parameter	Points number	Mean value (s)	Minimal value (s)	Maximal value (s)	Standard deviation (s)
Inter_time(34_1)	74 142	163.4263	0.00	12562.00	501.8184

3.6.2.1 Input Parameters

These are the parameters that were varied when the traffic sources were described:
burst_time_ is the averaged ON length.
idle_time_ is the averaged OFF length.
rate_ is the rate of information transmission during the ON period (bits per second)
packetSize_ is the packet size (constant frame size for the application in bytes).
shape_ is the shape parameter of Pareto distribution.

To describe Pareto traffic sources in more detail it is necessary to use the subsidiary variables, which can be defined as follows:

interval = **packetSize_** · **8/rate_** (the time interval for one packet transmission)
burstlen = **burst_time/interval** (in packets, defined in the parmeter **packetSize_**)
The variable **burstlen** determines the packet number that should be generated until the start of the ON period.

During each ON/OFF cycle two independent variables are calculated having Pareto distributions:

next_burstlen is the packet number that will be transmitted during the next ON period (e.g. the size of the next Web object in the packets with **packetSize_** length).
next_idle_time is the length of the next OFF period (e.g. the next waiting time).

3.6.2.2 Pareto Algorithm for the Traffic Generator

A brief description of the operation of Pareto traffic generator algorithm is presented below.

Step 1. The **next_burstlen** value is calculated using the ON period averaged length and shape parameter of Pareto distribution.
Step 2. All **next_burstlen** packets are transmitted.
Step 3. The **next_idle_time** is calculated using the averaged OFF period length (**idle_time_**) and the shape parameter of Pareto distribution (**shape_**).
Step 4. The expectation during the **next_idle_time** and transition to step 1.

3.6.2.3 Pareto Distribution

The random variable X with Pareto distribution (3.7) and mean value $M(X) = b\alpha/(\alpha - 1)$, if $\alpha > 1$, was modelled with the help of a special traffic generator. The shape parameter α in the expressions

$$\textbf{burstlen} = M(X) = b1\alpha/(\alpha - 1)$$
$$\textbf{idle\_time\_} = M(Y) = b2\alpha/(\alpha - 1)$$

was designated as **shape.** As a result,

$$b1 = \textbf{burstlen}(\alpha - 1)/\alpha$$
$$b2 = \textbf{idle\_time\_}(\alpha - 1)/\alpha$$

Ns2 has a built-in generator of the random numbers distributed according to the Pareto law, which have the shape and scaling parameters described as follows:

double pareto (double scale, double shape)

When the Pareto traffic generator calculates **next_burstlen**, the following operations are executed:

int **next_burstlen** = int(pareto(b1, α) + 0.5)

/* **next_burstlen** should have a minimum of one packet */

if(**next_burstlen** = = 0) **next_burstlen** = 1;

When the Pareto traffic generator calculates the next OFF period length, the following operations are executed by the generator:

double **next_idle_time** = pareto(b2, α)

3.6.3 Results of Separate Sources Modelling

The results of imitation modelling of the separate voice traffic sources in the ns2 system will be examined in accordance with the above-mentioned algorithm. Figure 3.31 shows the modelled traffic of the separate source in packets for a time resolution of 0.02 s. It is evident that the traffic represents the ON/OFF process, the period duration of which is determined by the model parameters and the modelling algorithm. ON periods in the modelled source correspond to the user's voice activity and OFF periods to the pauses between words or phrases. Since the sources without VAD usage do not expend the system channel resources effectively, the codecs using voice activity detectors were chosen as the voice sources. The pauses and activity average value of these codecs are less than the minimum by one or in some cases by two orders.

3.6.4 Results of Traffic Multiplexing for the Separate ON/OFF Sources

An analytical review will be made of the traffic obtained as a result of 500 ON/OFF sources multiplexing, each of which is considered as one of the parties engaged in a telephone

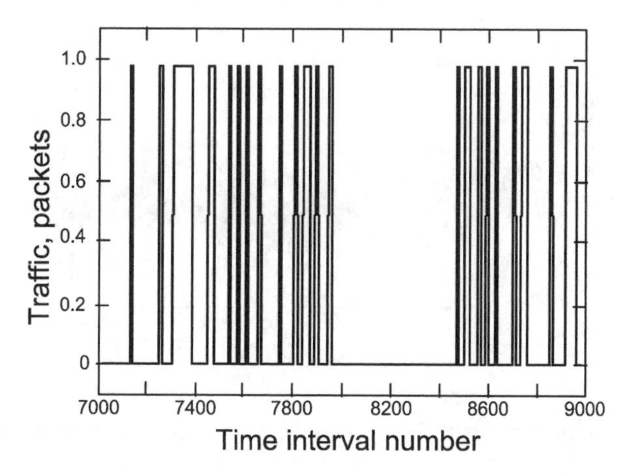

Figure 3.31 Modelled traffic of the single source in the ns2 system

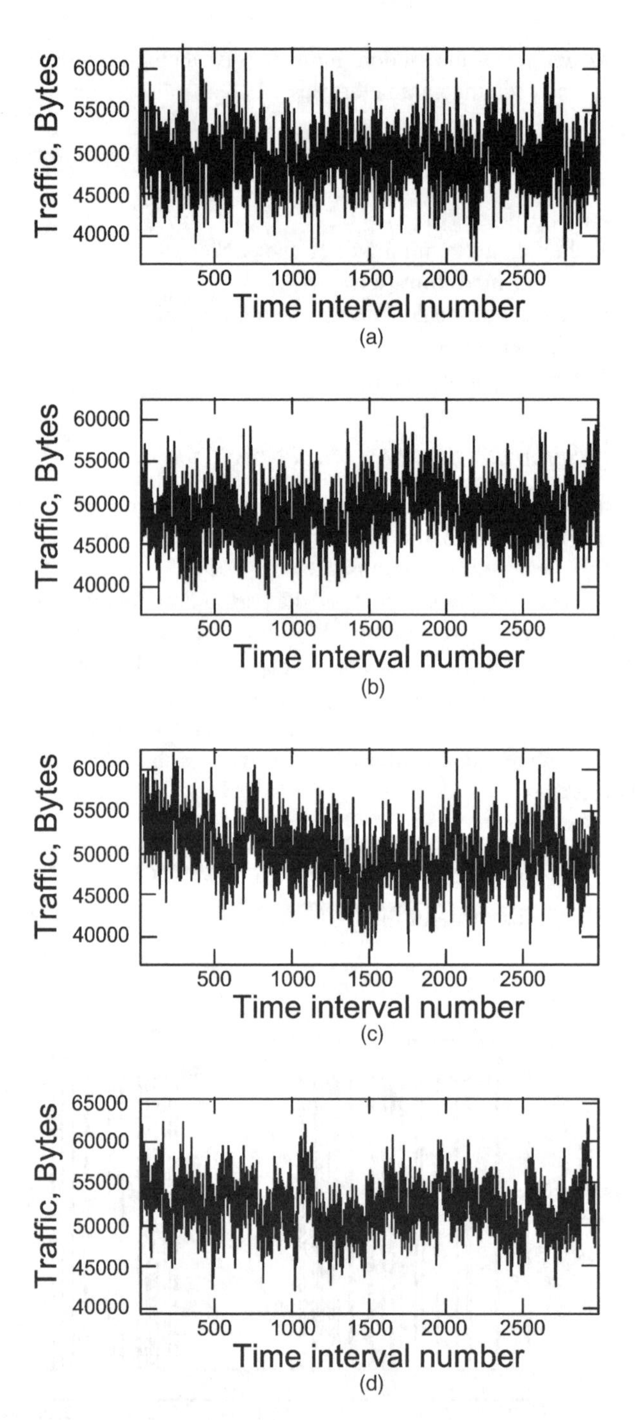

Figure 3.32 Voice traffic profiles (0.1 s time resolution) modelled in ns2 for multiplexing of 500 ON/OFF sources at various given shape parameters of ON/OFF state duration Pareto distribution: (a) $\alpha = 2.0$; (b) $\alpha = 1.7$; (c) $\alpha = 1.4$; (d) $\alpha = 1.1$

conversation. The traffic was obtained for a time resolution of 0.1 s, which ensures a high level of averaging, permitting reliable statistical estimates to be achieved. The main subject of the research is the study of the influence of change of the separate source properties on the properties of the multiplexed flow, which can be characterized by the shape parameter α of Pareto distribution for each source possessing values of 2,0; 1,7; 1,4 and 1,1 during the analysis.

On analysis will be made of how much the change of the separate source parameters will influence the Hurst exponent change for multiplexed flow. The preliminarily analysis will be carried out of the traffic generated by all the sources connected to the r5 router. The traffic is measured in the flow aggregation point, i.e. at the input of the FIFO queue.

To analyse the traffic the information presented in the traffic.txt file will be used. Figure 3.32 shows the traffic for the time resolution of 0.1 s in bytes for modelling with the parameters listed in Table 3.5. Figure 3.32 shows the traffic realizations at 0.1 s resolution for four different α parameter values. Table 3.6 demonstrates the basic statistics of the realizations shown in Figure 3.32. Having obtained the realizations presented in Figure 3.32, a Hurst exponent estimate is determined and a plot is made of the correlation coefficient versus delay.

The Hurst exponent estimate is conducted with the help of three widely spread approaches: the correlation function slope analysis in log-log scale, the variance–time plot analysis and R/S statistics. Figure 3.33 shows the histograms for the data of Figure 3.32, from which it can be seen that the analysed process distributions tend to Gaussian law. As an example, Figure 3.34 shows the Hurst exponent estimates for $\alpha = 1.4$ and the parameters used to obtain them. Table 3.7 summarizes the information concerning the Hurst exponent estimate carried out using the various approaches, on the basis of which it is shown that the shape parameter variation of the separate sources in the multiplexed flow influences the self-similarity exponent of the whole flow.

The approach based on the correlation function plot estimation in log-log scale demonstrates Hurst exponent growth as the shape parameter decreases, but with dynamics that do not correspond to the limiting theorem [11]. The variance–time analysis ensured an estimate close to the correlation coefficient plot, but the last realization showed Hurst exponent decay because in the case of large Hurst exponents this approach leads to their underestimation. The analysis of the R/S statistics data shows stable Hurst exponent growth as the distribution shape parameter decreases.

Table 3.5 Parameters of modelling

Parameter	Value	Description
r5_host_	100 sources	Pareto sources number connected to r5
r4_host_	100 sources	Pareto sources number connected to r4
r3_host_	100 sources	Pareto sources number connected to r3
r2_host_	100 sources	Pareto sources number connected to r2
r1_host_	100 sources	Pareto sources number connected to r1
bottleneck_speed	4.0 Mb/s	Carrying capacity of narrow channel in Mbits per second
fifo_buf	2000 packets	FIFO queue buffer size
modelling_time	300 s	Total modelling time in seconds
burst_time	500 ms	Average activity period
idle_time	1.5 s	Average pause interval

Table 3.6 Basic statistics of analysed traffic trace

Parameter for the single source	Point number	Mean (bytes)	Standard deviation (bytes)
$\alpha = 2.0$	2999	48 748.21	4653.07
$\alpha = 1.7$	2999	49 617.68	4552.17
$\alpha = 1.4$	2999	49 994.44	4214.92
$\alpha = 1.1$	2999	52 775.34	3238.39

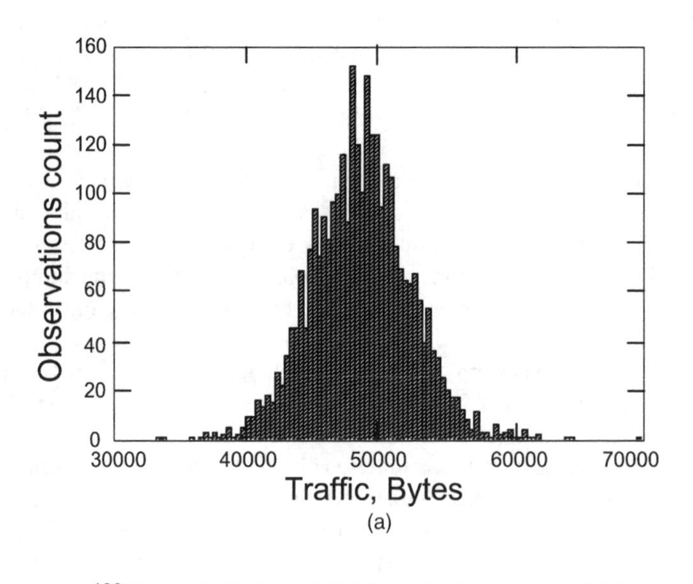

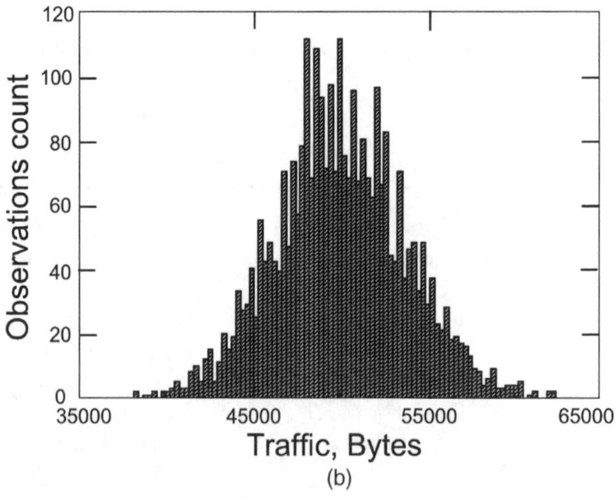

Figure 3.33 Voice traffic histograms (0.1 s time resolution) at multiplexing 500 ON/OFF sources for the various shape parameters of the Pareto distribution for ON/OFF states duration: (a) $\alpha = 2.0$; (b) $\alpha = 1.1$

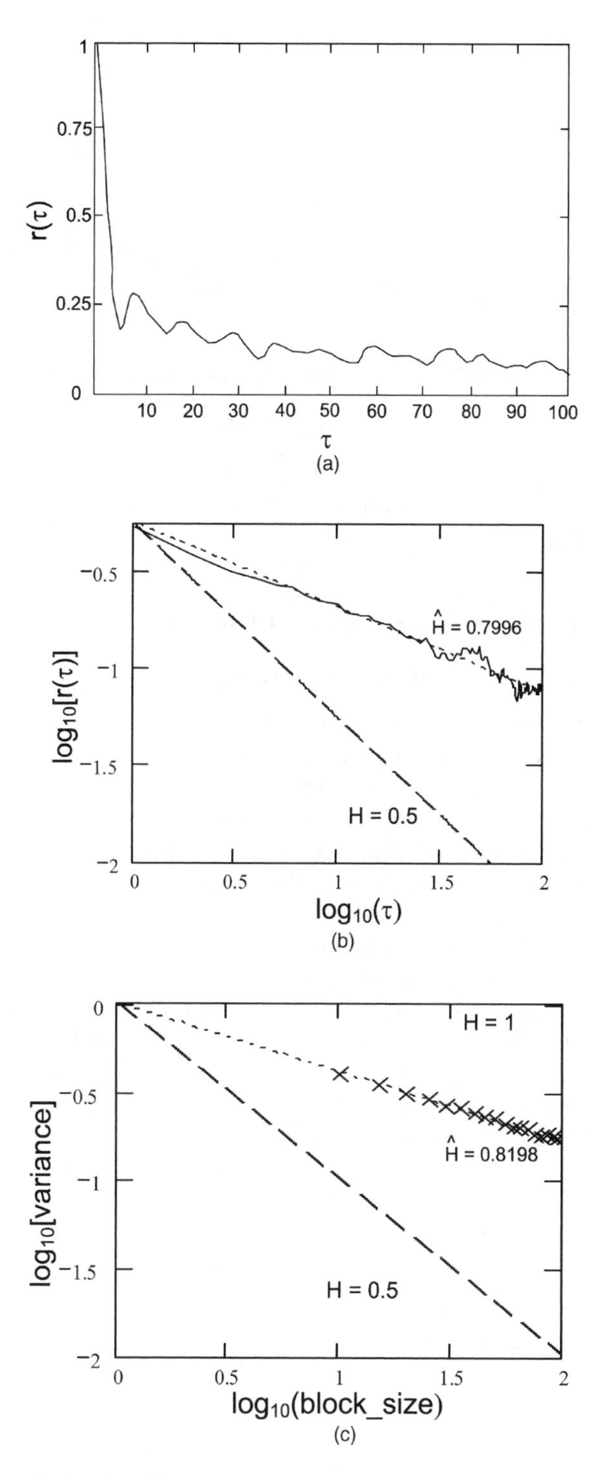

Figure 3.34 The multiplexed traffic estimates for $\alpha = 1.4$: (a) correlation function; (b) correlation function in log-log scale; (c) variance–time plot; (d) R/S statistics plot

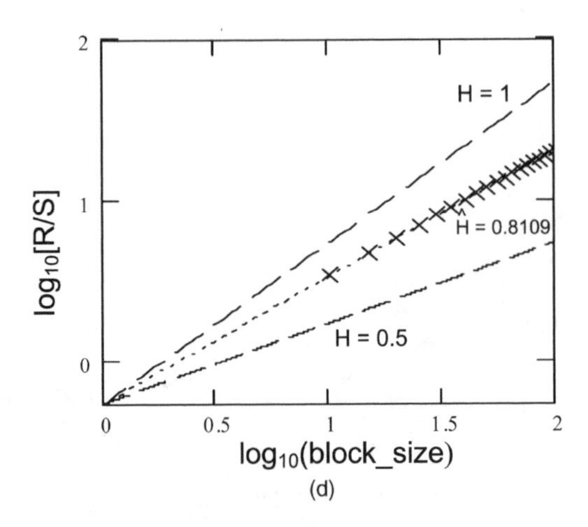

Figure 3.34 (*continued*)

3.7 Long-Range Dependence for the VBR Video

3.7.1 Distinguished Characteristics of Video Traffic

Video signals are the sequence of continuous spatially still pictures referred to as 'frames'. There are several physical reasons why the video traffic traces are special. Each fixed picture can be presented by the coding algorithm in the digital form and can then be compressed to decrease the bandwidth. The approach normally applied to decrease the bandwidth is to transmit the full initial frame and then to transmit the difference frames. This method of transmission is called *interframe coding*. Since the adjoining frames differ little from each other (for the motion is continuous), this leads to an essential correlation existence of the frames located nearby. To avoid transmission errors it is possible to ensure the periodic transmission of the full frame, where varying the frames no longer depends on the previous frames. In this case the functional correlation is over, which can put an end to the statistical correlation in frame sizes. As these variations require that the last frame should be fully transmitted, the scene duration has an effect on the trace character. For these and some other reasons video traffic differs from broadband data traffic and, therefore, the models and the conclusions revealed for video data cannot be used for other traffic types.

Table 3.7 Hurst exponent estimate

Shape parameter of the tested realization	Log–log correlation coefficient	Variance–time analysis	R/S statistics
$\alpha = 2.0$	$H = 0.592$	$H = 0.544$	$H = 0.672$
$\alpha = 1.7$	$H = 0.758$	$H = 0.794$	$H = 0.684$
$\alpha = 1.4$	$H = 0.7996$	$H = 0.8198$	$H = 0.811$
$\alpha = 1.1$	$H = 0.828$	$H = 0.745$	$H = 0.884$

When information is lost during transmission or the interarrival frame times are long or change significantly, the quality of video data is lowered. Video data reproduction can be controlled by limiting the buffer sizes, allowing the frames arriving with delay to be disregarded. The buffer size is often characterized by the time period before its underflowing (i.e. the maximal frame delay that can appear). The design goal is to avoid a delay of more than 100–200 ms. Since several buffers can be met between the source and the user, and there are also other delay sources (e.g. the signal propagation time), the value of 10 ms is often used as the maximal buffer size in certain studies.

3.7.2 Video Conferences

The main purpose of the development of the video signal compression methods is to decrease the transmission rate of the digital video flow to levels of 40 kb/s to 2 Mb/s. The H.261/H.263 group as part of H.230 recommendations contains recommendations for video and audio transmission with the rates of 46.4 kb/s and 16 kb/s in the B-channel (64 kb/s) together with the necessary overhead information required by H.221 recommendations.

Only the 'speaking heads' are shown during the video conference, so video conferences can be considered as the simplest type of video data for modelling. The signal in PAL (phase alternation line) as well as in NTSC (National Television System Committee) standards, converted further in the total standard of one of the two types CIF (common intermediate format) or QCIF (quarter CIF), can be the video data source.

The CIF standard provides a higher quality but requires a wider frequency band (when using the modern compression methods the recommended transmission rates are 384 kb/s and more). The achieved resolution is only twice as bad as in the NTSC system. The QCIF standard has a resolution twice lower than CIF for each measurement, i.e. the number of pixels is four times less, but the required transmission rate can be reduced to 64 kb/s.

The realization example typical for the video conference in H.263 standard is illustrated in Figure 3.35(a). To describe the entertainment video programs the models originating from them are used. The normalized correlation functions of the analysed sequences (Figure 3.35(b)) are the most widely distributed and available characteristics, and can be estimated from the results of the traffic experimental measurements.

Numerous fulfilled researches of the video sequences [7] permit the conclusion to be drawn that these sequences demonstrate long-range dependence, and the short-range correlations are also important when modelling the sources, as can be seen from the correlation coefficient plots.

To describe the video conference traffic one-dimensional distribution the negative binomial and the gamma distribution are widely used. They are completely defined by two parameters, which can easily be estimated using the moments method of selective estimates of real traffic estimation values and the variance. As a result, these two moments and the correlation coefficient r are required only to describe the VBR teleconference traffic properties.

3.7.3 Video Broadcasting

Video sequences typical for movies, news, sports and entertainment programs are more dynamic. VBR video broadcasting differs from VBR video conferences by the flow rate. Therefore video conference sequences consist mainly of 'head–shoulders' pictures with slight panning or without it, while video broadcasting can be characterized by a changing stationary

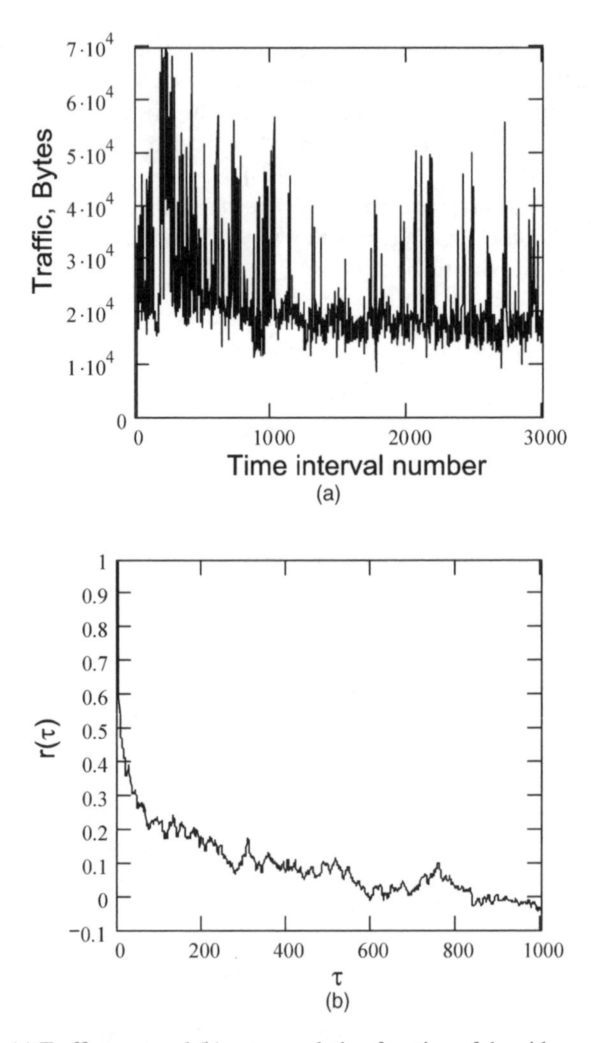

Figure 3.35 (a) Traffic trace and (b) autocorrelation function of the video conference trace

scene. Because of interframe coding, it is evident that as the scene changes more bits will be required than for the intrascene frames, a feature that distinguishes video broadcasting from video conferences [12]. There are also other differences described in the publications [13, 14], in which the DPCM (differential pulse code modulation) algorithm was used, and it was also shown that the bit number per frame for video broadcasting has a different correlation function from those for video conferences and video telephony. For the latter two processes the correlation functions are similar to each other and decay to zero geometrically. For video broadcasting the correlation function does not decay to zero. Moreover, the first frame after scene changing contains many more bits than the other frames in the scene. It has already been mentioned that the correlation function for a small delay decays faster than for a large delay. As a result, the time series can be described by the pseudo-Markovian process, which is determined by the bit intensity of various scene types (and by the scene changing condition) [15]. As a result, the simple models describing a video conference are insufficient for video broadcasting.

The modelling strategy of the scene changing recognition method and the further model formation for scene length and cell sum in the scene changing frame is in the development stage. There are some data sets determining the bit sum per frame for the sequences coded by the in-field/interframe scheme of DPCM coding without DCT (discrete cosine transform) usage or motion compensation.

The maximal X_{max}, the average X_{av} bit intensities as well as their ratios can be included in the list of the basic statistical characteristics of the video sequences. As a rule, the ratios of the maximal value to the mean value for video broadcasting change from 1.3 to 2.4 [7]. For comparison, the ratio of the maximal to the mean for a video conference with the same codec is equal to 3.2. It is noteworthy that the large ratios of maximal to mean are related to the decrease in the average intensity. Sequences having the reduced average intensity and high ratios of the maximal to the mean are typical for the various TV programs recorded from the cable TV network. The sequences with low ratios of the maximal to the mean (such as football, sport, news, etc.) are typical for high quality video broadcasting.

3.7.3.1 Scene Changing Determination

Figure 3.36 shows two sections of the film trace. It can be visually observed that there are several sharp hits that might appear due to the scene changing. If the duration is simply fixed instead of modelling these duration distributions, the multiplexed sources with different initial conditions will sometimes not have the coincident hits, which can lead to underestimation of the traffic characteristics.

In the case where ATM (asynchronous transfer mode) technology is used for video data transmission, it is assumed that the scene changing occurs when the frame contains unusually large cell numbers compared to its neighbours. For a quantitative estimate X_i is considered as the cell number in the i frame. When changing the scene, the second-order difference $\Delta X_i^2 = (X_{i+1} - X_i) - (X_i - X_{i-1})$ will be large in value and negative by sign. To define what 'large' means, the second-order difference is divided by the average value of several previous frames. A fixed value was chosen as the critical value, its choice being absolutely subjective.

The correlation function plots show that the scene lengths are noncorrelated. Thus the main problem when modelling them consists of finding a description of the frame sum distribution in the scene.

Experimental researches show [16] that the scene length corresponds, as a rule, to the single-humped distribution. The following three distributions have the widest circulation for the scene length description:
Gamma distribution:

$$w_G(x) = \frac{\lambda(\lambda x)^\beta}{\Gamma(\beta+1)} e^{-\lambda x}, \quad x > 0$$

Weibull distribution with an additional distribution function:

$$1 - F(x) = 1 - e^{-\lambda x^\beta}, \quad x, \lambda, \beta > 0 \tag{3.8}$$

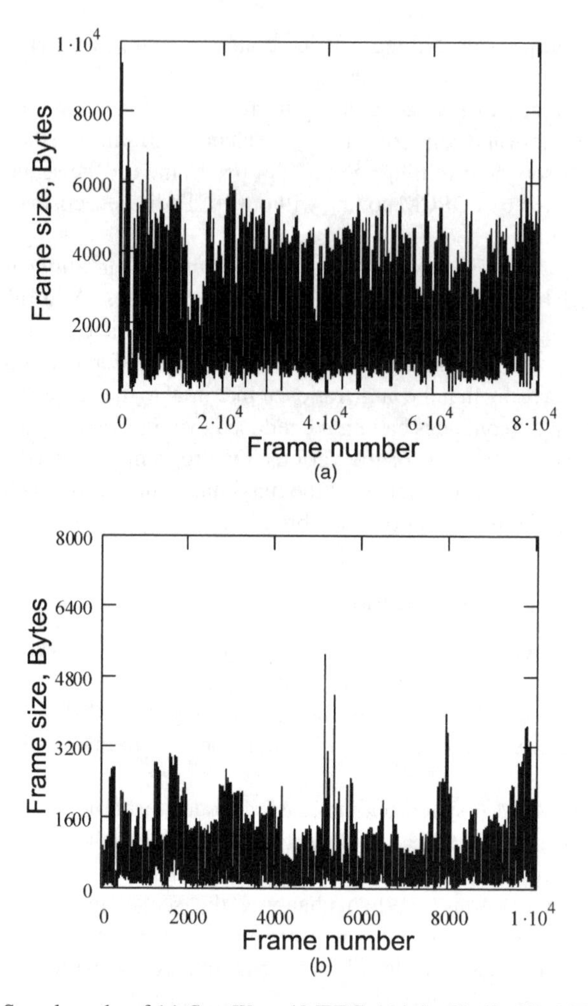

Figure 3.36 Sample paths of (a) *Star Wars 4* MPEG-4 high; (b) *Aladdin* MPEG-4 medium

Generalized Pareto distribution:

$$w_{\mathrm{P}}(x) = \frac{\Gamma(\beta + k)\lambda^{b}x^{k-1}}{\Gamma(\beta)\Gamma(k)(\lambda + x)^{k+\beta}}, \quad x, \beta, \lambda, k > 0 \tag{3.9}$$

It should be noted that the classical Pareto distribution is the specific case for $k = 1$.

In all mentioned distributions λ is the scaling parameter and β is the shape parameter; Pareto distribution has the second shape parameter k.

3.7.3.2 Intrascene Frames

The bit intensity plot analysis [7] shows that scene changing influences two frames and that after scene changing the first frame is very large. If Z_n is considered as the cell number in frame n of the scene changing and Y_n as the cell number in the next $(n+1)$ frame, then

$$Y_{n+1} = a + bZ_n + \varepsilon_n \tag{3.10}$$

where ε_n are the independent and identically distributed normal random variables with zero mean value. The correlation function and PDF are the important statistical characteristics in other intrascene frames.

Researches show that the correlation function of the TV signal scenes does not decay to zero geometrically as it does, for example, for video conferences. As a result, other models are required for the scene changing frames rather than for intrascene frames.

A more accurate estimate $r(i)$ can be derived by means of dividing the data to the scenes, the experimental correlation function calculation for each scene and averaging over all scenes. Determining the correlations $r_s(i)$ for delay i in scene s, the correlation function estimate can be found as

$$r(i) = \frac{1}{S}\sum_{s=1}^{S} r_s(i)\ 1_s(I), \quad i = 1, 2, \ldots \tag{3.11}$$

where S is the scene number in the sequence. The indicated function $1_s(i)$ equals unity if scene s contains at least $i+3$ frames (note that the first two frames are considered separately, which is necessary at least for $i+1$ observations for the correlation estimates for i delay). Otherwise, $1_s(i) = 0$.

The wide variety of possible scenes implies that in-scene frame distribution will be well described by the combination of single-humped distributions.

3.7.4 MPEG Video Traffic

The choice of how the data traces should be presented using the statistical models is the main problem for source modelling. The source model is chosen so as to consider it as the input process of the analysed system. It is evident that the source model is acceptable if it adequately describes the trace in the created system model. It is understood that when the source model is used in the system model, the performance values concerned, obtained during modelling, are ‘close enough’ to the values given by the real trace. The definition of the sufficient closeness can depend on which model is used as the basis.

It is well known that the video data are the frames sequence. Each frame is coded and compressed in accordance with the applied algorithm. For example, the MPEG (Motion Pictures Experts Group) standard was developed for the moving images and is well adapted to the ‘real’ data. This approach implies compression with losses; to increase the compression it uses limitations on the spatial and temporal resolution of the eye. In addition, MPEG uses the image property characterized by large frame numbers in the video sequence being similar to their nearest neighbours. Therefore, transmitting the information concerning the changed pixels between the adjacent frames only, it is possible to achieve the essential increase in the compression factor.

The MPEG standard uses three modes for frame coding. They are called intraframe (I), predicted (P) and interpolating (B). The I frame is the JPEG (Joint Photography Experts Group) coding of a separate frame (i.e. without using time redundancy). It is the most inefficient among the three modes, but is capable of correcting any error introduced by the B and P frames. In the common version each twelfth frame is an I frame, i.e. the following coding sequence is used:

...IBBPBBPBBPBBIBBPBBP... P frames are generated starting from the nearest I or P frames if the conformity between the frames is close enough. The P frame is coded into a block of 8×8 DCT coefficients and if these values are close to the previous coefficients, for this block the difference information is transmitted. Therefore, the P frame increases the I frame compression if there is a time correlation between adjacent I and P frames. B frames are always generated using the difference information and the direct and inverse correlations. The encoder compares the previous and future I and P frames and uses those frames that are closest to the B frame.

If the initial video data contain nonessential motion or scene changing, MPEG should give very high compression coefficients. This is the main usage for such applications as video telephony. On the other hand, if the video sequence contains an abrupt scene or plan change, the compression coefficient will not be very high and 'artificiality' may arise due to errors in the B and P frames.

During coding the so-called coding template is usually given, i.e. an accurate sequence determining the time moments of the full frame arrival. This template is referred to as the GOP (group of pictures) and appears to be self-sufficient for decoding the frame sequence.

From the video sequence correlation function shown in Figure 3.37(a) it can be seen that MPEG coding introduces strict periodicity. To avoid this periodicity MPEG data can be grouped into blocks of 12 frames, called the GOP. The correlation function of GOP data is shown in Figure 3.37(b).

Since the frames inside this template do not differ very much from each other (only the difference between them is transmitted), this leads to the existence of an essential correlation of their sizes. During transmission of the next full frame the correlation between them practically disappears. For this reason the video traffic differs considerably from the usual traffic of the telecommunication network. Therefore, the conclusions and the models obtained for the usual network traffic cannot be applied for the analysis and modelling of video traffic.

Figure 3.38 shows the trace fragment of the MPEG sequence (the first 3000 frames). For further analysis and modelling of MPEG sequences three separate sequences formed by I, B and P frames are of interest, but not a complete sequence. These sequences are shown in Figures 3.39(a), (b) and (c) respectively.

3.7.4.1 Intergroup Characteristics

The intergroup traffic character is well described in the literature [17–20]. It can be described by the first-order and second-order statistical characteristics of I frame processes. Figure 3.39 shows examples of I, P, B and IPB process distributions. Gamma distribution is a good approximation of the I process size PDF:

$$F_{X_{\mathrm{I}}}(r) = \frac{r^{m_{\mathrm{I}}-1}}{\Gamma(m_{\mathrm{I}}) {\ell_{\mathrm{I}}}^{m_{\mathrm{I}}}} \mathrm{e}^{-r/\ell_{\mathrm{I}}}, \quad \forall r > 0 \tag{3.12}$$

where m_{I} is the shape parameter and ℓ_{I} is the scaling coefficient. They are related by the μ value and the variance σ_{I}^2 of the I frame trace using the following relations:

$$m_{\mathrm{I}} = \frac{\sigma_{\mathrm{I}}^2}{\mu_{\mathrm{I}}^2} \quad \text{and} \quad \ell_{\mathrm{I}} = \frac{\sigma_{\mathrm{I}}^2}{\mu_{\mathrm{I}}} \tag{3.13}$$

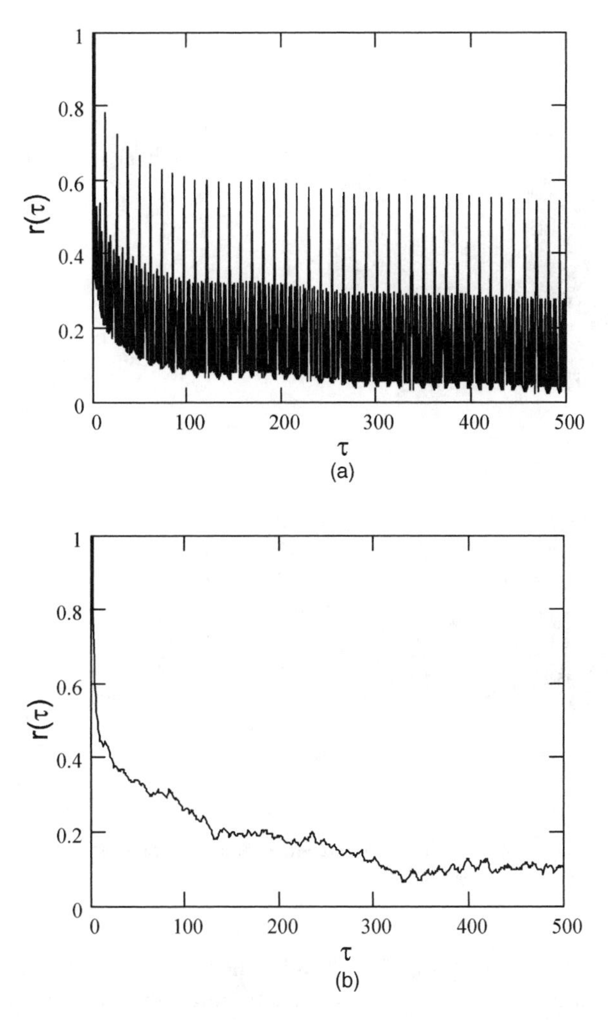

Figure 3.37 The first 500 values of the correlation function of (a) the real video sequence and (b) the GOP for *Aladdin* MPEG-4 medium

The I frame trace has self-similar properties and can be characterized by the SRD parameter λ_{I}, the LRD parameter H_{I} and the 'boundary parameter' K_{I}. Thus the autocorrelation function calculated as has the form

$$R_{x_{\mathrm{I}}x_{\mathrm{I}}}(m) = \begin{cases} \mathrm{e}^{-\lambda_{\mathrm{I}} m}, & m \leq K_{\mathrm{I}} \\ Lm^{-\beta_{\mathrm{I}}}, & m > K_{\mathrm{I}} \end{cases} \tag{3.14}$$

where $\beta_{\mathrm{I}} = 2 - 2H_{\mathrm{I}}$. The same procedure can be used to describe P and B frame distributions.

Figure 3.40 shows examples of the I process autocorrelation function (ACF) for I, P, B and IPB for various types of analysed sequences. In References [17] and [19] it is shown that the ACF of I processes for analysed sequences has two different characteristics: the self-similar

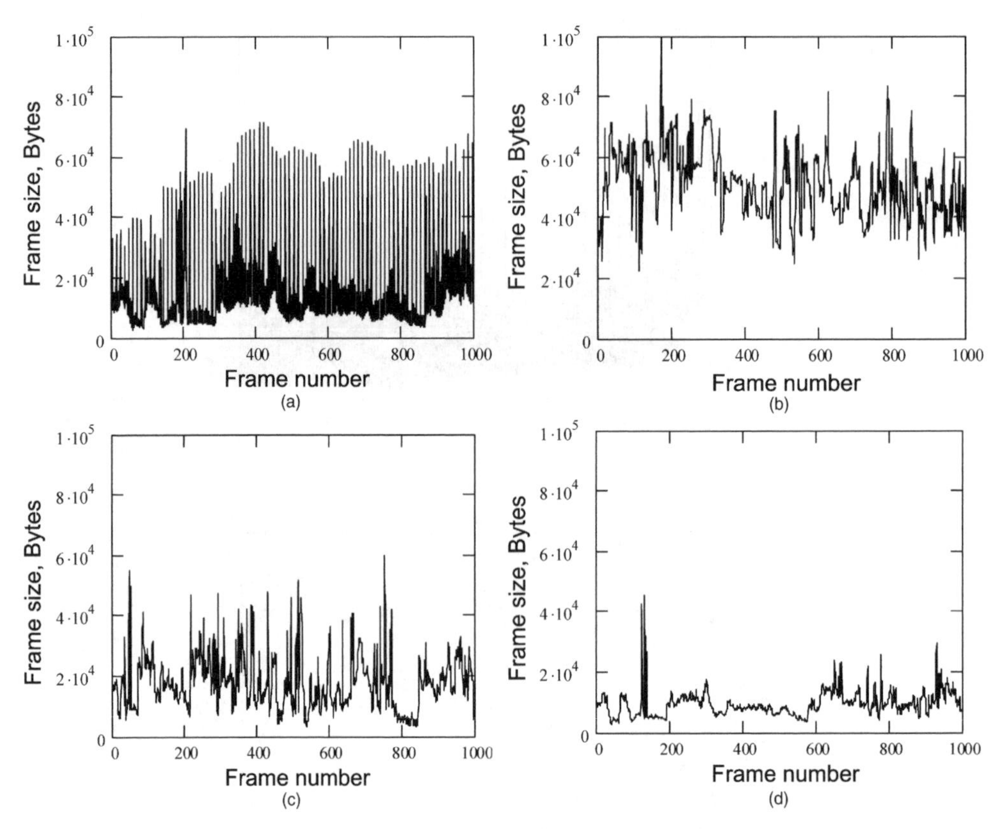

Figure 3.38 The bit intensity of MPEG-coded video data (the first 3000 frames): (a) IPB frames; (b) I frames; (c) P frames; (d) B frames

character (long-range dependence), described by the Hurst exponent H_{I}, and exponential decay similar to the function $e^{-\lambda_{\mathrm{I}} x}$ over short time intervals. Two regions are divided by coefficient K_{I} characterizing the boundary. For example, in the case of the cartoon the exponent $H_{\mathrm{I}} = 0.873$, $\lambda_{\mathrm{I}} = 0.891$ and coefficient $K_{\mathrm{I}} = 30$ frames. The same character is typical for the correlation functions of B and P frames.

Taking into account that the gamma distribution is fully characterized by the mean and the variance, it is necessary to analyse the statistical characteristics of the processes reflecting the B and P frame size distributions for the given I frame sizes. These characteristics are designated as $M\{\mathrm{B}|\mathrm{I}\}, \sigma^2_{\mathrm{B}|\mathrm{I}}$ and $M\{\mathrm{P}|\mathrm{I}\}, \sigma^2_{\mathrm{P}|\mathrm{I}}$ respectively. The mean values of B and P processes and their variances depend to a large extent on the same I frame values belonging to one GOP, in accordance with the linear law. Due to this correlation, the PDF of the total sequence of the B frame sequence (further B sequence) and of the P frame sequence (further P sequence) are similar to that of the I sequence. Moreover, the second-order statistical characteristics of all these processes are similar. It should be noted that the whole sequence as well as B and P processes are self-similar with the Hurst exponent, the value of which is close to the Hurst exponent for the I process. For example, for the cartoon, $H_{\mathrm{I}} = 0.884, H_{\mathrm{B}} = 0.821$ and $H_{\mathrm{P}} = 0.878$.

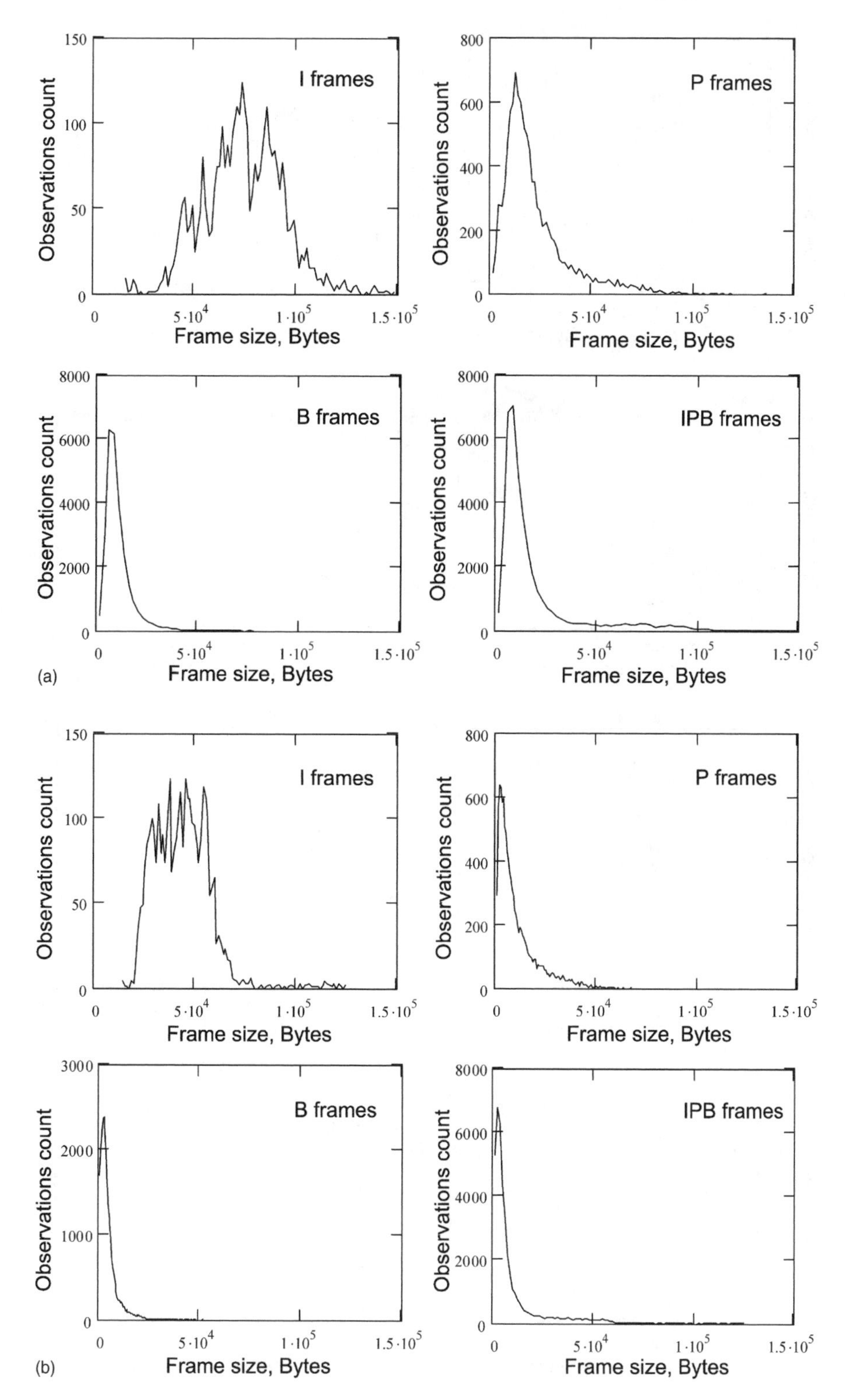

Figure 3.39 PDF plots for four video sequences of MPEG-1 (I, P, B, IPB frames): (a) the cartoon *Simpsons*; (b) the movie *Star Wars*; (c) the video record of a football match; (d) the video record of Formula 1

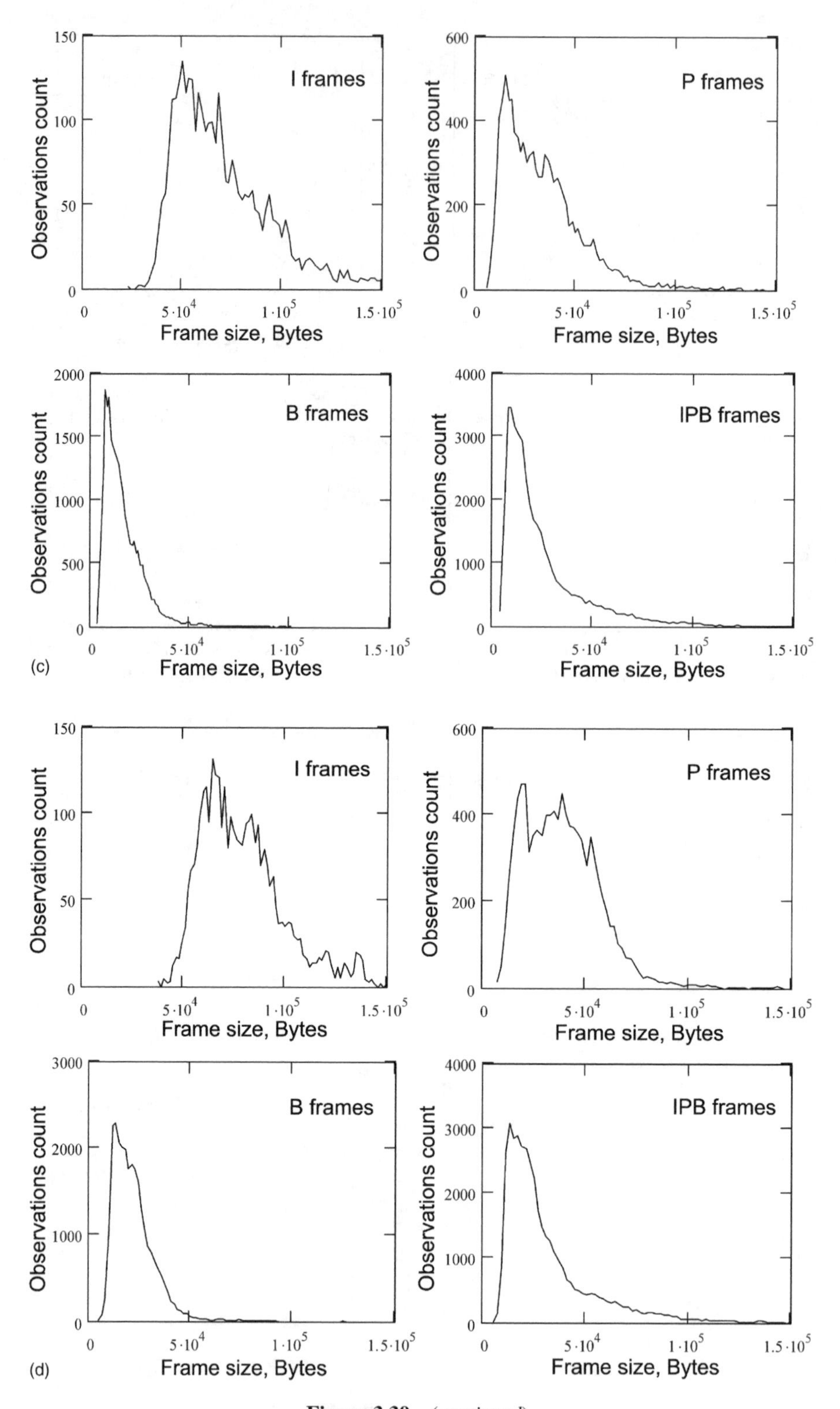

Figure 3.39 (*continued*)

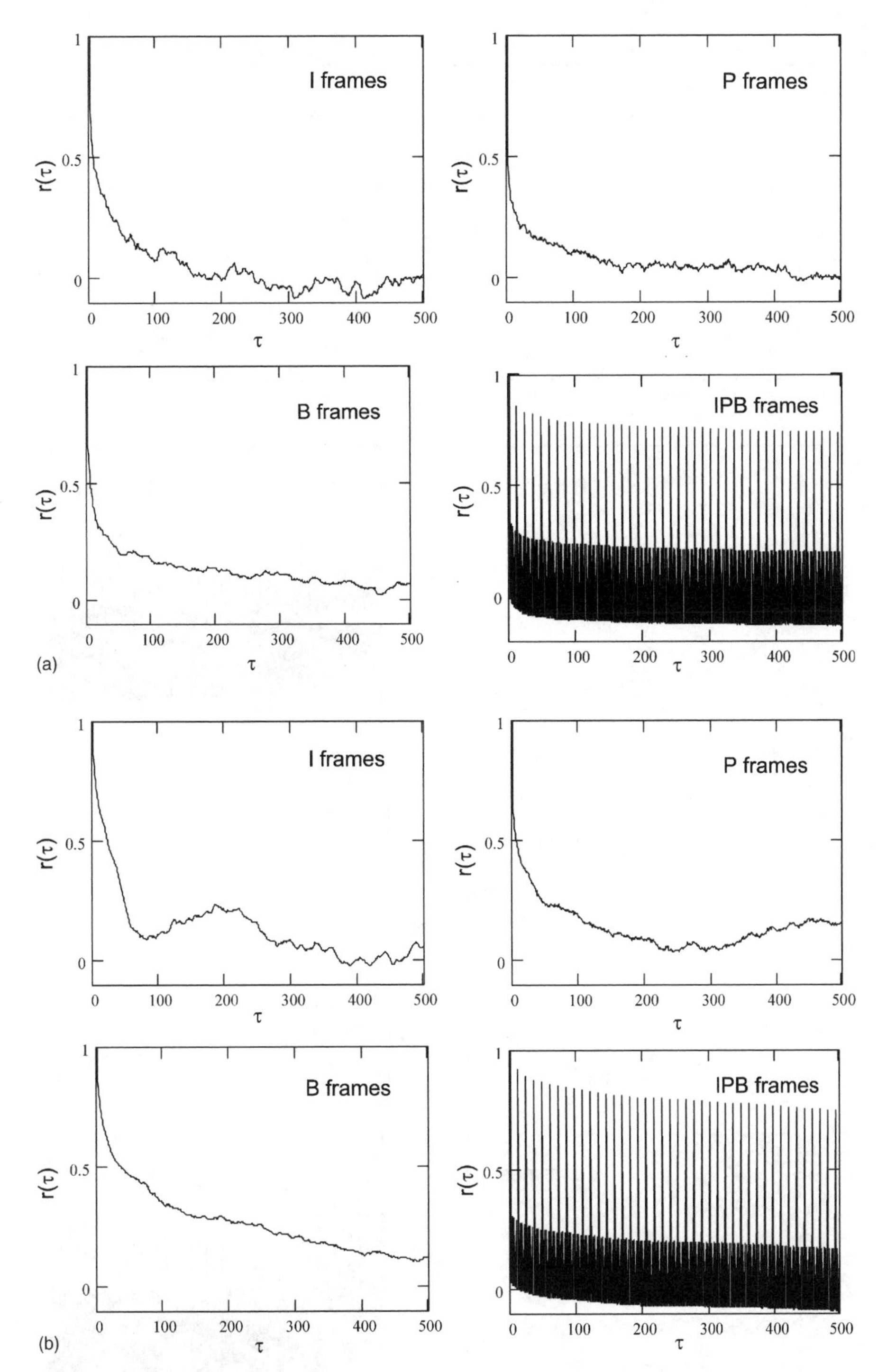

Figure 3.40 ACF plots for the analysed video sequences MPEG-1 (I, P, B, IPB frames): (a) the cartoon *Simpsons*; (b) the movie *Star Wars*; (c) the video record of a football match; (d) the video record of Formula 1

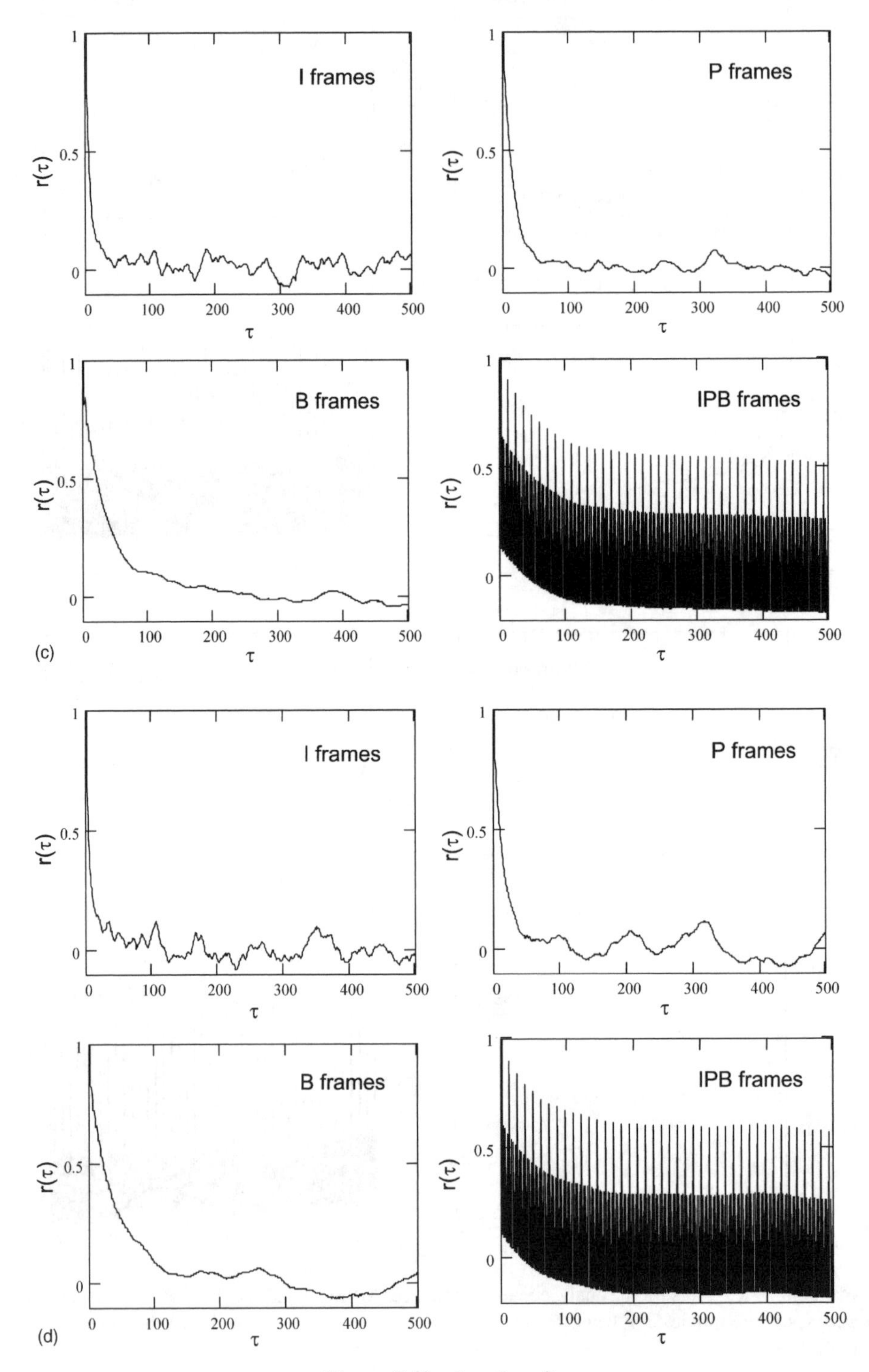

Figure 3.40 *(continued)*

A large number of models are known to have non-MPEG video sources. Most of them are based on Markovian models and therefore it is impossible to adapt them to MPEG traffic, which, as shown in Reference [21], is self-similar. That is why the classical models based on the Markovian processes cannot be the basis for the VBR traffic models. The first step in the analysis and modelling of video traffic was undertaken in Reference [22], where gamma/Pareto distribution was used.

This model reflects two main statistical characteristics of the real video sequence: the quotient distribution with the heavy tail and the long-range part of the autocorrelation function. Since this model does not approximate the SRD accurately enough, it is suitable for modelling the set of a large number of video sources. In this case the quotient distributions are close to Gaussian and the singular correlation effects are random over the short time spaces.

From the above-mentioned information, it is possible to draw a conclusion about the importance of the analytical model formation for MPEG traffic emulation. First of all, it is necessary to calculate the network capacity (the multiplexers in the first place) and the quality of the video traffic transmission in telecommunication networks.

3.7.5 Nonstationarity of VBR Video Traffic

It was proved in publications [23] and [24] that VBR video traffic can belong to the class of long-range dependent processes:

- The correlation of r_k demonstrates the hyperbolic decay for large delays k: $r_k \to c_0 k^{-\beta}$, as $k \to \infty$.
- The power spectral density $S(\omega)$ for small frequency values ω corresponds to the law $S(\omega) \to c_1 \omega^{\beta-1}$, as $\omega \to \infty$.
- The variance σ_n^2 of the sample mean value decreases slower than the inverse sample size n: $\sigma_n^2 = \sigma^2(\bar{X}_n) \to c_2 n^{-\beta}$, as $n \to \infty$ ($\bar{X}_n = \sum_{i=1}^{n} X_i/n$ for several constants $c_0,\ c_1,\ c_2$).

The constant value $\beta \in [0; 2]$ reflects the function type, $0 \leq \beta < 1$ indicates the long-range dependence and $1 < \beta \leq 2$ demonstrates the short-range data dependence. (The persistence degree is often expressed with the help of the Hurst exponent $H = 1 - \beta/2$.) The long-range dependence is defined within the limits of the weak stationarity structure [23, 25], i.e. the stationarity in the wide sense.

The stationarity and the ergodicity allow statistical estimates such as the mean value and the variance or other model parameters to be found from each separate data sample, or in this case from the separate time series. If the assumptions of stationarity and ergodicity do not hold, certain measures such as the mean value and the variance may be without meaning. In reality the mean value of the VBR video time series converges very slowly, which can be caused by nonstationarity and not necessarily by long-range dependence.

3.7.5.1 Testing for Stationarity

Let $X(n), n = 0, 1, 2, \ldots$, be the stochastic process with the power spectral density $S(\omega)$. This process periodogram can be estimated in the form

$$\hat{I}_N(\omega) = \left(\frac{1}{2\pi N}\right) \left| \sum_{n=0}^{N-1} [X(n) - \bar{X}] \mathrm{e}^{\mathrm{j}\omega n} \right|^2$$

where $\bar{X}$ is the sample mean value which converges to $\frac{1}{2}S(\omega)\chi_2^2$ (see, for example, Reference [26]) for $\omega \neq 0;\ \pm\pi;\ \pm 2\pi;\ldots$. This implies that $\hat{I}_N(\omega)$ for large N is an unbiased but ungrounded estimate, as $\lim_{N\to\infty} \sigma^2 \hat{I}_N(\omega) = S^2(\omega)$. Nevertheless, it is true that for two fixed frequencies, ω_1 and ω_2, the periodogram ordinates $\hat{I}_N(\omega_1)$ and $\hat{I}_N(\omega_2)$ are approximately noncorrelated. These properties are also correct for long-range dependent processes [25]. The application of the *spectral window* $\Lambda(\omega)$ gives the consistent estimate [26]

$$\overline{I_N}(\omega) = \int_{-\pi}^{+\pi} \hat{I}_N(\omega)\Lambda(\Theta - \omega)\mathrm{d}\omega \tag{3.15}$$

Choosing the Bartlett–Priestley spectral window [26] gives the following expression $\sigma^2\left[\overline{I_N}(\omega)\right] \approx [(6M)/(5M)]S^2(\omega)$ for the variance. The variance still depends on the power spectral density itself. To avoid this functional dependence the logarithmic *variance stabilizing transform* can be used [27].

For the first accuracy order,

$$M[\log(\overline{I_N})] \approx \log(I_N) \tag{3.16}$$

$$\sigma^2[\log(\overline{I_N})] \approx \frac{2\pi}{N}\int_{-\pi}^{+\pi} \Lambda^2(\Theta)\mathrm{d}\Theta \tag{3.17}$$

where $\omega \neq 0; \pm\pi; \ldots$. Thus, the estimate $\log(\overline{I_N})$ is closer to the normal value than the nontransformed estimate. To prove (or to negate) the assumption of weak nonstationarity, the X process is divided into I segments, each of which is centred by time t_i and has the length N. For each ith segment the power spectral density $\overline{I_{N,i}}(\omega)$ is calculated in accordance with Equation (3.15). The discretization of the smoothed periodogram (3.15) is carried out by frequencies $\omega_i = \pi j/N (j = j_0 + k\Delta j, k = 0, 1, \ldots, J)$, and taking a logarithm gives the two-dimensional random variable $Y_{ij} = \log[\overline{I_{N,i}}(\omega_j)]$. If the frequencies ω_i, like the times t_i, have a wide enough dispersion, the random variable Y_{ij} is distributed approximately normally and is noncorrelated [28]. The assumption of Y_{ij} approximate normality and lack of correlation in both measurements implies Y_{ij} approximate independence. Therefore, to define the structure of the basic random process the method of *variance analysis* can be used [27,28]:

$$Y_{ij} = \mu + a(t_i) + b(\omega_i) + c(t_i, \omega_i) + \eta_{ij} \tag{3.18}$$

where η_{ij} is the independent and identically distributed normal random variable with zero mean value and variance σ^2, defined by the relation (3.17). The presence of $c(t_j, \omega_j)$ and $a(t_i)$ can be checked using the variables

$$S_{I+R} = \sum_{i=1}^{I}\sum_{j=1}^{J}(Y_{ij} - Y_{\cdot j} - Y_{i\cdot} + Y_{\cdot\cdot})^2 \tag{3.19}$$

$$S_T = J\sum_{i=1}^{I}(Y_{i\cdot} - Y_{\cdot\cdot})^2 \tag{3.20}$$

where the dot shows the mean value over the index for which it substitutes: e.g. $Y_{\cdot j} = \sum_{i=1}^{I} Y_{ij}/I$. In the stationary process the terms $c(t_j, \omega_j)$ and $a(t_i)$ can be expected to disappear. In this case the variables S_{I+R}/σ^2 and S_T/σ^2 are χ^2-distributed with $(I-1)(J-1)$ and $(I-1)$ degrees of freedom respectively. The stationarity hypothesis is rejected if one of the statistical tests exceeds 1% of the quantile of the appropriate χ^2 distribution.

This test cannot be used in the case of long-range dependence because the noise is not normally distributed and correlated. In this case the series method described in Chapter 1 should be applied.

3.8 Self-Similarity Analysis of Video Traffic

The features of estimation of video traffic self-similarity will be considered using different methods and concrete examples.

3.8.1 Video Broadcasting Wavelet Analysis

The video traffic trace for the *Star Wars* movie was chosen. MPEG-1 was used as the coding algorithm. The trace length was 40 000 frames, taking into account the frame frequency of 25 fps. As a result, the length of the analysed sequence is approximately estimated as 26 minutes and 40 seconds. The analysed trace realization is presented in Figure 3.41. Although the data have long been recorded, it is a well-known reference test set, useful for a fractality and LRD check for video traffic.

As repeatedly mentioned, many methods have been used to find a Hurst self-similarity exponent estimate, such as R/S analysis, variance–time plots, the periodogram analysis and the Whittle analysis. However, video traffic is strongly correlated and demonstrates a long-range dependence. The long-range dependence property leads to a serious estimate displacement and difficulties in making a convergence estimate. An estimate will be made of video traffic self-similar properties over large intervals using the wavelet analysis as an example.

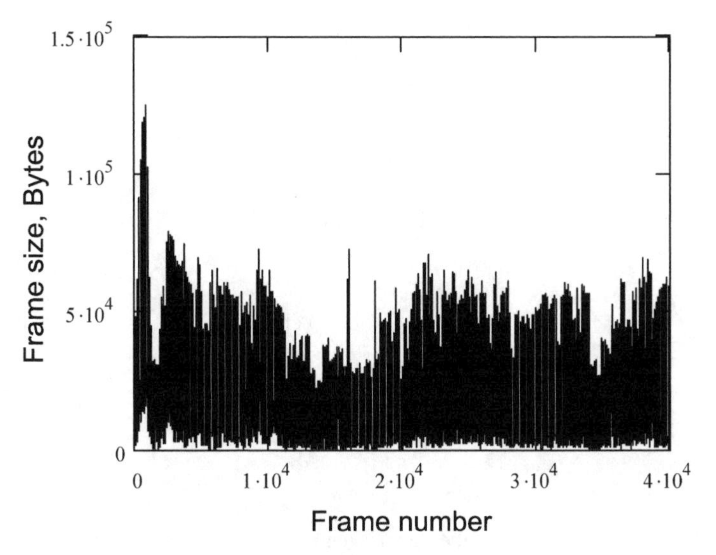

Figure 3.41 Video traffic trace for the *Star Wars* movie

The advantages of the wavelet analysis result from the fact that the wavelet functions themselves demonstrate the scaling property and therefore form the optimal 'coordinates system', from which the scaling phenomena can be traced. This analysis provides steady detection of the scaling behaviour, its type and an accurate measurement of the parameters in order to describe this scaling behaviour.

Using this approach, the wavelet analysis will be carried out by means of the expansion of the successive sample $X(t)$: $\{x(t_0), x(t_1), \ldots, x(t_{N-1})\}$ of volume $n_0 = 2^J_{\max} (n_0 \leq N)$ into the refining functions of a varying scale. Here $J_{\max} = [\log_2 N]$ is the maximal number of expansion scales and $[\log_2 N]$ is the integer part of the $[\log_2 N]$ number. The scaling index value $j = 0$ corresponds to the maximal resolution case, which is the most accurate approximation, equal to the initial series $X(t)$, consisting of n_0 counts. The transition to a rougher resolution occurs as $j (0 < j \leq J_{\max})$ increases.

In accordance with the wavelet analysis regulations described in Chapter 1, the time series $X(t)$ is presented in the form

$$X(t) = X_J(t) + \sum_{j=1}^{J} D_j(t) \tag{3.21}$$

where $X_J(t) = \sum_{k=0}^{n_0/2^J - 1} s_{J,k} \varphi_{J,k}(t)$ is the initial approximation function corresponding to the scale $J (J \leq J_{\max})$; $s_{J,k} = \langle X(t), \varphi_{J,k} \rangle$ is the scaling coefficient equal to the scalar product of the initial series $X(t)$ and the scaling function of the 'roughest' scale J, displaced by k scale units to the right from the origin of coordinates; $D_j(t) = \sum_{k=0}^{n_0/2^j - 1} d_{j,k} \psi_{j,k}(t)$ is the refining function of the jth scale; and $d_{j,k} = \langle X(t), \psi_{j,k} \rangle$ is the wavelet coefficient for scale j, equal to the scalar product of the initial series $X(t)$ and the wavelet with scale j, displaced by k scale units to the right from the origin of coordinates.

The normalized wavelet and scaling functions of the Haar system give good results for the discrete time series analysis (Figure 3.42). If

$$\varphi(t) = \begin{cases} 1 & \text{for } 1 \leq t < 0 \\ 0 & \text{otherwise} \end{cases}$$

and

$$\psi(t) = \begin{cases} 1, & \text{as } 1 \leq t < \frac{1}{2} \\ -1, & \text{as } \frac{1}{2} \leq t < 1 \\ 0, & \text{otherwise} \end{cases}$$

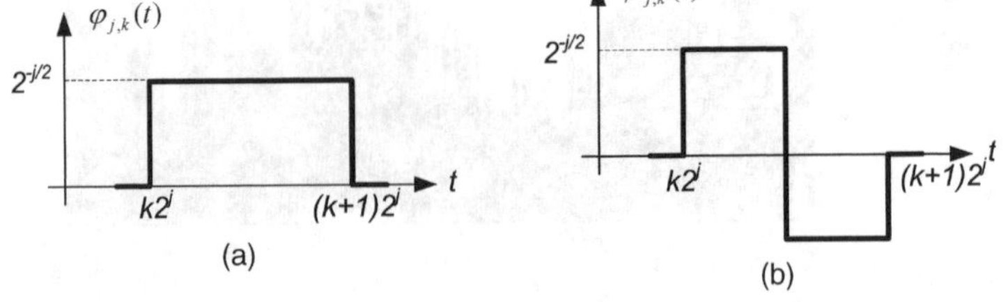

$\varphi_{j,k}(t)$ $2^{-j/2}$ $k2^j$ $(k+1)2^j$ t

(a)

$\psi_{j,k}(t)$ $2^{-j/2}$ $k2^j$ $(k+1)2^j$ t

(b)

Figure 3.42 Normalized functions of the Haar system: (a) the scaling function $\varphi_{j,k}(t)$; (b) the wavelet function $\psi_{j,k}(t)$

ψ is the orthonormal wavelet in $L^2(R)$ space. It is called the Haar wavelet and $\{\psi_{j,k} : j, k \in Z\}$ is the orthonormal system in $L^2(R)$.

The relation of the wavelet coefficients found for the time series expansion over the wavelet functions basis and the Hurst exponent H is determined from the equation

$$\begin{aligned}\log_2 \mu_j &\approx \log_2\left(\frac{1}{n_j}\sum_{k=1}^{n_j} |d_x(j,k)|^2\right) \sim (2H-1)j + C_W \\ &= \log_2\left(\frac{1}{K_j}\sum_{k=0}^{K_j-1} |d_{j,k}|^2\right) = \alpha j + C_W\end{aligned} \tag{3.22}$$

where $K_j = n_0/2^j$ is the wavelet coefficient number for the scale $j, C_W = c_f C(\alpha, \psi)$ is the parameter that does not depend on scale j and $\alpha = 2H - 1$.

The number of wavelet coefficients decreases as the scale increases. Formula (3.22) is used for the Hurst exponent estimate of the LRD video sequences. This means that if X is the LRD process with the Hurst exponent H, the plot of function j, referred to as the logarithmic diagram (LD), should have the linear slope $2H - 1$, and demonstrates that the scaling exponent $(2H - 1)$ can be obtained from the plot slope estimate of the function $\log_2((1/K_j)\sum_{k=0}^{K_j-1} |d_{j,k}|^2)$ of j. Therefore, the Hurst exponent estimate can be found by means of the choice of the approximated curve equation using the weighted least squares (WLS) method.

The logarithm of this variable will be the estimate of $\log_2 \mu_j$, but will be displaced as the logarithm nonlinearity shows that $M \log_2(\bar{d}_j^2) \neq \log_2(M\bar{d}_j^2) = j\alpha + \log_2 C_W$. As shown in References [29] to [31], the regression analysis problem is reduced to considering the equation $My_j = ja + \log_2 C_W$. The estimation of slope $\widehat{\alpha}$ can be obtained by carrying out the weighted linear regression, in which $x_j = j$ and $\sigma_j^2 = \mathrm{Var}(y_j)$. Determining the quantities $S = \sum_{j=j_1}^{j_2} 1/\sigma_j^2$, $S_1 = \sum_{j=j_1}^{j_2} j/\sigma_j^2$ and $S_2 = \sum_{j=j_1}^{j_2} j^2/\sigma_j^2$, the weighted estimate $\widehat{\alpha}$ can be obtained for α as

$$\widehat{\alpha} = \frac{\sum_{j=j_1}^{j_2} y_j(S_j - S_1)/\sigma_j^2}{SS_2 - S_1^2} = \sum_{j=j_1}^{j_2} w_j y_j \tag{3.23}$$

which is unbaised over the interval $[j_1; j_2]$. In addition,

$$\log_2 C_W = \frac{\sum y_j(S_2 - S_1 j)/\sigma_j^2}{SS_2 - S_1^2}$$

Assuming a weak correlation between wavelet coefficients in the case when $d_{j,k}$ are Gaussian values, the variance σ_j^2 can be estimated by the expression

$$\sigma_j^2 = \frac{\varsigma(2, n_j/2)}{\ln^2 2} \sim \frac{2}{n_j \ln^2 2}$$

where

$$\varsigma(2, z) = \sum_{n=0}^{\infty} \frac{1}{(z+n)^2}$$

is the generalized Rieman zeta function.

In view of the above, we can suggest the following algorithm of the Hurst exponent estimate for video traffic [32]:

Step 1. Determine the initial value of the scaling coefficients $s_{0,k}$, corresponding to the $j = 0$ scale, which are equal to the values of the initial time series $X(t)$:

(a) set $j = 0$;
(b) $s_{0,0} = x(t_0), s_{0,1} = x(t_1), s_{0,2} = x(t_2), \ldots, s_{0,n_0-1} = x(t_{n_0-1})$.

Step 2. For transition to the $j + 1$ scale: (a) set $j = j + 1$; (b) determine the scale and wavelet coefficients $s_{j,k}, d_{j,k}$ for scale j with the help of the recurrence formulas:

$$s_{j,k} = \frac{1}{\sqrt{2}}\left[s_{j-1,2k} + s_{j-1,2k+1}\right], \qquad d_{j,k} = \frac{1}{\sqrt{2}}\left[s_{j-1,2k} - s_{j-1,2k+1}\right] \tag{3.24}$$

where $k = 0, 1, \ldots, K_j - 1$; $K_j = n_0/2^j$.

Step 3. Calculate the second-order initial moment $m_2^{(j)}$ of wavelet coefficients of scale j using the formula $\mu_{j=m} = m_2^{(j)} = \sum_{k=0}^{K_j-1} |d_{j,k}|^2/K_j$.

Step 4. Check the condition $j < J$, where $J = \log_2(n_0)$. If this condition is fulfilled, return to step 2. If not, then complete the transition to step 5.

Step 5. Determine the approximated straight line equation on the basis of the WLS method in the form $\log_2(\mathrm{m}_2^{(j)}) = \alpha j + C_W$, where $j = l, 2, \ldots, J$. Each point on the ordinate axis corresponding to the $\log_2(\mathrm{m}_2^{(j)})$ value is given the weight $W_j = K_j$.

Step 6. Determine the point estimate for the Hurst exponent H from the equation $2H - 1 = \alpha$. This completes the calculation.

3.8.2 Numerical Results

Self-similarity degree wavelet analysis results carried out on the example of the *Star Wars* movie trace are shown in Figure 3.43. For each H estimate it is possible to indicate the octave range in which the linear regression was chosen. These ranges are chosen with the help of the visual analysis of the logarithmic diagrams and identification of the linear region in it. The approximately linear interval of the logarithmic plot can be observed for the octaves $5 \leq j \leq 14$. Larger values of j were not considered due to the restricted data volume and also because the wavelet coefficients set at the large scaling level demonstrate several values only, which do not ensure a stable approximation. (These restrictions are also considered in the scale analysis of other flows.) The fit of the linear function over this interval gives the estimate of the LRD exponent as $H = 0.86$ to 0.996. The slope for the large time scales is constant, indicating that the real traffic is self-similar (monofractal) over large time scales. However, over small time scales the slope value has an essential difference, which indicates that the traffic is a multifractal process. The linearity of the logarithmic diagrams for the roughest octaves shows that the spectral densities for the traces demonstrate the power character near zero with the exponent $\alpha > 1$, which contradicts the assumption about LRD, because for LRD processes $\alpha < 1$.

Hurst exponent estimation results obtained using other approaches are listed in Table 3.8. It can be seen that the periodogram and the variance approaches give results close to the wavelet method. The results that differ the most are given by the Whittle estimate.

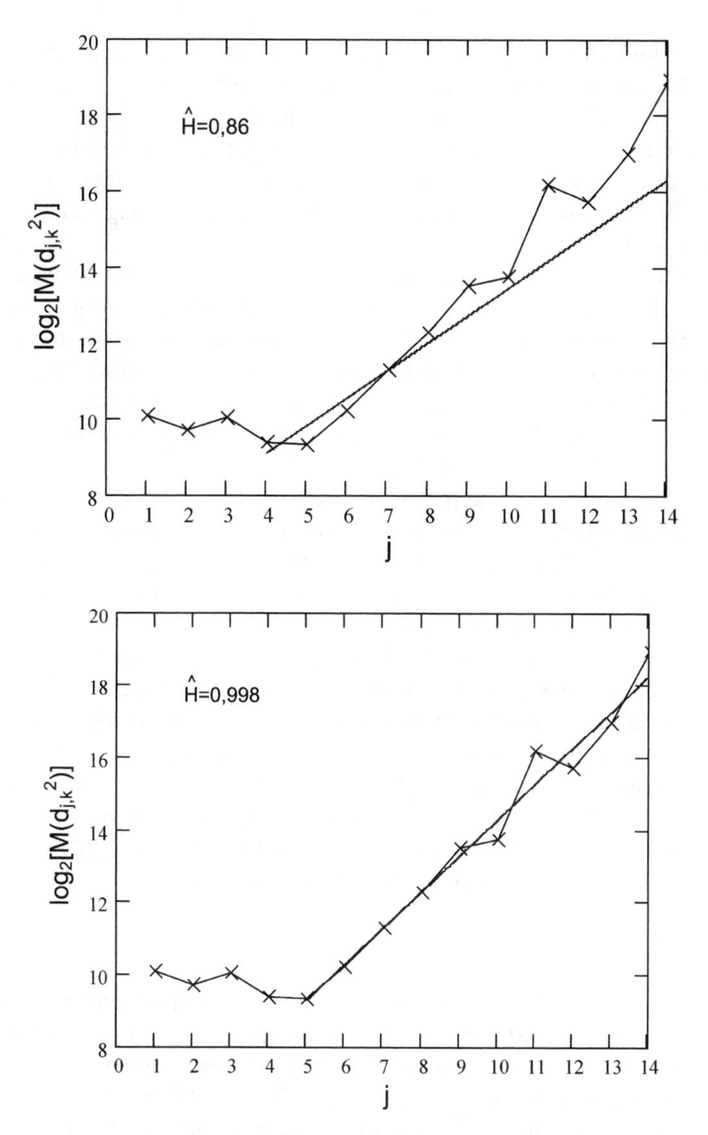

Figure 3.43 Wavelet estimates of the *Star Wars* trace

Table 3.8 Hurst exponent estimation for the *Star Wars* movie trace using different techniques

Aggregate variance	$H = 0.819$
R/S statistics	$H = 0.789$
Periodogram	$H = 0.993$
Variance of residuals	$H = 0.992$
Abry–Veitch estimator	$H = 0.929$
Whittle estimator	$H = 0.599$

Since the plot looks 'broken', the trace probably does not demonstrate strict scaling of the second order. As the video sequences consist of various scenes, the video frames representing a separate scene are similar due to the identical or analogous background present in this scene. This implies similar sizes of adjacent frames, which proves a strong positive correlation for small delays. All this testifies to the fact that the multifractal wavelet estimate often gives unreliable results when used for processes with strong SRD and strong LRD components.

Note that it is not known *a priori* which property of the scale invariance and for which scales can exist if it exists at all. In practice, the decision can be made for lower octave thresholds, for each presupposed scaling mode. This decision is made with the help of the logarithmic diagram investigation by confidence intervals. The useful heuristics consists in the regression line restricting (or nearly doing it) each confidence interval in the chosen range $[j_1; j_2]$. This approach can be normalized by the chi-square fitting criterion. The scaling modes of three octaves in width only should be presumed to be possible, except for cases where the confidence intervals are very small, as it is possible that they may be 'formed in line' even if the scaling is absent. It is especially necessary to pay attention to the scaling behaviour, starting from the straightforwardness at small scales (small j, high frequency), for when n_j is large the confidence intervals are small (i.e. a large data volume is present). Therefore, the regression line, which seems to approximate the points at lower scales well, can really be a very bad approximation once the known weights are taken into account.

The calculating advantages of the logarithmic diagram (LD) are part of its usefulness as a statistical tool. An LD based on the discrete wavelet transform can be realized with the help of the fast pyramidal algorithm of the filter bank with the very low calculating complexity of order $O(n)$, where n is the length of the time series. This algorithm also has advantages from the point of view of memory usage when dividing data into blocks, analysed and recovered with minimal calculating expenses.

The logarithmic diagram demonstrates not only the long-range dependence of the traffic but also shows the second-order statistical characteristics over each time scale. This wavelet analysis property can be used to develop an effective algorithm for predicting the queuing character.

3.8.2.1 Video Conferences

The results will be analysed of the self-similarity estimate of video conference traffic on the examples of three realizations H.263, 64 kbit/s and VBR [5]. Figure 3.44 shows traces of video traffic (H.263, 64 kbit/s) for different types of content and its Hurst exponent wavelet estimation. Figure 3.45 shows the trace of video traffic (H.263, VBR) for different types of content and its Hurst exponent wavelet estimation. For each H estimate the octave range is shown where the linear regression was fitted. These ranges are chosen with the help of the visual analysis of the logarithmic diagrams and identification of the linear region in it. Table 3.9 gives the results of the self-similarity degree estimate using various known approaches. It is seen that the wavelet analysis results correspond well with the periodogram, variance of residuals, and Whittle estimator and worse with the aggregate variance and R/S statistics.

The multifractal estimate is the extension of the monofractal wavelet estimate. In addition to the second moments (variances) for the wavelet coefficients this estimate may also account for the higher-order moments

$$S_q(j) = \frac{1}{n_j}\sum_{k=1}^{n_j} |d_{j,k}|^q \tag{3.25}$$

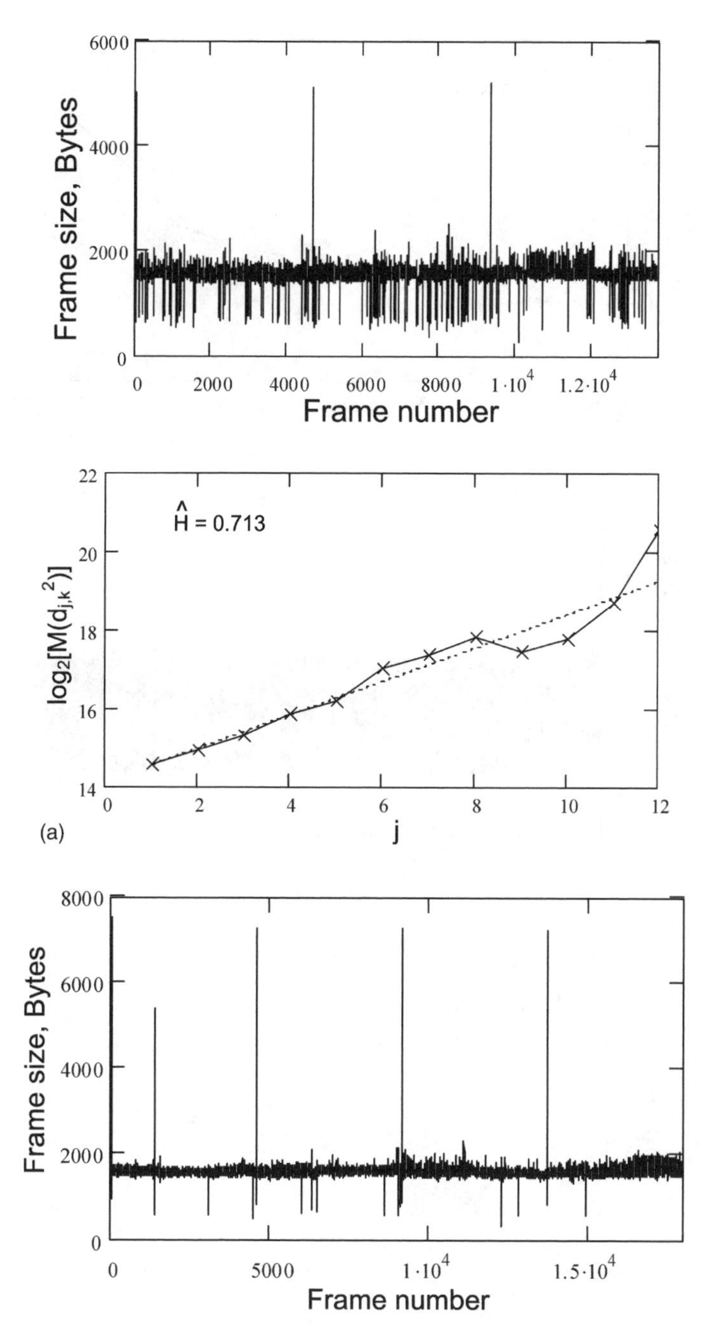

Figure 3.44 Traces of video traffic (H.263, 64 kbit/s) and its Hurst exponent wavelet estimation: (a) office cam; (b) parking cam; (c) lecture cam

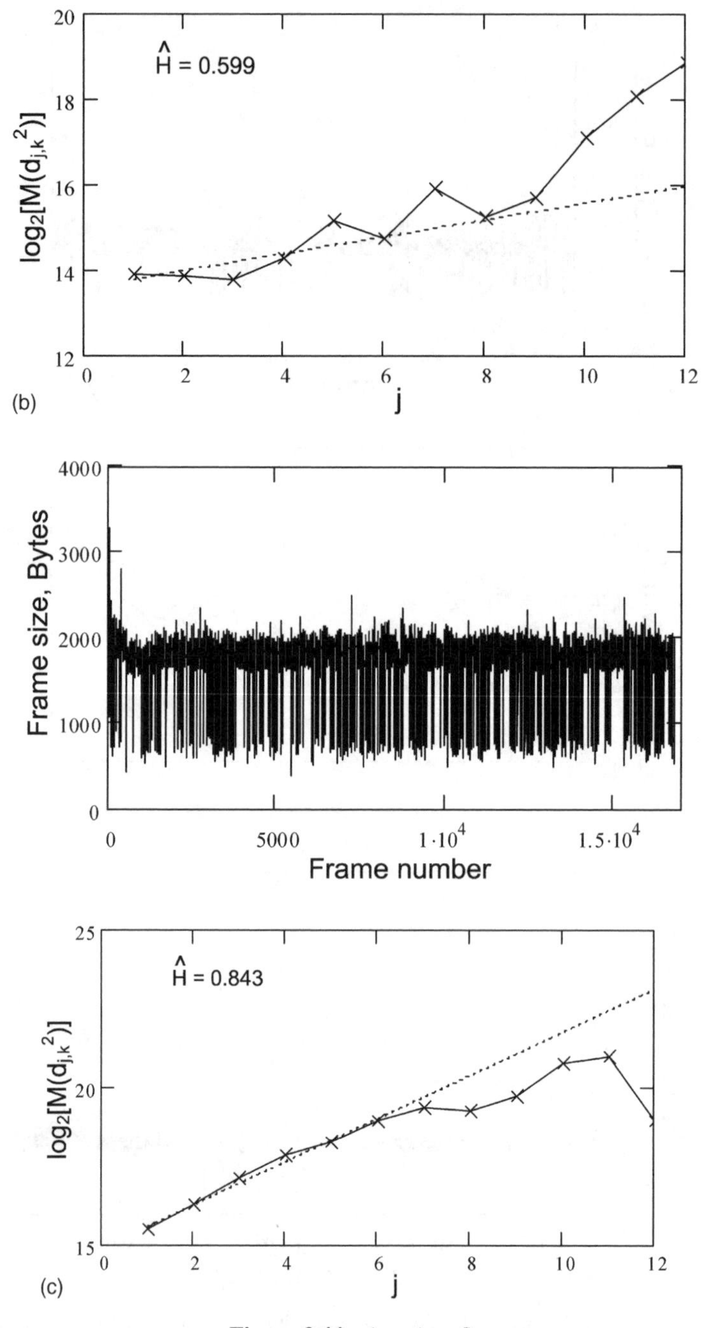

Figure 3.44 (*continued*)

Table 3.9 Hurst exponent estimation using different techniques for traces shown in Figures 3.44 and 3.45

Realization type	Office cam, H		Parking cam, H		Lecture cam, H	
estimate method	CBR	VBR	CBR	VBR	CBR	VBR
Aggregate variance	0.547	0.719	0.719	0.894	0.569	0.908
R/S statistics	0.331	0.591	0.346	0.634	0.428	0.640
Periodogram	0.756	1.164	0.755	0.98	0.756	1.193
Variance of residuals	0.797	1.036	0.872	1.186	0.810	1.104
Abry–Veitch estimator	0.768	1.225	0.697	0.917	0.867	1.134
Whittle estimator	0.723	0.999	0.595	0.896	0.881	0.999

The multifractal estimate evaluates slope α_q by carrying out the linear regression for $S_q(j)$ in the given j range. The Hurst exponent H is calculated using the expression $H = 0.5 + \alpha_q/q$, which is similar to the monofractal case $H = 0.5(1 + \alpha)$, by considering the calculation at the moment order q.

Averaged values $M|d_{j,k}|^q$ of the coefficients for different q values (for the trace *Star Wars* mpeg1, IBP frames) are shown in Figure 3.46. The Hurst exponent estimate for different q moment values for this trace is shown in Figure 3.47.

3.8.3 Multifractal Analysis

The video traffic multifractality can be estimated using the Legendre spectrum in accordance with the approach described in Section 3.4. These results are shown in Figure 3.48 for the trace presented in Figure 3.41. It can be seen from the plot of the multifractal Legendre spectrum as well as from the nonlinear character of the functions $c(q)$ and $\tau(q)$ that video traffic contains the essential multifractal scaling.

The protocol type, the image quality and the transmission rate seriously influence the multifractal traffic character. Figure 3.49 illustrates this for the examples of the ‘action’ traces, a cartoon for the low and high image quality and the transmission CBR (constant bit rate) and VBR (variable bit rate) modes on the H.263 protocol.

It can be seen that the largest multifractal spectrum width is typical for the VBR sequence and the lowest width is typical for the monofractal case of the CBR (constant bit rate) sequence. As the image quality is improved the Legendre spectrum width increases.

3.8.3.1 Multifractal Spectrum Estimation Based on Wavelets

Although the multifractal spectrum $f(\alpha)$ contains significant information concerning the character of singularities in Y, unfortunately it is difficult to calculate it. The $f(\alpha)$ function can be interpreted as the velocity function using the large deviation principle. In other words, f measures how often (in k units) the observed process $(1/j)\log|d_{j,k}|$ deviates from the ‘expected value’ α_0 in scale j. This corresponds to the scaling behaviour analysis of the wavelet coefficient moments.

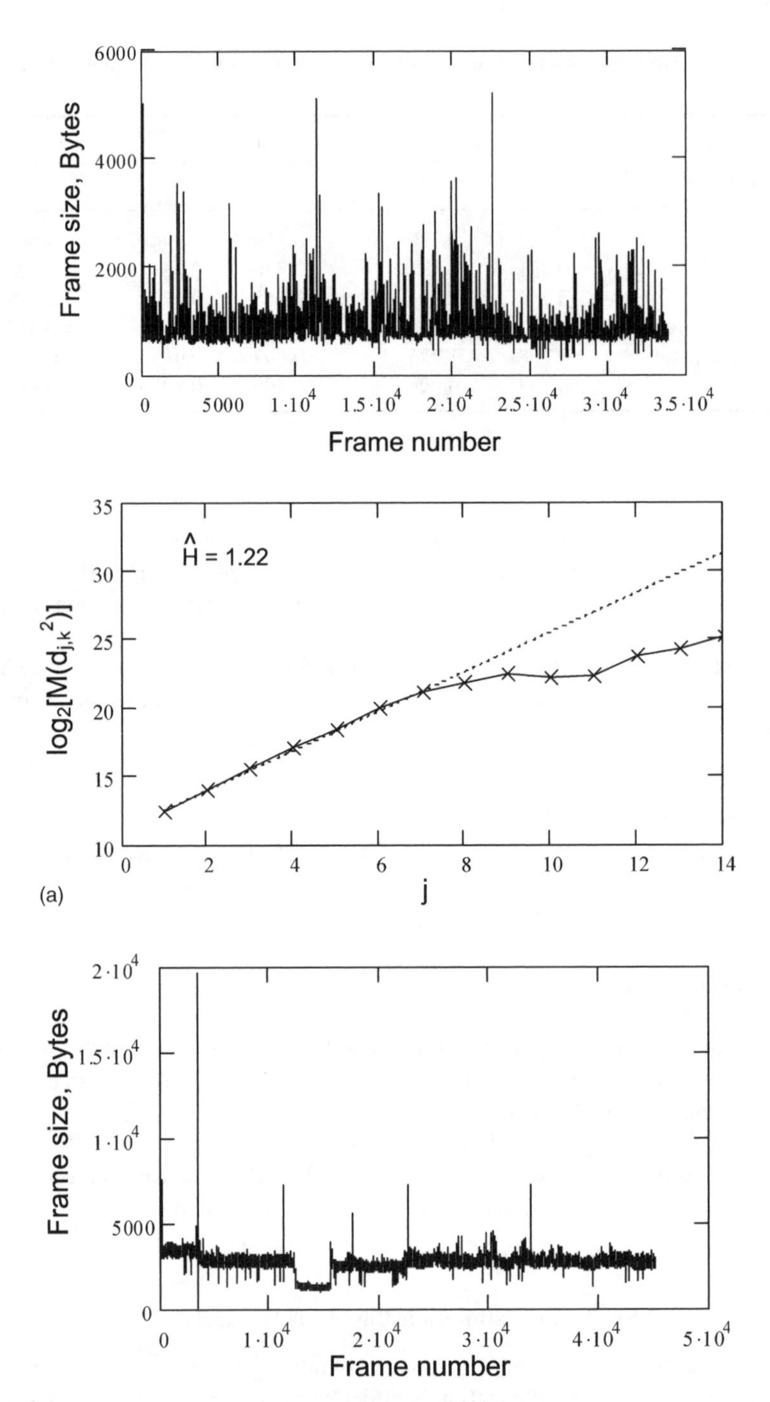

Figure 3.45 Traces of video traffic (H.263, VBR) and its Hurst exponents wavelet estimation: a) office cam; b) parking cam; c) lecture cam.

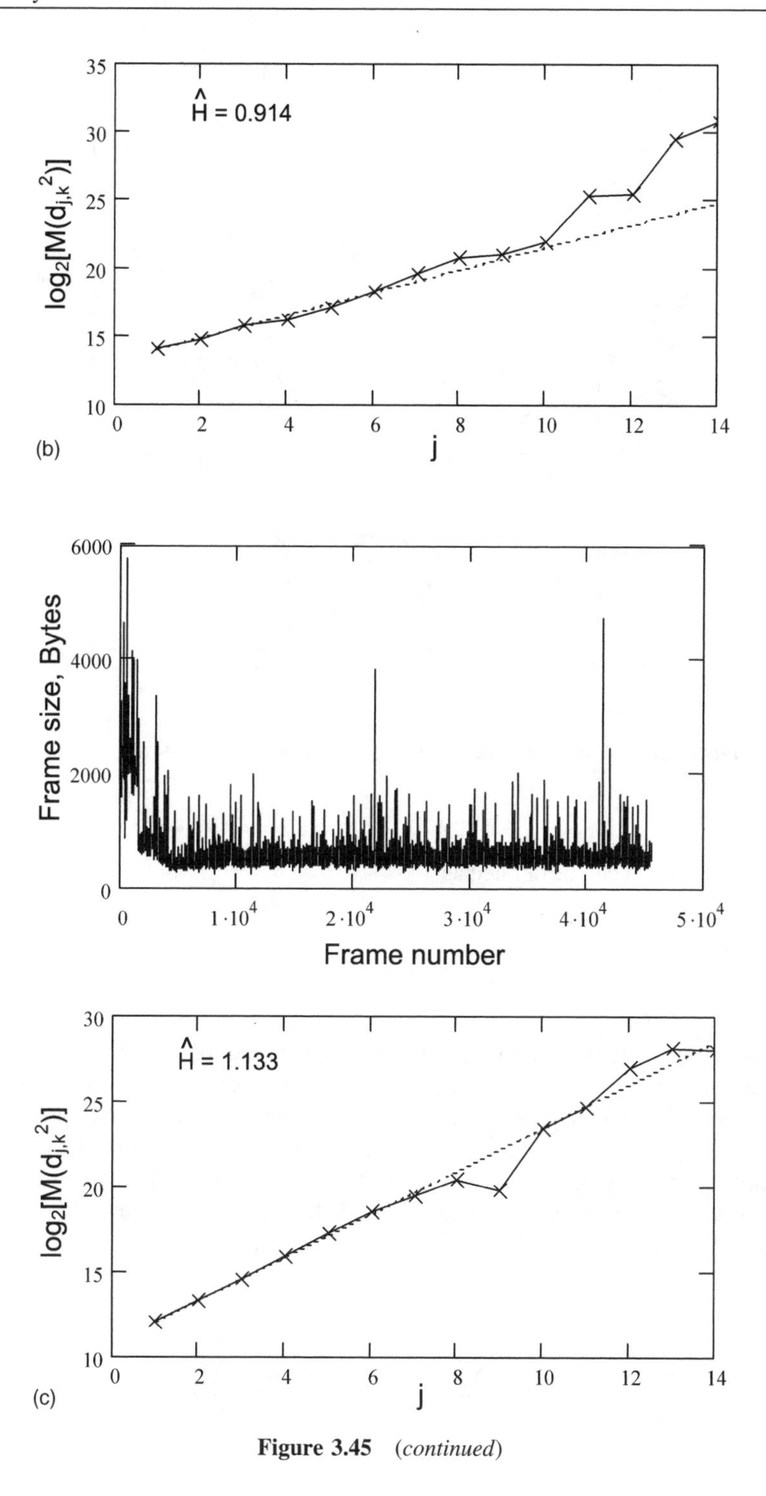

Figure 3.45 (*continued*)

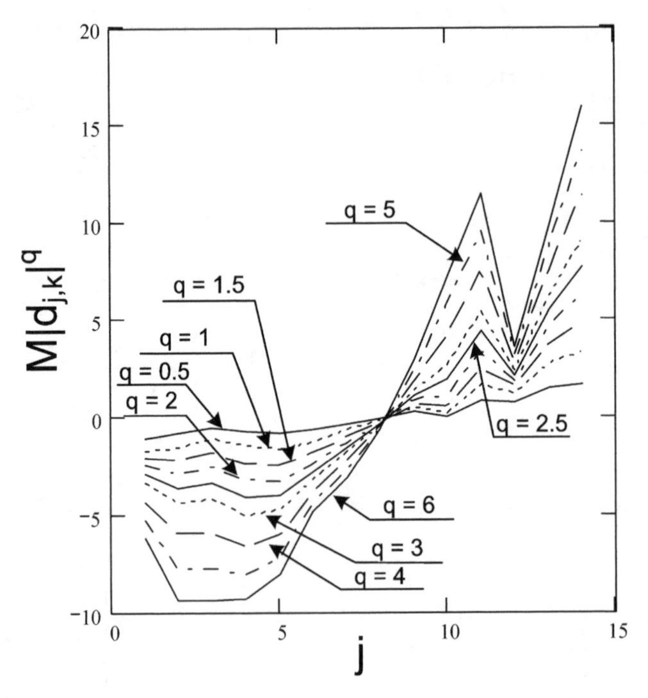

Figure 3.46 Averaged values $M|d_{j,k}|^q$ of the coefficients for different q values

The scaling function can be defined as

$$\tau(q) = \lim_{j \to -\infty} \log_2 M|d_{j,k}|^q$$

which measures the moment scaling in the higher moment functions of the wavelet coefficients, including the singularity structure in the process. To estimate the multifractal spectrum for N given samples of the separate process realizations use is made of the advantages of the wavelet coefficients $d_{j,k}$: $j = 1, \ldots, \log_2 N, k = 0, \ldots, N2^{-j} - 1$ stationarity within the scale limits. A rough sample estimate is made for the wavelet coefficient moments for $q > -1$. It is assumed that the wavelet coefficients are Gaussian: $d_{j,k} \sim N(0, \sigma_\Psi 2^{2jH})$, where σ_ψ is a constant value depending on the mother wavelet ψ. As a result, the partition function

$$\hat{S}_j(q) := \frac{1}{N2^{-j}} \sum_{k=0}^{N2^{-j}-1} |d_{j,k}|^q = 2^{jqH} \frac{1}{N2^{-j}} \sum_{k=0}^{N2^{-j}-1} |d_{0,k}|^q$$

converges for $q > -1$ only.

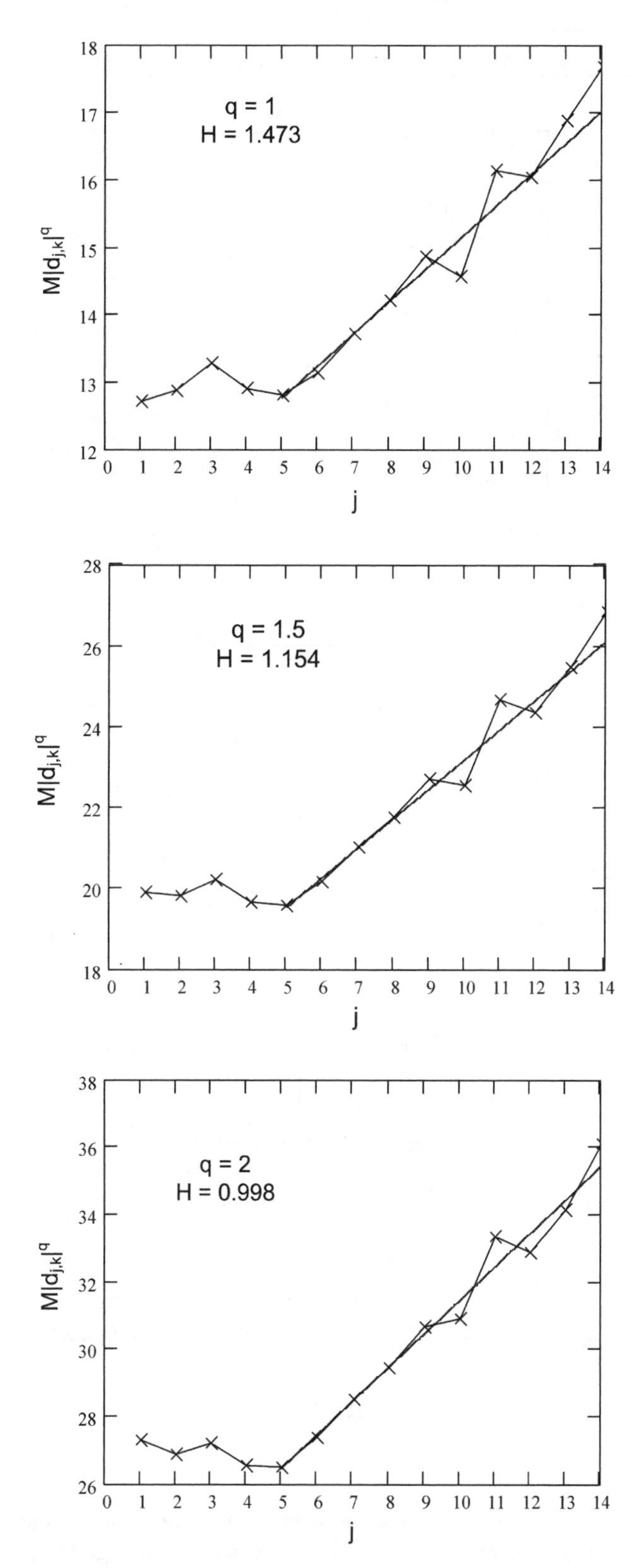

Figure 3.47 Estimation of the Hurst exponent for different moment values q (*Star Wars* MPEG-1, IBP frames)

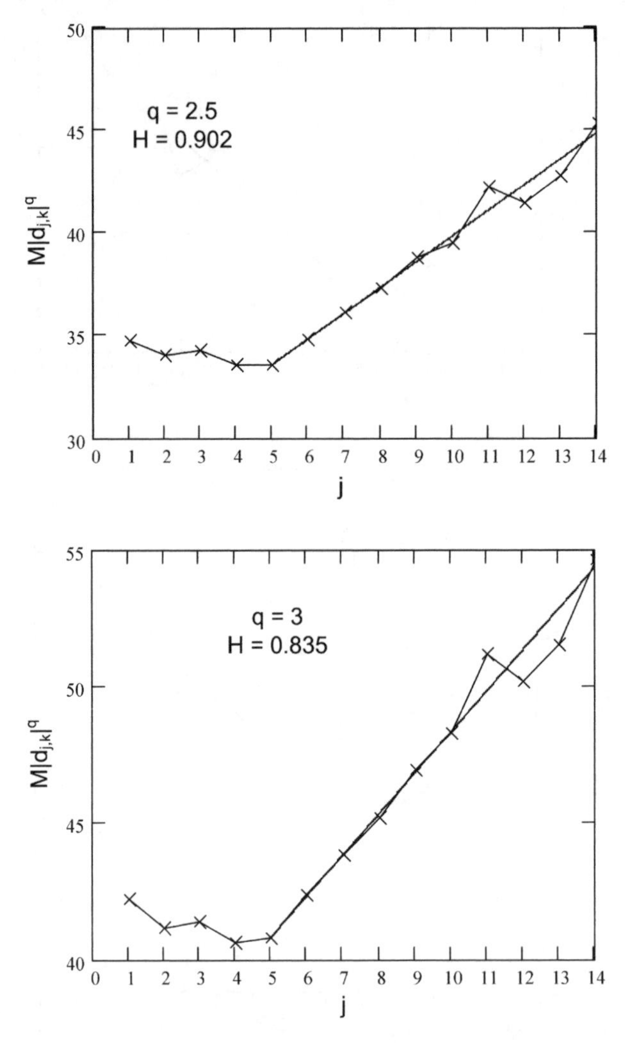

Figure 3.47 *(continued)*

The scaling function will be estimated as the exponent for variation $q > -1$ with scale 2^j. In practice, the linear regression for $\log_2 \hat{S}_j(q)$ with respect to j between scales j_1 and j_2 is used, so that

$$\hat{\tau}(q) := \sum_{j=j_1}^{j_2} a_j \log_2 \hat{S}_j(q)$$

It can be shown that $\hat{S}_j(q)$ is the unbiased estimate of the sample mean value for the random variable $|d_{j,k}|^q$. In addition, using the scaling property $d_{j,k} \overset{d}{=} 2^{jH} d_{0,k}$ gives $\hat{S}_j(q)$ as asymptoti-

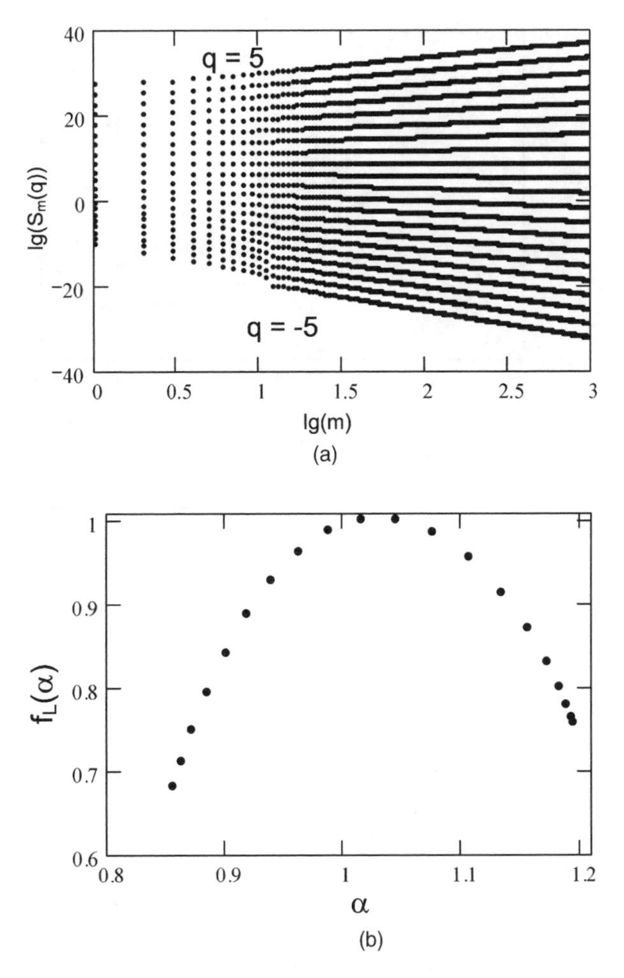

Figure 3.48 Analysis of multifractal scaling with the help of: (a) partition sum $Sm(q)$ with respect to m in the double logarithmic scale (if viewed top-down, q changes from 10 to -10 with iteration 1); (b) multifractal Legendre spectrum; (c) $c(q)$ function; (d) $\tau(q)$ function

cally normal with the mean value $\mu = 2^{j\tau(q_0)}M|d_{0,0}|^q$ and the variance $\sigma_{j,q}^2 = 2^{2j\tau(q)} \times \sigma^2|d_{0,0}|^q/2^{-jN}$. The regression weights in the equation for $\hat{\tau}(q)$ should satisfy two requirements: $\sum_j a_j = 0$ and $\sum_j ja_j = 1$. Using the fact that the Legendre transform can be used for differentiable functions such as $\tau(q)$, it is possible to estimate the Legendre spectrum $f_L(\alpha)$ by means of the local slope estimation for $\hat{\tau}(q)$ [33]:

$$\hat{a}(q_i) = \frac{[\hat{\tau}(q_{i+1}) - \hat{\tau}(q_i)]}{q_0}, \quad q_i = iq_0$$

$$\hat{f}_L(\alpha(q_i)) = q_i\hat{\alpha}(q_i) - \hat{\tau}(q_i)$$

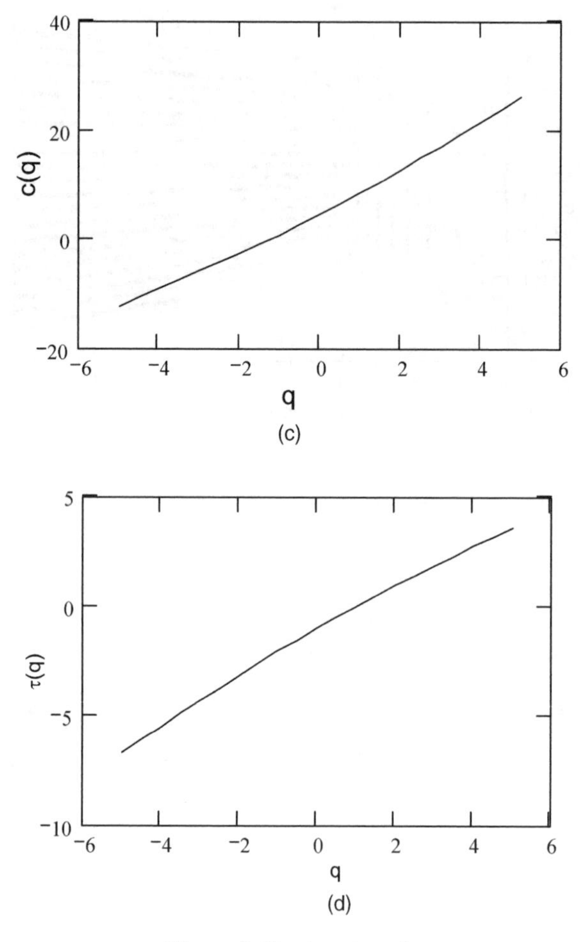

Figure 3.48 *(continued)*

The estimate $\hat{\tau}(q)$ for the partition sum is the linear combination of the (independent) asymptotically normally distributed random variables. Therefore, $\hat{\tau}(q)$ is asymptotically unbiased. However, for the sample of finite size $N < \infty$, $\hat{\tau}(q)$ behaves instead as $M \log_2 \hat{S}_j(q) = j\hat{\tau}(q) + g(j)$ with $\lim_{N2^{-j} \to \infty} [g(j)/j] = 0$. At the same time, in practice, the scaled-dependent displacement $g(j)$ introduces the appropriate displacement in $\hat{\tau}(q)$ which should be estimated additionally.

3.9 Models and Modelling of Video Sequences

3.9.1 Nonstationarity Types for VBR Video Traffic

The analysis of the scene and video traffic models shows that these models have similar structures. Therefore the scene model is nonstationary over the mean value, i.e. the data intensity demonstrates the jumps. The VBR traffic check-up over a shorter time scale

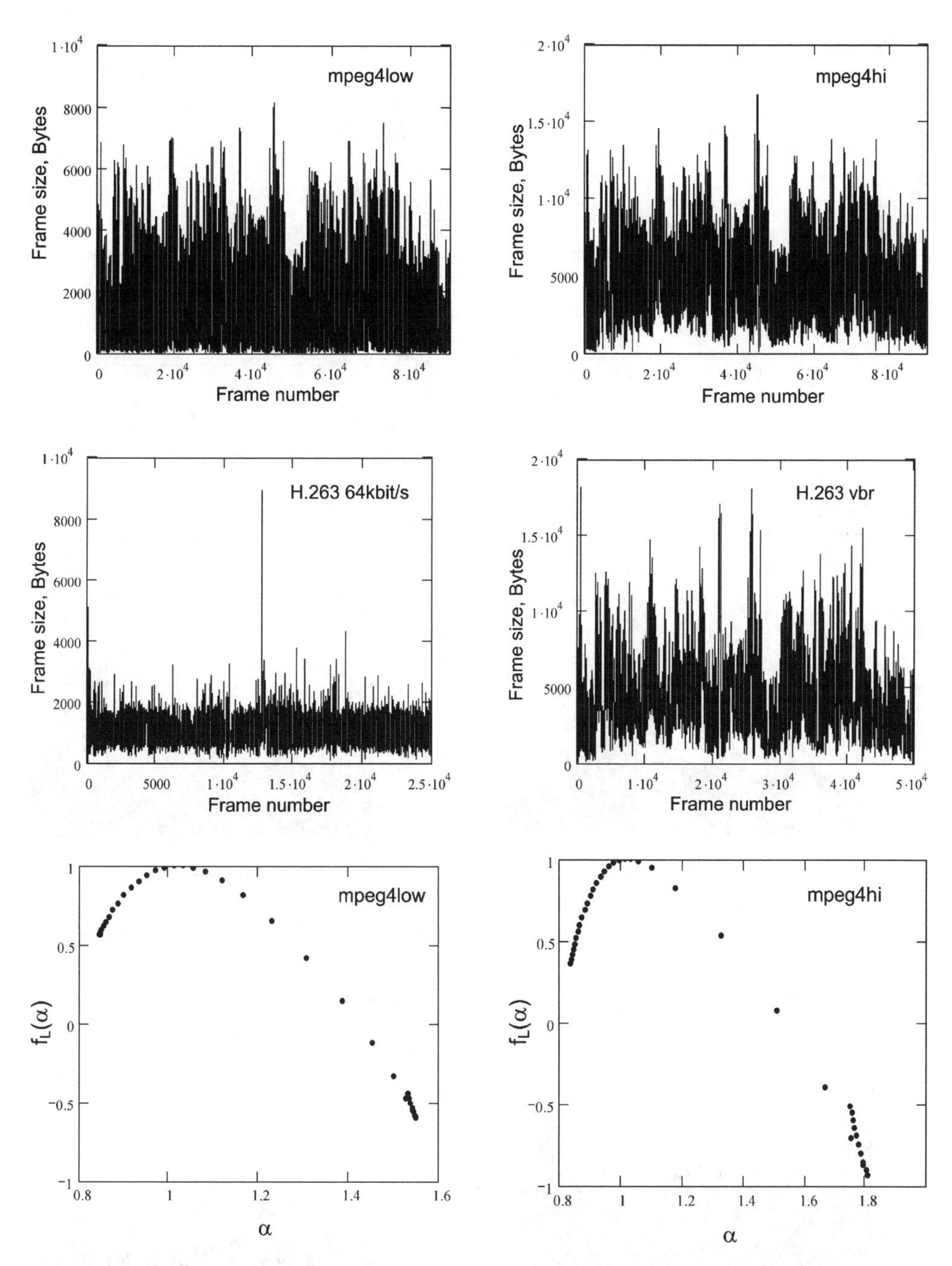

Figure 3.49 (a) Multifractal analysis of frame sequence for the *Jurassic Park* movie for different coding cases: (a, a1) MPEG-4 low; (b,b1) MPEG-4 high; (c,c1) 4.263, 64 kbit/s; (d,d1) H.263 VBR. Multifractal analysis of frame sequence for the *aladdin* cartoon for different coding cases: (e,e1) MPEG-4 low; (f,f1) MPEG-4 high; (g,g1) 4.263, 64 kbit/s; (h,h1) H.263 VBR

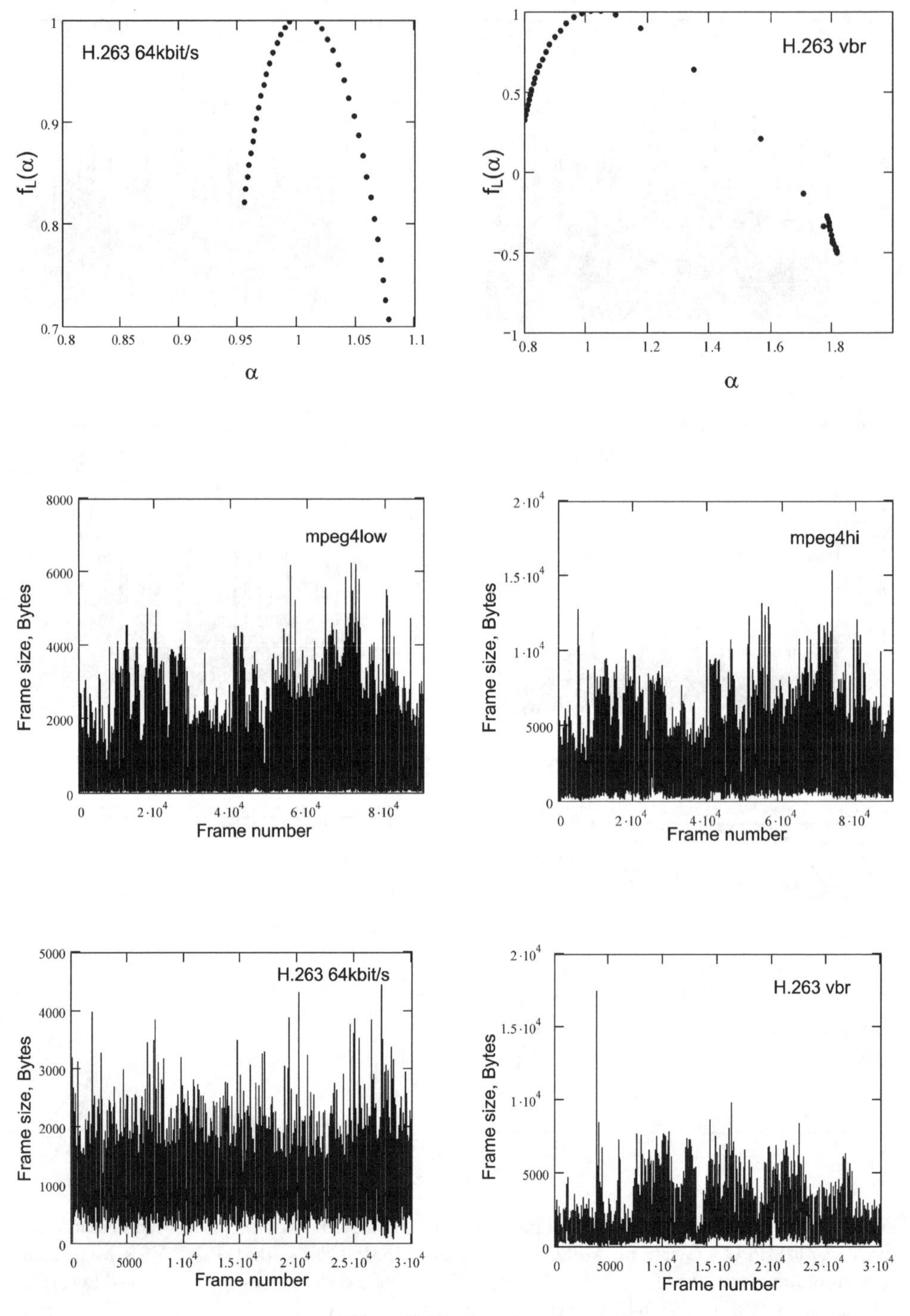

Figure 3.49 (*continued*)

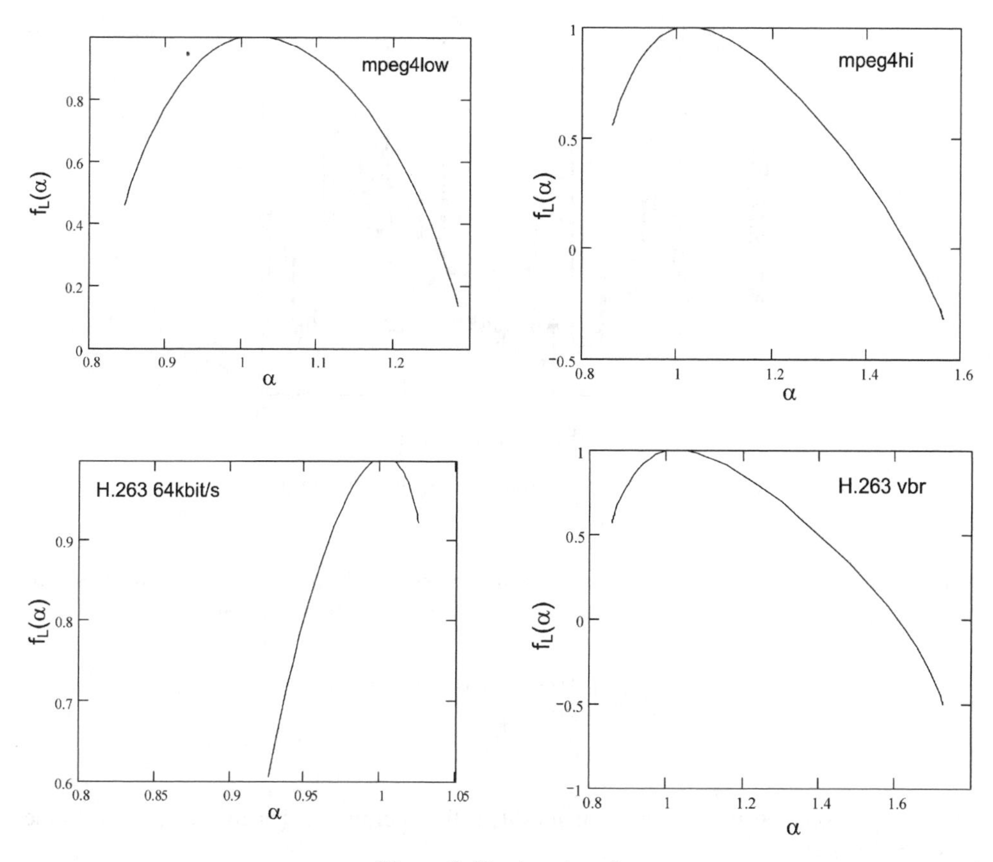

Figure 3.49 (*continued*)

demonstrates the same behaviour (Figure 3.50). It shows that the convenient assumption of weak stationarity should be discharged and the long-range dependence should be regarded as artificial nonstationarity. At the same time, this argument is confirmed by publication [34] in which the analysed process is referred to as a *shifting level process* (SLP).

Definition

Let Y_i be the independent and identically distributed random variable with the mean value m and the variance σ_Y^2. *Existence of the* group moments of the higher order $m'_{Y,r}$ should be possible. Let $\Delta T_i := t_{i+1} - t_i$ (periods) be independent and identically distributed variables with the PDF w_t and the mean value m_t. Then the stochastic process $X(t) = Y_i$ for $t_i \leq t < t_{i+1} (i = 1, 2, \ldots)$ is called the shifting level process (SLP).

The SLP was first introduced in economics by Mandelbrot, who called it the recovering process [35]. The application to video traffic was fulfilled in Reference [36]. These processes were introduced as the alternative explanation of long-range dependence observed in hydrology [37]. Their relation to the Hurst exponent was investigated. The behaviour of the moment estimates for these processes was considered in Reference [38].

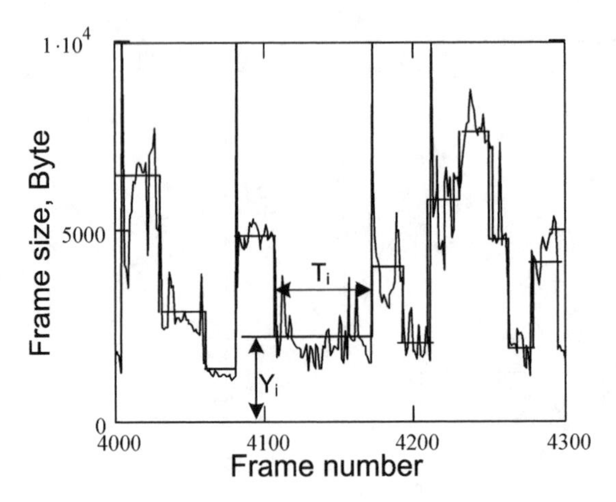

Figure 3.50 VBR video traffic. Musical videoclip (H.263)

These processes are *asymptotically weakly stationary*, which ensures an acceptable basis for practical applications. It is assumed that the periods have Pareto distribution:

$$w_t(t) = \begin{cases} f(t) & \text{for } 0 < k \leq t < t_0 \\ \dfrac{\Theta t_1^{\Theta}}{t^{\Theta+1}} & \text{for } t \geq t_0 \end{cases} \tag{3.26}$$

where $f(t)$ is some positive function such as w_t is the appropriate density and t_0, t_1 are some positive standards.

It can be shown that the following theorem is valid.

Theorem 3.1

The shifting level process has the period distribution defined by Equation (3.26), if and only if the appropriate correlation function demonstrates the long-range dependence on the Hurst exponent $H = (3 - \Theta)/2$.

Mandelbrot also proved [35] that the power spectral density near zero corresponds to the power law for the SLP with the distribution (3.26). This testifies to the fact that nonstationarity for the mean value can be the reason for long-range dependence. Moreover, Equation (3.18) determines the sufficient flexibility for short-range behaviour modelling as well as for the $w(t)$ description.

Thus, it is necessary to be very accurate when describing network traffic in order not to confuse the real nonstationarities with the stationary fractal behaviour. These effects can give similar results for many statistical tests. It is noteworthy that there are very promising methods that can be used to differenciate the nonstationarity and the long-range dependence [25] or to estimate the Hurst exponent in the presence of several types of nonstationarities [39].

In some cases in practice it is possible to speak about the local stationarity and it is important to define accurately the appropriate time scales for the stationary fractal behaviour. It should be noted that sometimes, besides the statistical proof of the fractal behaviour, a physical description of the traffic generation mechanism can also help in the choice of fractal models.

3.9.2 Model of the Video Traffic Scene Changing Based on the Shifting Level Process

The analysis of the video traffic experimental traces shows that, as in the case of voice traffic, the video sequence can be presented in the form of two component aggregates:

- the process describing the video information scene varying;
- the process within the limits of the specific scene.

The shifting level process will be used to describe the scene changing. In this case the traffic model for the source, which changes the rate in the given time moment, follows two independent and identically distributed processes: S_i in the scene limits (the arrival intensity in the scene limits) and T_i for the scene duration.

Let $\{S_n: n = 0, 1, 2, 3, \ldots\}$ be independent and identically distributed variables with the state space $\{0, 1, 2, \ldots, i, \ldots, M\}$. The PDF, the mean value and the variance of S_n variables will be designated as $w_S(\cdot), \mu_S$ and σ_S^2 respectively. The delayed recovery process $0 = t_0 < t_1 < t_2 < \cdots$ has the inter-arrival time $T_n = t_n - t_{n-1}, n = 1, 2, 3, \ldots$, where $\{T_n: n = 2, 3, 4, \ldots\}$ are the independent and identically distributed variables with the distribution function $F_T(\cdot)$, density $w_T(\cdot)$ and mean value $\mu_T \cdot T_1$ corresponds to $F^e(t)$, which is the distribution of the residuary life time for T [40]:

$$F_T^e(t) = \frac{1}{\mu_T} \int_0^t [1 - F_T(x)]\mathrm{d}x \tag{3.27}$$

The shifting level process $\{X(t)\}$ is then the fluid model, for which the arrival intensity in time moment t is determined by i, if $S_{N(t)} = i$,

$$X(t) = \sum_{n=0}^{\infty} S_n 1_{\{t_n \le t \le t_{n+1}\}} \tag{3.28}$$

It is evident that $\{X(t)\}$ is the stationary process with the mean value $\mu_X = \mu_S$ and the variance $\sigma_X^2 = \sigma_S^2$.

The relation between the correlation function

$$R(t) = \frac{M([X(\tau) - \mu_X][X(\tau + t) - \mu_X])}{\sigma^2[X(t)]}$$

and the scene duration distribution $\mathrm{F}_T(\cdot)$ will be found. The correlation function of the SLP is defined as [40]

$$R(t) = \frac{1}{\mu_T} \int_t^{\infty} [1 - F_T(y)]\mathrm{d}y \tag{3.29}$$

The differentiation of this equation gives the simple relations between the PDF for T and the correlation function

$$F_T(t) = 1 + \mu_T \frac{\mathrm{d}R}{\mathrm{d}t} \tag{3.30}$$

and

$$w_T(t) = \mu_T \frac{d^2 R(t)}{dt^2} \tag{3.31}$$

The SLP will now be applied to video traffic modelling. In the general case, the traffic structure of the coded video traces depends on the change of the visual information inherent in them and on the coding algorithm used for data compression. In particular, the MPEG coder generates periodic bursts as a result of operation of its coding algorithm for images. The characteristics peculiar to VBR video traffic will be without considering the video coding aspects. In other words, the traffic periodicity corresponds to an account of the envelope of the video traffic correlation function only. A similar assumption corresponds to the video traffic model at the scene level.

In the analysed case the scene size, i.e. the source intensity within the limits of the scene, corresponds to S_i and the scene duration corresponds to T_i. In the general case, the process of the scene sizes may fail to be independent and identically distributed, and the actual correlation depends on the process of the scene determination. It was shown (Section 3.5.3) that for voice traffic modelling on the call layer the correlation is modelled by the discrete Markovian chain. However, the large parameter number in the transition matrix for semi-Markovian processes makes it difficult to select the matrix procedure. Moreover, the productivity estimate is very sensitive to state determination. Therefore, semi-Markovian models cannot cover the PDF (first-order statistical characteristics) of the experimental traces with the restricted state number.

However, the most important advantage of the SLP, from a practical point of view, is that the experimental statistics of scene changing are practically inaccessible because they are only rarely met. This also concerns the statistics of this or that scene duration. As a result, it is often impossible to obtain the transit probability matrix and the distribution function of the appropriate state duration from the experimental data. The video sequence correlation properties are the only available information in practice.

On the other hand, when the SLP is used, it is assumed that the scene size process is also the recovery process. The SLP has the property that the marginal distribution and the correlation function are unambiguously determined with the help of S_i and T_i respectively. Therefore, it is easy to select the PDF and the correlation function of the model for these experimental traces.

In Reference [41] it was shown that the experimentally found correlation function of the video traffic is close to the exponential function (SRD) at small delays and to the hyperbolic function (LRD) at large delays. Therefore, the shifting level process with the compound correlation (SLCC) of the exponent and the hyperbola is considered in the following way:

$$R(t) = \begin{cases} R_e(t) = e^{-t/\tau}, & \text{for } 0 < t < t_0 \\ R_h(t) = c_0(t + t_1)^{-\beta}, & \text{for } t_0 < t \end{cases} \tag{3.32}$$

Thus, in the SLP the distribution the $w_s(\cdot)$ histogram is defined by the scene size S_i and the correlation function $R(t)$ is defined by the scene duration T_i.

As a result, it is possible to offer the following algorithm for the SLCC parameter selection for the real video traffic statistical characteristics:

Stage 1. Since the experimental statistics concerning the scene changing is absent, the statement is used that $\{S_n\colon n = 0, 1, 2, 3, \ldots, M\}$ are the independent and identically distributed random variables with the state space $\{0, 1, 2, \ldots, i, \ldots, M\}$. Using the results of Reference [42], it can be affirmed that the PDF of scene sizes $(w_S(0), w_S(1), \ldots, w_S(M-1), w_S(M))$ are well described by the following negative binomial distribution:

$$w_S(i) = \binom{-r}{i} p^r (-q)^i = \binom{i+r-1}{i} p^r q^i \tag{3.33}$$

$(i = 0, 1, 2, \ldots, M-1)$ and $w_S(M) = 1 - \Sigma_{i<M} \omega_S(i)$, where M is the peak rate in the frame.

The mean value and the variance of this distribution are determined as

$$M[X(t)] = \frac{r(1-p)}{p} \quad \text{and} \quad \sigma^2[X(t)] = \frac{r(1-p)}{p^2} \tag{3.34}$$

respectively. Here $0 < p < 1, q = 1 - p$ and $r > 0$. Therefore, the parameters can be estimated as

$$p = \frac{M[X(t)]}{\sigma^2[X(t)]} \quad \text{and} \quad r = \frac{M[X(t)]^2}{\sigma^2[X(t)] - M[X(t)]} \tag{3.35}$$

As a result, for $\{S_n : n = 0, 1, 2, 3, \ldots, M\}$ modelling it is necessary to simulate the random variable following $w_S(i)$ distribution (3.33) with p and r parameters according to Equation (3.15).

Stage 2. Consider the selection procedure for the correlation function for the shifting level process. To approximate the correlation function of the real video sequence with the help of $R(t)$, given in Equation (3.32), with the experimental data usage, it is necessary to determine five parameters τ, β, t_0, t_1 and c_0 included in Equation (3.32).

The distribution function will be found for the combined correlation function model, using Equation (3.30), for the generation of the random numbers corresponding to the scene duration

$$F_T(t) = \begin{cases} 1 - \mathrm{e}^{-t/\tau}, & \text{for } t < t_0 \\ 1 - \tau\beta c_0 (t + t_1)^{-(\beta+1)}, & \text{for } t_0 \leq t \end{cases} \tag{3.36}$$

When modelling the random functions by the inverse function method, it is necessary to know the $F_T(\cdot)$ inversion, defined as

$$F_T^{-1}(y) = \begin{cases} -\tau \ln(1-y), & \text{for } y < 1 - \mathrm{e}^{-\beta + t_1/\tau} \\ \left(\frac{\beta c_0 \tau}{1-y}\right)^{1/(\beta+1)} - t_1, & \text{for } y \geq 1 - \mathrm{e}^{-\beta + t_1/\tau} \end{cases} \tag{3.37}$$

From this expression the advantage of using the given approach to describe the correlation function (3.32) can be seen. It therefore became possible to find the explicit formulation of the inverse PDF.

Note that to describe the correlation function over small and large correlation intervals it is also possible to use other approaches. For example, the description of exponential type functions in the form of the sum of exponential functions is widely used. However, it is impossible to calculate the inverse PDF analytically using these functions. The above models can be regarded as the basis for various trace modellings: JPEG [42] and smoothed GOP for MPEG [43,44].

3.9.3 Video Traffic Models in the Limits of the Separate Scene

As mentioned above, the video data containing LRD have short-range as well as long-range correlation structures. Therefore, it is necessary to take this feature into account when developing the video traffic model. One suggestion is to take the existing model on the basis of the AR process of p-order as the main model, which can be presented in the following form [45]:

$$x[i] = -\sum_{k=1}^{p} a[k]x[i-k]| + u[i] \tag{3.38}$$

where $\{u[i]\}$ is the sequence of the independent and identically distributed Gaussian variables.

3.9.3.1 The Choice of AR Model Order

There are many criteria that can be used to choose the order of the AR model, which are in a sense objective functions. Two criteria were offered by Akaike [46]. The first one is the *final error of prediction* (FEP). In accordance with this criterion, the order choice for the AR process is made in such a way as to minimize the error average variance at each prediction step. The prediction error is considered as the power sum in the nonpredictable part of the analysed process and as the variable characterizing the estimation inaccuracy for AR parameters. The FEP for the AR process can be defined by the expression

$$\mathrm{FEP}[k] = \hat{p}_k \left[\frac{N + (k+1)}{N - (k+1)} \right] \tag{3.39}$$

where N is the number of the data samples, p is the order of the AR process and $\hat{p}_k$ is the assessed value of the white noise variance (which will be used as the linear prediction error). Note that it is assumed in Equation (3.39) that the sample mean value is subtracted from the data. The term in square brackets increases with the order growth, thereby characterizing the increase of the estimate uncertainty $\hat{p}_k$ for the prediction error of the variance. The order value has been chosen for which the FEP value is minimal.

The second Akaike criterion is based on the maximum likelihood method and is referred to as the *informational Akaike criterion* (IAC). In accordance with this criterion, the model order is defined by minimization of some information–theoretical functions. If it is assumed that the analysed AR process has Gaussian statistics, the IAC will be determined by the following expression:

$$IAC(k) = N \ln \hat{p}_k + 2k \tag{3.40}$$

where $\hat{p}_k$ is the noise variance estimate for the AR model of the kth order. The term $2k$ in Equation (3.20) characterizes the cost of the use of the additional coefficients. This does not lead, however, to an essential decrease in the prediction error variance. As a result, the model order is chosen that minimizes the IAC value. As $N \to \infty$, the first and second Akaike criteria are asymptotically equivalent. As for the FEP criterion, many researchers mentioned that the model order, chosen in accordance with the FEP criterion in the case of data that does not correspond to the AR processes, is very often underestimated.

For a further discussion the IAC criterion will be used to estimate the model order. The experimental estimates show [45], for example, that for MPEG data the twelve-order model caused by the GOP sequence structure fits well.

3.9.3.2 Estimation of a[k] Coefficients

To determine the $a[k]$ coefficients the Yule–Walker [47] criterion is used. To do this it is necessary to solve the linear equation system in the form

$$\begin{pmatrix} \hat{R}_0 & \hat{R}_1 & \dots & \hat{R}_{p-1} \\ \hat{R}_1 & \hat{R}_2 & \dots & \hat{R}_{p-2} \\ \vdots & \vdots & \ddots & \vdots \\ \hat{R}_{p-1} & \hat{R}_{p-2} & \dots & \hat{R}_0 \end{pmatrix} \begin{pmatrix} \hat{a}_1 \\ \hat{a}_2 \\ \vdots \\ \hat{a}_p \end{pmatrix} = - \begin{pmatrix} \hat{R}_1 \\ \hat{R}_2 \\ \vdots \\ \hat{R}_p \end{pmatrix} \tag{3.41}$$

where the matrix elements $\hat{R}_\tau$ are found from the displaced estimate of the correlation function $\hat{R}_\tau \equiv (1/N) \sum_{t=\tau+1}^{N} y_t y_{t-\tau}$. Thus the matrix of $\hat{R}_\tau$ elements is the so-called Toplitz matrix, the elements of which are equal along each diagonal.

To simplify the process of finding the $a[k]$ coefficients as the elements of the Toplitz matrix, the first 12 samples of the covariational function in Equation (3.21) were found with the help of standard tools of the mathematical packet Matlab.

The samples from three movies of different genre, an action film (a thriller), a comedy and a cartoon, were chosen in the MPEG format as the analysed sequences. Sample profiles for the above-mentioned sequences are shown in Figures 3.51 (a),(b) and (c).

Figure 3.52 shows the variance–time plots for the size increase of the aggregation block for the samples from the real video sequences. For this, the minimal size chosen was equal to 5 and the maximal size equal to 8000. When executing this test the following Hurst exponents were obtained for three considered video sequences: 0.83, 0.8253 and 0.8808.

The fulfilled test results illustrate the self-similar structure of the video traffic. The correlation function of the analysed sequences does not converge to zero, even for large delay values, which confirms the presence of long-range dependence in the sequences.

The $a[k]$ coefficient values were calculated at the next stage of video sequences modelling in accordance with the algorithm (3.38). The first 13 values of the real sequence correlation function were taken for this calculation as the matrix elements.

A fragment of the generated sequence with white Gaussian noise at the input in accordance with the algorithm (3.18) is shown in Figure 3.53. The obtained sequence proves the absence of self-similar properties, which demonstrates the necessity to modify the model of (3.38).

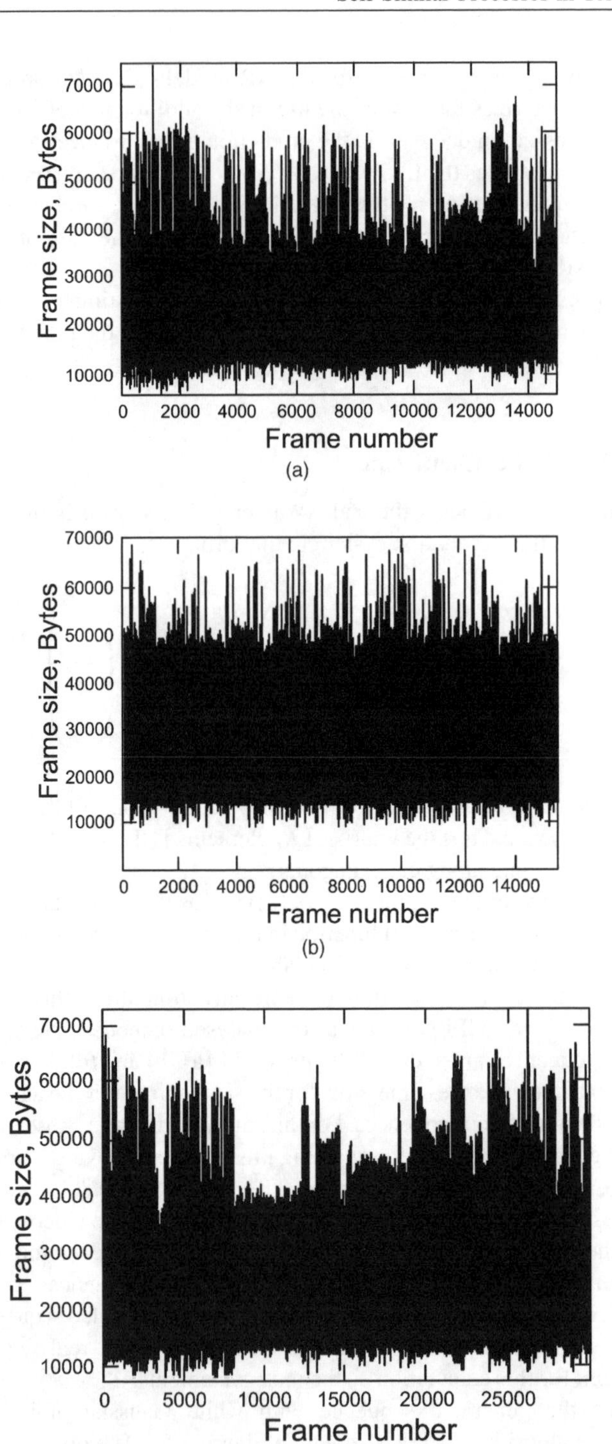

Figure 3.51 Sample profiles of the real video sequences in the MPEG format: (a) 1, an action; (b) 2, a comedy; (c) 3, a cartoon

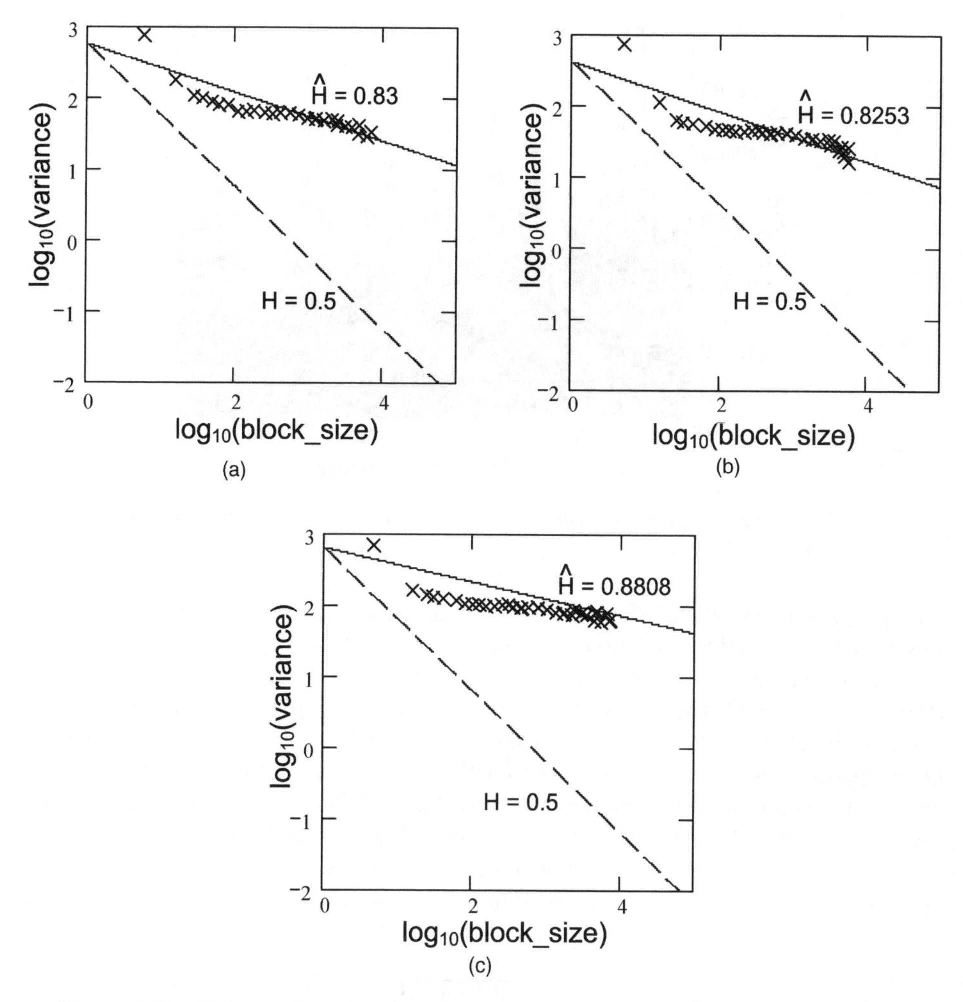

Figure 3.52 Variance–time plots for the samples from video sequences: (a) 1; (b) 2; (c) 3

3.9.4 Fractal Autoregressive Models of p-Order

The independent Gaussian noise source was initially used in the model of (3.38), which does not show the fractal properties. Some FGN source can be used instead of it as an experiment. The software described in Reference [48] can be applied as a source. The algorithm used in Reference [48] is based on the fast Fourier transform (FFT) and is called the FFT algorithm. This approach allows the generation of approximate self-similar processes based on the FFT and on the process known as fractional Gaussian noise.

As a modified model of video traffic the following algorithm is offered:

$$x[i] = -\sum_{k=1}^{p} a[k]x[i-k] + G_H[i] \qquad (3.42)$$

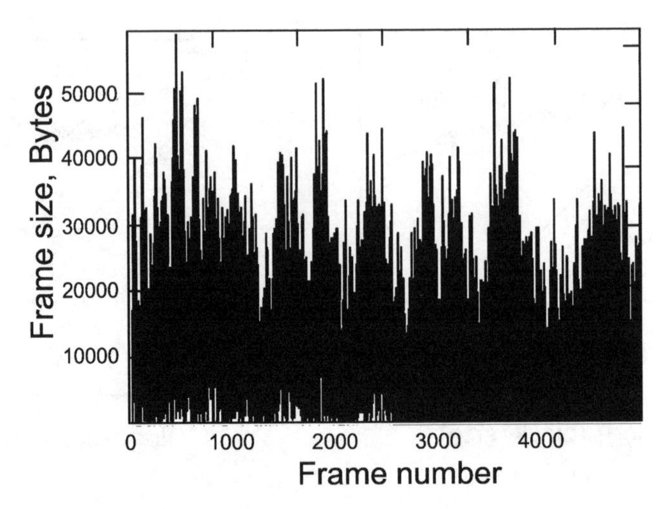

Figure 3.53 The fragment of the generated sequence with the usual (white) Gaussian noise

where $G_H(i)$ is fractional Gaussian noise (FGN) with the Hurst exponent H, the mean value m, r.m.s. deviation σ and the correlation function $R(k) = (\sigma^2/2)[(k+1)^{2H} - 2k^{2H} + (k-1)^{2H}]$.

A fragment of the sequence obtained as a result of the sequence algorithm (3.42) realization is shown in Figure 3.54. The sequence tested using the variance-time method for increasing the aggregated block size shows the close fit of the results with the experimental data (Figure 3.55). For comparison, the calculation of these coefficients was carried out on the basis of the correlation coefficients of two other sequences. Figure 3.56 shows the sequence fragments generated with the help of the model (3.42), using $a[k]$ values found from Equation (3.41) and the fractional Gaussian noise generator. Figure 3.57 shows the variance–time plots at an aggregated block size increasing for the samples from the real video sequences. The minimal

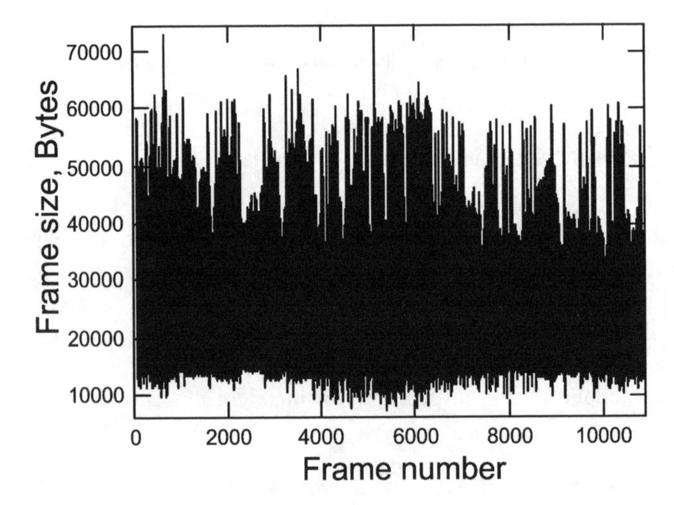

Figure 3.54 Sequence 1 fragment with fractional Gaussian noise

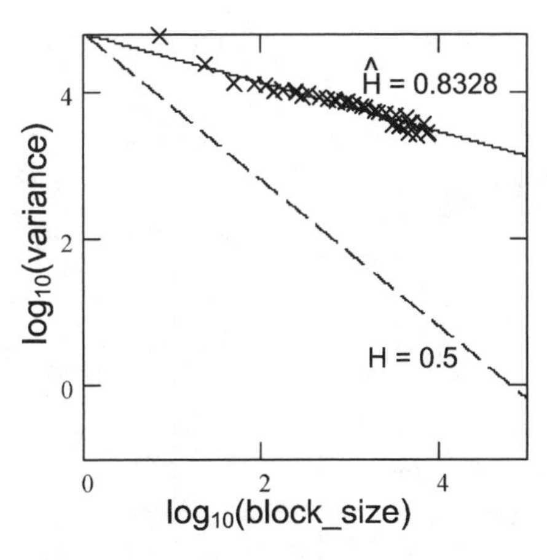

Figure 3.55 Variance–time plot of the video sequence with fractional Gaussian noise

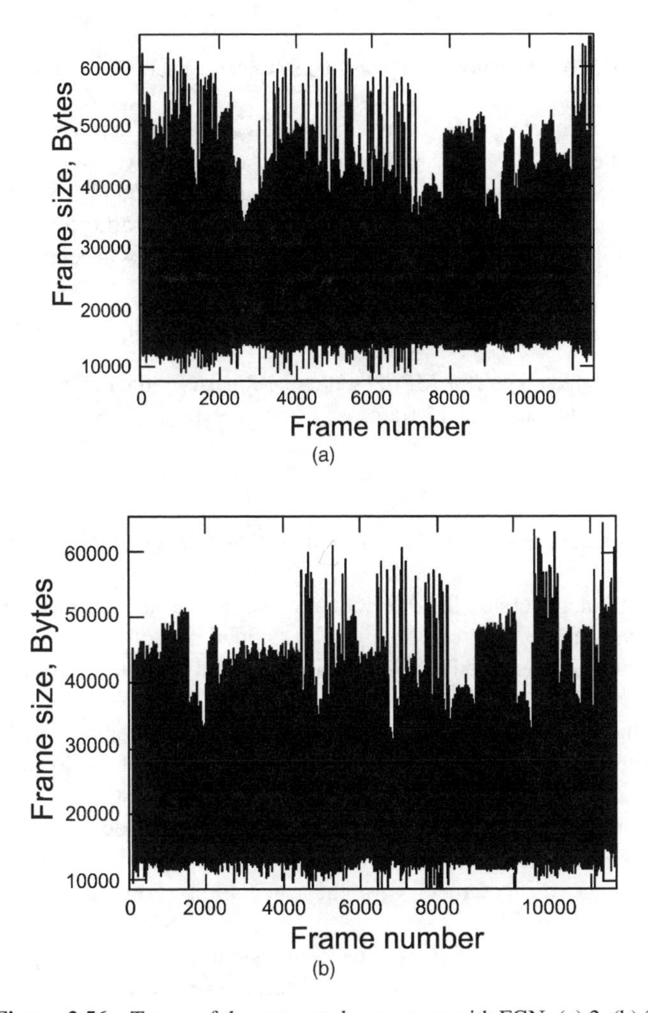

Figure 3.56 Traces of the generated sequences with FGN: (a) 2; (b) 3

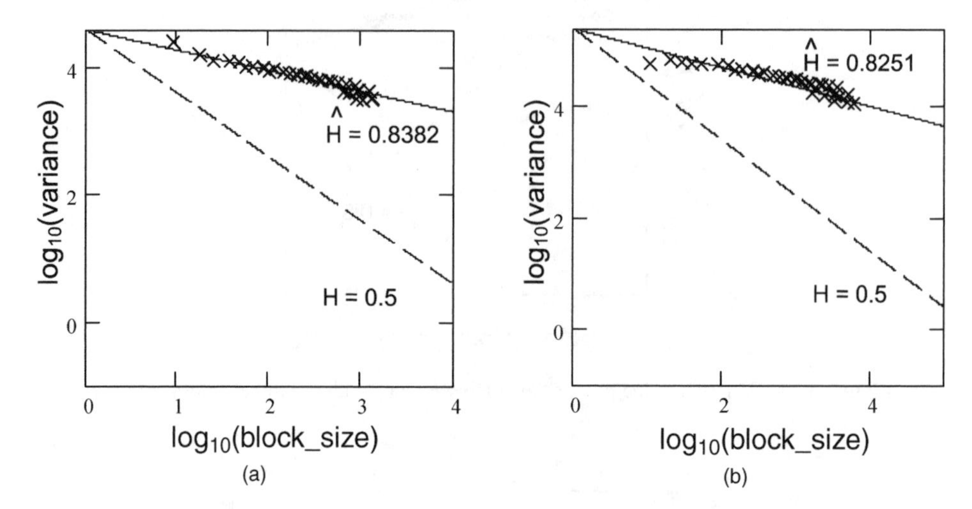

Figure 3.57 Variance–time plots for the sequences with FGN: (a) 2; (b) 3

block size chosen was equal to 5 and the maximal size was 2000. When carrying out this test, the values of the Hurst exponent were 0.8382 and 0.8251 respectively.

It should be noted that when matched with MPEG data the model was able to generate negative traffic values, which should be truncated when real traffic flows are obtained. In general, the algorithm (3.38) modification to the form (3.42) gives acceptable results.

3.9.5 MPEG Data Modelling Using I, P and B Frame Statistics

As repeatedly mentioned above, the MPEG sequence consists of the image group sequence (i.e. GOP). Let the hth GOP consist of 12 frames:

$$\mathrm{I}^{(h)}, \mathrm{B}_1^{(h)}, \mathrm{B}_2^{(h)}, \mathrm{P}_1^{(h)}, \mathrm{B}_3^{(h)}, \mathrm{B}_4^{(h)}, \mathrm{P}_2^{(h)}, \mathrm{B}_5^{(h)}, \mathrm{B}_6^{(h)}, \mathrm{P}_3^{(h)}, \mathrm{B}_7^{(h)}, \mathrm{B}_8^{(h)}$$

The statistical characteristics of the B and P frames do not depend on their position inside the group.

The algorithm that generates the MPEG frame sequence and presupposes the relations between the frames in one GOP can be logically divided into two parts. The first is I frame sequence generation. The second is the generation of B and P frame sequences and all MPEG sequences.

The mean value μ_{I} and the variance $\sigma^2_{X_{\mathrm{I}}}$, estimated from Equation (3.13), are the parameters, which can be considered as the input data for the PDF of the I process $F_{\mathrm{I}}(x)$ (3.12) (the process characterizing I frame size distribution). The Hurst estimate H_{I}, which describes the process long-range dependence properties, the exponent λ_{I}, defining the process short-range dependence properties, and the parameter K_{I}, determining the boundary between SRD and LRD regions, are the parameters for the calculation of the autocorrelation function $R_{X_{\mathrm{I}}X_{\mathrm{I}}}(m)$.

Thus, the offered algorithm generates three independent I, B and P subsequences using three different PDFs for three empiric I, B and P sequences. However, none of these models account for the correlation between the frames belonging to one image group (the GOP).

3.9.6 ON/OFF Model of the Video Sequences

Let the ON/OFF model generate the frames $X(t)$, each p bits in size or 0 bits in accordance with the output value of the random number generator (RNG) and with the threshold value a, i.e.

$$X(t) = \begin{cases} p, & \text{if RNG}(t) \geq a \\ 0, & \text{if RNG}(t) < a \end{cases} \tag{3.43}$$

The frame statistical characteristics are $M(X) = ap$ and $\sigma^2(X) = a(1-a)p^2$. The model parameters a and p for the JPEG, found in Reference [45], are $a = 0.952$ and $p = 29\,198$. Since MPEG data are too bursty, in order to describe this by the ON/OFF model, the given model was applied to GOP data only. The calculations show that the model parameters are $a = 0.880$ and $p = 16\,999$ [45].

The ON/OFF model generates limited arrivals because it presupposes only two values for the frame size, but one of them (namely zero) is impossible to come across in practice.

3.9.7 Self-Similar Norros Model

The Norros model was introduced in Chapter 2. Using this model it is possible to generate the discrete flow of arrivals (in this case in bits/frame) in accordance with the algorithm $A[i] = m + kX_H[i]$, where m is the mean value, k is some scale constant and $X_H[i], i = 0, 1, 2, \ldots$, is the FGN process. To estimate the k value, it is necessary:

1. To find the size, the mean value, the variance and the Hurst exponent for each video data set.
2. To generate the FGN trace with the Hurst exponent and the length in accordance with these data.
3. To calculate the FGN trace variance and use it as well as the video data variance.

For example, the Norros model for MPEG data, obtained in Reference [45], has the following parameters: $N = 171\,000, m = 18\,697.7, H = 0.801$ and $k = 18\,104.0$.

3.9.8 Hurst Exponent Dependence on N

The Hurst exponent estimate variation $\hat{H}_N$ depending on the sample size N for the MPEG and for the modelled data sets using the aggregation layers 1, 12 and 144 will be analysed. For nonaggregated MPEG data $\hat{H}_N$ changes by more than 50 %, and the variation decreases with the aggregation growth. After aggregation the value $\hat{H}_N$ was estimated to be in the range 0.80–0.85 for any N. This result is evidently explained by the fact that SRD influences the Hurst exponent estimate method when N is not large, but when N rises, SRD becomes less important, while the LRD influence becomes greater. This is reflected in the fact that $\hat{H}_N$ is small for N that is not large, but increases with the growth of N.

Estimations of H_N for averaged MPEG data would be more uniform for variations in N because SRD is partially truncated at process aggregation. In all cases the Hurst exponent estimate lay in the range of 0.75–0.90.

It is interesting that the AR model describes $\hat{H}_N$ behaviour only for nonaggregated MPEG data. The FARI (fractional autoregressive integrated) model also has increasing $\hat{H}_N$ values as N rises due to SRD and LRD that are too strong. It can also be noted that ON/OFF and AR models have low values of $\hat{H}_N$ for all N and all aggregation layers. In some cases $\hat{H}_N$ exceeds 0.6. This shows that $\hat{H}_N$ estimates are not accurate (in the limits of ± 0.1). The Norros model is close to the expected value of 0.8 for all aggregation layers and block sizes, which is correct for MPEG, JPEG and Norros data.

References

[1] A.M. Kondoz, '*Digital Speech: Coding for Low Bit Rate Communication Systems*', John Wiley & Sons, Ltd, 1999, p. 442.

[2] D. J. Wright, '*Voice Over Packet Networks*', John Wiley & Sons, Ltd, 2001, p.252.

[3] A.V. Osin, 'The influence of voice traffic self-similarity on quality of service in telecommunication networks', PhD Thesis, Moscow Power Engineering Institute (Technical University), Moscow, Russian Federation, 2005.

[4] O.I. Sheluhin, A.V. Osin, I.A. Nevstruev and G.A. Urev, 'Comparative study of the evaluation techniques for self-similarity processes stationarity' (in Russian), *Electrotekhnicheskie i Informacionnie Kompleksi i Sistemi*, **2**(1), 2006, 55–61.

[5] O.I. Sheluhin and A.V. Osin, 'Multifractal properties of the real-time traffic' (in Russian), *Electrotekhnicheskie i Informacionnie Kompleksi i Sistemi*, **2**(3), 2006, 36–44.

[6] T.D. Dong, B. Sonkoly and S. Molnar, 'Fractal analysis and modelling of VoIP traffic', in NETWORKS2004, Vienna, Austria, 13–16 June 2004.

[7] K. Park and W. Willinger (eds), '*Self-Similar Network Traffic and Performance Evaluation*', John Wiley & Sons, Ltd, 2000.

[8] O.I. Sheluhin, A.V. Pruginin, A.V. Osin and G.A. Urev, 'Mathematical models and imitation modelling of VoIP traffic aggregation' (in Russian), *Electrotekhnicheskie i Informacionnie Kompleksi i Sistemi*, **2**(1), 2006, 32–38.

[9] O.I. Sheluhin, A.V. Osin and G.A. Urev, 'Voice traffic experimental study in VoIP networks' (in Russian), *Electrotekhnicheskie i Informacionnie Kompleksi i Sistemi*, **2**(2), 2006, 54–59.

[10] O.I. Sheluhin and A.V. Osin, 'Speech traffic self-similarity impact on the QoS parameter optimization in the telecommunication network' (in Russian), *Nelineinii Mir*, **4**(3), 2006, 116–121.

[11] M.S. Taqqu, W. Willinger and R. Sherman, 'Proof of fundamental result in self-similar traffic modeling', *Computer Communication Review*, **27**, 1997, 5–23.

[12] Y. Yasuda, H. Yasuda, H. Ohta and F. Kishino, 'Packet video transmission through ATM networks', in Proceedings of IEEE Globecom, 1989, pp. 25.1.1–25.1.5.

[13] W. Verbiest and L. Pinnoo, 'A variable bit rate video codec for asynchronous transfer mode networks', *IEEE Journal on Selected Areas in Communications*, **7**, 1989, 761–770.

[14] W. Verbiest, L. Pinnoo and B. Vosten, 'The impact of the ATM concept on video coding', *IEEE Journal on Selected Areas in Communications*, **6**, 1988, 1623–1632.

[15] G. Ramamurthy and B. Sengupta, 'Modeling and analysis of a variable bit rate video multiplexor', in Proceedings of INFOCOM '92, Florence, Italy, 1992, pp. 817–827.

[16] D.P. Heyman and T. V. Lakshman, 'Long-range dependence and queueing effects for VBR video', in '*Self-Similar Network Traffic Analysis and Performance Evaluation*, (eds K. Park and W. Willinger), Wiley–Interscience, New York, 1999.

[17] M.W. Garett and W. Willinger, 'Analysis, modeling and generation of self-similar VBR video traffic', in Proceedings of ACM SIGCOMM'94, London, 1994.

[18] C. Huag, M. Devetsikiotis, I. Lambadaris and A.R. Kaye, 'Modeling and simulation of self-similar VBR compressed video: a unified approach', in Proceedings of ACM SIGCOM'95, Cambridge, MA, January 1995.

[19] R. Grunenfelder, J.P. Cosmos, S. Manthrope and A. Odinma-Okafor, 'Characterization of video codecs as autoregressive moving average processes and related queuing system performance', *IEEE Journal on Selected Areas in Communications*, **9**, June 1989.

[20] P.R. Jelencovic and A.A. Lazar, 'The effect of multiple time scales and subexponentiality in MPEG video streams on queuing behavior', *IEEE Journal on Selected Areas in Communications*, **15**, August 1997.

[21] A. Lombardo, G. Morabito and G. Schembra, 'An accurate and treatable markov model of MPEG-video traffic', in Proceedings of IEEE Infocom'98, San Francisco, Ca, April 1998.
[22] M. Krunz and H. Hughes, 'A traffic model for MPEG coded VBR streams', Michigan State University, Department of Electrical Engineering, Technical Report, 1997.
[23] J. Beran, R. Sherman, M.S. Taqqu and W. Willinger, 'Long-range dependence in variable-bit-rate video traffic', *IEEE Transactions on Communications*, **43**, 1995, 1566–1579
[24] M. Grasse, M.R. Frater and J.F. Arnold, 'Statistics of variable bit rate video coders with and without motion compensation', In 6th International Workshop on '*Packet Video*', Portland, OR, 26–27 September 1994.
[25] J. Beran, '*Statistics for Long-Memory Processes*', Chapman & Hall, New York, 1994.
[26] M.B. Priestly, '*Spectral Analysis and Time Series*', Volume 1, Academic Press, London, 1981.
[27] M.G. Kendall and A. Stuart, '*The Advanced Theory of Statistics*', Vol. 3, 2nd edn, Charles Griffin & Company Ltd, London, 1968.
[28] M.B. Priestly and T.S. Rao, 'A test for non-stationarity of time-series', *Journal of the Royal Statistical Society, Series B*, **31**, 1969, 140–149.
[29] P. Abry and D. Veitch, 'Wavelet analysis of long-range-dependent traffic', *IEEE Transactions on Information Theory*, **44**(1), 1998, 2–15.
[30] P. Abry, D. Veitch and P. Flandrin, 'Long range dependence: revisiting aggregation with wavelets', *Journal of Time Series Analysis*, **19**(3), 1998, 253–266.
[31] D. Veitchand and P. Abry, 'A wavelet based joint estimator for the parameters of long-range dependence', *IEEE Transactions on Information Theory*, **45**(3), 1999, 878–897.
[32] O.I. Sheluhin, A.V. Osin and R.R. Ahmetshin, 'Telecommunication traffic self-similarity estimation by wavelets' (in Russian), Electrotekhnicheskie i Informacionnie Kompleksi i Sistemi, **3**(2), 2006, pp. 29–36.
[33] P. Goncalves, R. Riedi and R. Baraniuk, 'A simple statistical analysis of wavelet-based multifractal spectrum estimation', in Asilomar Conference on '*Signals, Systems, and Computers*', Volume 1, Pacific Grove, CA, November 1998, pp. 287–291.
[34] V. Klemens, 'The Hurst phenomenon: a puzzle?', *Water Resources Research*, **10**, 1974, 675–688.
[35] B. Mandelbrot, 'Some noise with $1/f$ spectrum: a bridge between direct current and white noise', *IEEE Transactions on Information Theory*, **13**(2), April 1967, 289–298.
[36] M. Grasse, M.R. Frater and J.F. Arnold, 'Origins of long-range dependence in variable bit rate video traffic', in Proceedings of ITC-15, Washington DC, 23–27 June, 1997, pp. 1397–1388.
[37] D.C. Boes and J. D. Salas, 'Nonstationarity of the mean and the Hurst phenomenon', *Water Resources Research*, **14**(1), 1978.
[38] D.B.H. Cline, 'limit theorems for the shifting level process', *Journal of Applied Probability*, **20**(2), 1983.
[39] M. Roughan and D. Veitch, 'Measuring long-range dependence under changing traffic conditions', in Proceedings of IEEE INFOCOM'99, New York, March 1999, pp. 338–341.
[40] S.M. Rose, '*Stochastic Process*', 2nd edn, John Wiley & Sons, Ltd, 1996.
[41] H. Ahn, J.-K. Kim, S. Chong, B. Kim and B. D. Choi, 'A video traffic model based on the shifting-level process: the effects of SRD and LRD on queueing behavior', in INFOCOM 2000, Tel aviv, Israel, 2000, pp. 1036–1045.
[42] M.W. Garrett and W. Willinger, 'Analysis, modeling and generation of self-similar VBR video traffic', in Proceedings of ACM SIGCOM'94, London, August, 1994, pp. 269–280.
[43] H. Ahn, 'The effects of multiple time scale burstiness and and long-range dependence in VBR video traffic on traffic control in multimedia networks', PhD Dissertation, Department of Electrical Engineering, KAIST (Korea Advanced Institute of Science and Technology), Taejon, Korea, February, 2000.
[44] P.R. Jelenkovic, A.A. Lazar and N. Semret, 'The effect of multiple time scales and subexponentiality in MPEG video streams on the queueing behavior', *IEEE Journal on Selected Areas in Communications* **15**(6), August 1997, 1052–1071.
[45] S. Bates, 'Traffic characterization and modelling for call admission control schemes on asynchronous transfer mode networks', Thesis submitted for the Degree of Doctor of Philosophy, The University of Edinburgh, 1997.
[46] H. Akaike, 'Power Spectrum Estimation through Autoregression Model Filting', *Ann. Inst. Stat. Math.*, **21**, 1969, 407–419.
[47] M. Kay, '*Modern Spectral Estimation: Theory and Application*', Prentice-Hall, Inc., Englewood Clipp, NJ, 1988.
[48] V. Paxson, 'Fast, approximate synthesis of fractional Gaussian noise for generating self-similar network traffic', *Computer Communication Review*, **27**, October 1997, 5–18.

4

Self-Similarity of Telecommunication Networks Traffic

4.1 Problem Statement

Modern investigations of traffic measurements executed at high resolution in a large number of real telecommunication networks [1–4] prove that the network traffic is *self-similar* or *fractal* in its structure, i.e. bursty within wide range of the time scale. Since the self-similarity presupposes a great influence on the network characteristics [5,6], understanding the reasons and effects of traffic self-similarity is an important problem.

It will be illustrated how the self-similar structure of the network traffic on the macroscopic level (i.e. the aggregated traffic generated by all active hosts in the network) provides for a new understanding of the traffic dynamics on the microscopic level (i.e. the traffic structure generated by the separate hosts). For this purpose two of the most frequently occurring area network environment will be considered: local networks (LAN) and wide area networks (WAN).

LANs (local area networks) were introduced in the middle of the 1970s for the interconnection of equipment for data processing (central computers, file servers, printers etc.) in offices, research and development areas or within the limits of university departments. One of the most popular LAN technologies is the Ethernet.

Self-similarity of LAN traffic leads to structural models that can be reduced to ON/OFF sources (known also as the packet trains) with the typical property that their ON and/or OFF periods follow HTD with infinite variance. As repeatedly mentioned above, the superposition of ON/OFF processes defines the direct connection between the self-similarity characteristics on the macroscopic level and the heavy tails phenomenon observed on the microscopic level, i.e. between the aggregated traffic flow and the traffic structure that is typical for the separate pairs of *source–recipient*.

WAN (wide area network), in contrast to LAN, provides the interconnection between the users (for instance, the central computers for different LANs), which are located, as a rule, in various geographical regions. The most well-known WAN is the Internet (global network

Self-Similar Processes in Telecommunications O. I. Sheluhin, S. M. Smolskiy and A.V. Osin
© 2007 John Wiley & Sons, Ltd

connecting more that ten million hosts and users). The proof of WAN traffic self-similarity is presented in the publications [3] and [4], in which the nonadequacy of the exponential (Poisson) traffic traditional models for key feature descriptions of WAN traffic behaviour is shown on the basis of the various WAN traffic traces analysis.

The interconnection between file sizes and self-similar traffic was studied in Reference [7] and was also confirmed by Reference [8], where it was shown that self-similarity in World Wide Web traffic can occur due to heavy tails distribution of the file sizes attending at the Web. Later attempts were made to find structural models for WAN traffic including the same one reducing to ON/OFF models for separate pairs of source–recipient and a WAN traffic description at separate application levels, e.g. Telnet, FTP and HTTP.

It was shown in Reference [9] that self-similar traffic can occur as a result of the complex system high-level state in which the sizes of the files transmitted through the network have heavy tails distribution. The superposition of similar transfer sets of the client/server type in the network environment causes self-similar traffic. This causal mechanism extensively relates to network resource variations (the critical carrying capacity and the buffer capacity), the topology, interference of the outside traffic with dissimilar traffic characteristics and the variation in the interarrival time distribution of the file request.

Therefore, for LAN as well as for WAN, self-similarity of the aggregated network traffic is a direct result of the structural models that iterate the network dynamics at low network levels and define the traffic characteristics at these levels, which is confirmed by real traffic measurements with high resolution. The found models are simple and effective, giving the possibility of ensuring that the required properties would be stable in permanently changing network conditions.

At the same time, some publications [10] mention that not everything can be interpreted so unambiguously. The real processes, e.g. in the Internet network, are nonstationary, which is extremely important and should be taken into account in a traffic description. In this research the authors suggest adding nonstationarity as one of the fundamental characteristics of WAN traffic in spite of the long-range estimation and marginal distributions.

The growth and wide distribution of multimedia traffic has only complicated the issue, causing the appearance of self-similar network traffic.

4.2 Self-Similarity and ‘Heavy Tails’ in LAN Traffic

The Ethernet is a broad band system of multiple access for local network implementation with distributed control. In the Ethernet network a certain number of stations is connected to the separate channel. When the station is about to transfer a packet the carrier control method is used in such a way that the station waits for the channel to become free (the absence of the other transmitting stations). If two or more stations make the decision that the channel is free and begin to transmit simultaneously it may lead to conflict [11]. These mechanisms illustrate the Ethernet protocol of the random access and can be referred to as the method of shared access, with carriage recognition and collision detection, or CSMA (carrier-sense multiple access)/CD.

Ethernet networks have already been used for more than 100 years and remain among the most popular and successful local network technologies (LAN) at present. The attractive properties of Ethernet networks include: the convenience of network reconfiguring and the high level of fault tolerance (due to the absence of a central control device). A 10 Mbps Ethernet is a multiple access system for local computer networks with distributed control, which was and has remained the LAN technology base. The past 10 years were marked by the abrupt increase in LAN quantity, which illustrates the necessity to provide user interconnections and common resource provisions for them, such as file servers and printers.

Using actual instrumental equipment that provides monitoring, the time of arrival and heading information can be recorded for each (full) packet that falls into the tracked Ethernet cable, connected to any host. Such types of measurements of Ethernet LAN traffic with high resolution during week duration periods were carried out in Reference [12] where the 'standard' daily traffic volume consisted of nearly 20–30 million packets or nearly 2 Gb of useful data.

The first investigation of traffic fractal dynamics was published in References [1] and [2] by researchers from Bellcore. On the basis of the all-purpose measurements fulfilled in the local Ethernet network, they came to the conclusion that the greater the loading to the Ethernet, the higher the estimated Hurst exponent H for the traffic or, which is equivalent, the higher the degree of self-similarity. This result is extremely important because it is true at high loadings when the effectiveness problems become more evident.

4.2.1 Experimental Investigations of the Ethernet Traffic Self-Similar Structure

An investigation will be illustrated of real network traffic collected in a small commercial network, the main part of which is the examination of web-pages in the Internet network and in a small household local network, where the traffic was created by typical applications used in the networks of such types (files transfer, chat, netmeeting, games, etc.). The diagram in Figure 4.1 shows that the topology of the examined network is of the star type. The twisted-pair cable UTP-5 is used as the connecting cable (100-BaseT). The cybercafe on the diagram is shown as a separate recipient, which has its own local network and is a subnetwork of the main network. The measurements in this network were carried out using a traffic analyser which was mounted on a proxy-server (PROXY), thus catching all information running on the dedicated Internet channel (rate of 2 Mbps).

To investigate the traffic structure and to illustrate its fractal (self-similar) character special software was used in which the appropriate tests were fulfilled. Examples of the appropriate profiles of byte intensity (W) and packet intensity (N) per time unit equal to 0.5 s are shown in Figure 4.2 (in th figure k is the index of the appropriate time interval). Figure 4.3 shows the 'tails' of distribution functions for the samples discussed above. For comparison, the curve of the Gaussian distribution function with zero expectation and unit variance is shown in the same plot.

4.2.2 Estimation of Testing Results

The test results will be discussed of self-similarity discovery for the series reflecting the network loading in bytes per time unit. It is obvious that for practical aims information on the degree of network loading is more useful. Information about the packet number per time unit may mislead the researcher, since many small control packets not carrying useful information and minimal length packets, which create hits not coinciding with byte rate hits, may be present in the network.

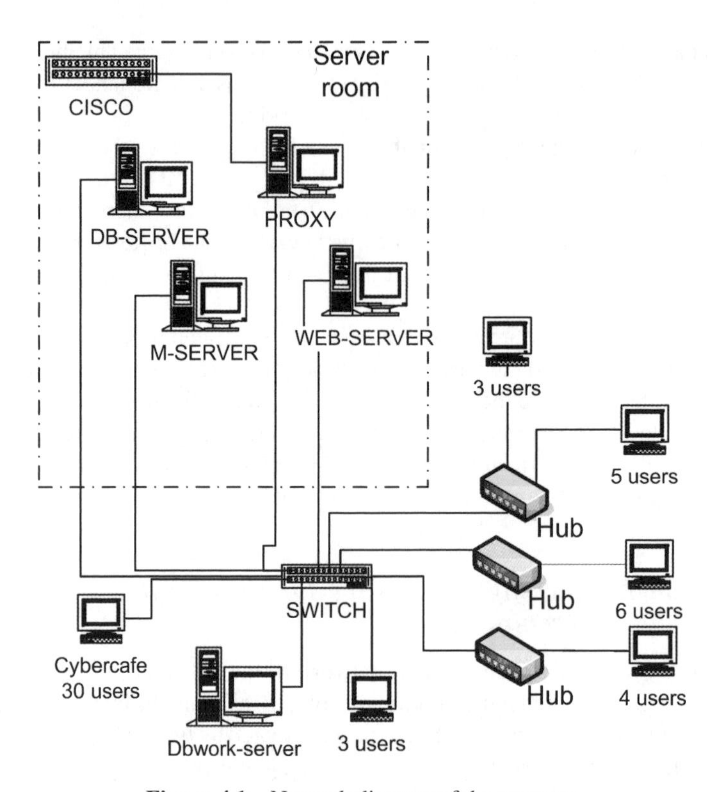

Figure 4.1 Network diagram of the star type

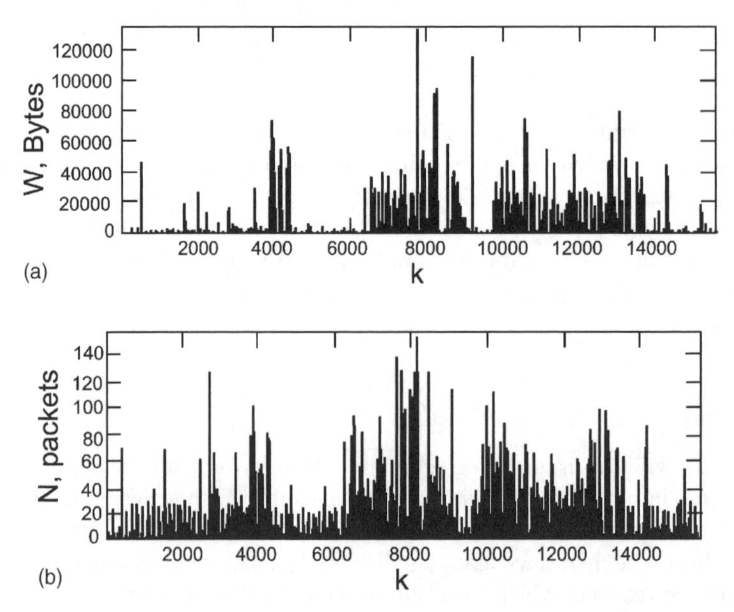

Figure 4.2 Traffic trace: (a) bytes intensity; (b) packets intensity

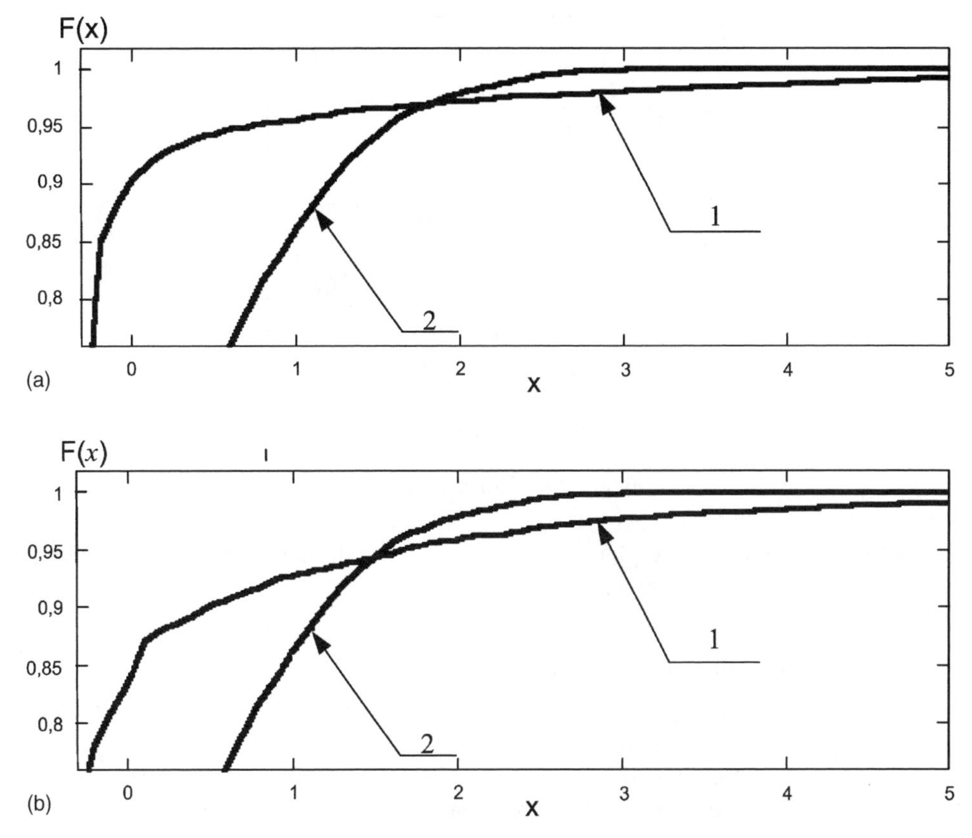

Figure 4.3 'Tails' of the distribution function for the Gaussian case and for measured data: (a) bytes loading; (b) packets intensity. Curve 1 is the distribution function for normalized experimental data; curve 2 is the distribution function for Gaussian random numbers

4.2.2.1 Variance–Time Plot

Figure 4.4 shows the results of a variance variation character test for the aggregation block size growth. The minimal and maximal block sizes were chosen equal to 10 and 400 discrete points for all measurements. All measurements of the examined series were taken with a time resolution of 0.5 s. Starting from these results it is obvious that the examined traffic has self-similar properties as an evident deflection in value of the Hurst exponent $H = 0.5$ is observed (the diagonal dotted curve).

4.2.2.2 R/S Statistics

Figure 4.5 shows the results of the test examining R/S statistics for the tested block size growth. For all measurements the minimal block size was chosen equal to 50 and maximal to 400 discrete points. The time resolution is 0.5 s.

The results obtained for R/S statistics, once again, confirm the heuristic character of the examined tests since the numerical results for R/S statistics essentially differ from those

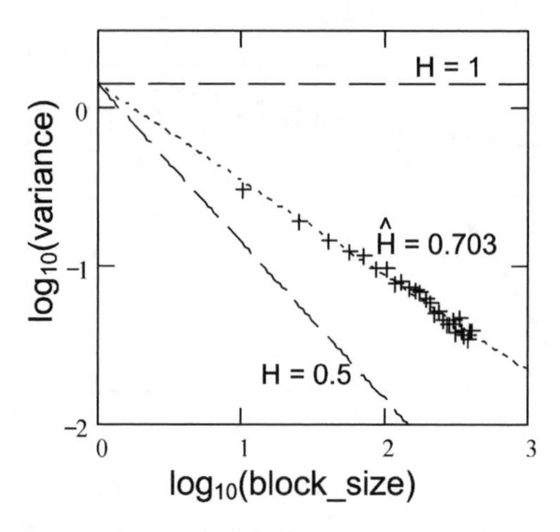

Figure 4.4 Variance–time plot

obtained by the variance–time plot. However, both tests illustrate that the true H for the examined data is within the 0.69–0.71 band.

4.2.2.3 Correlation Function Estimation

It is known that self-similar processes with H from 0.5 to 1 have the LRD property, i.e. their autocorrelation function at infinity does not tend to zero. Having estimated the degree of correlation function decay, it is possible to estimate H. When plotting the correlation function the first 100 delays are considered. The correlation function plot is shown in Figure 4.6.

Even for 100 delays this function does not tend to zero but shows power decay, which confirms the correlation function plot in the log–log scale. The estimation of the correlation function plot is also the heuristic method and consequently the results obtained with its help confirm qualitatively the presence of the self-similar structure in the examined data.

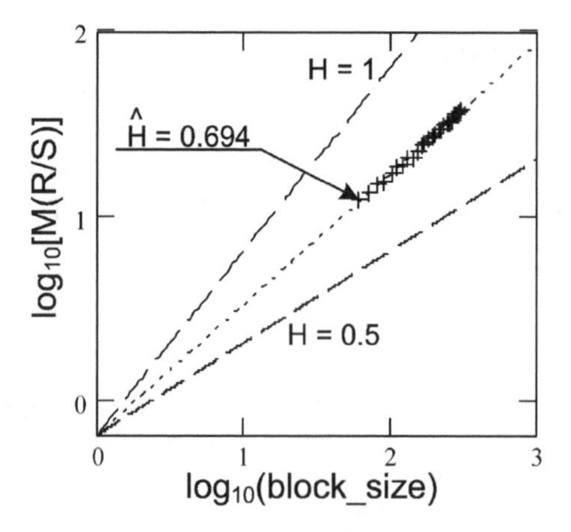

Figure 4.5 R/S statistics plot

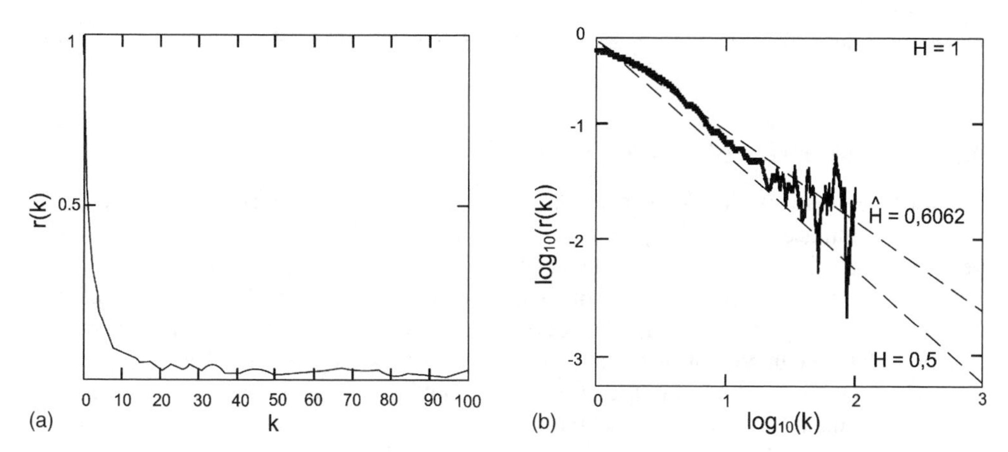

Figure 4.6 Examination of the correlation function: (a) correlation coefficient *r* versus the delay *k*; (b) log–log *r* versus the delay *k*

Processing the recorded Ethernet packets as black boxes and using the information on the packet time of arrival only, in References [1] and [2] it was shown that the measured aggregated Ethernet LAN traffic (i.e. the number of packets or bytes transmitted with the help of the Ethernet by all active hosts per time unit) with a mean value subtraction is the statistically self-similar process of the second order. In other words, Ethernet LAN statistical characteristics measured over microseconds and seconds have the identical statistical characteristics of the second order similar to Ethernet LAN traffic measured over minutes and even higher time scales. The scaling invariance of the measured Ethernet LAN traffic becomes apparent in the absence of the typical peak length. Ethernet traffic is bursty in all (or in the wide limit) time scales and various time scales, drawn on the plot, result in similar pictures and have a typical 'peak within peak' structure (see Reference [2]).

To explain this self-similarity, experimentally observed, the structural modelling approach was offered in Reference [2] and considered in detail in Reference [13]. The concept of the studies consists in the following. Using the information on the packet arrival time and also on the address of the Ethernet source and recipient contained in the recorded heading information for each packet, available in the Ethernet, the aggregated traffic can be divided into separate components appropriate to traffic flow between each active pair of the computer hosts or the pair of source–recipients. The simple traffic models such as ON/OFF sources or packet series models are popular at the level of the separate pair of source-recipients. For easy understanding, these models presuppose that the source interlaces between the 'active' state (ON period) and the 'passive' state (OFF period). During the ON period the packets are transmitted at a fixed rate but during the OFF period the packets are not transmitted. The packet group transmitted during the ON period is referred to as a 'series' and the calm period between two series (i.e. OFF period) is referred to the as the 'interseries interval'. The ON periods following each other as well as the OFF periods following each other are considered independent of each other and identically distributed. Therefore, the only distributions that determine the duration of ON and OFF periods are the stochastic elements describing ON/OFF sources. It was shown in Reference [14] that the superposition of such sets of ON/OFF sources will show the experimentally observed self-similar structure of the measured aggregated Ethernet LAN traffic in the case when distribution of the ON/OFF periods of the separate pair of source recipients has infinite variance.

4.3 Self-Similarity of WAN Traffic

The wide area networks (or WANs) appeared in the 1970s for ensuring computer interconnections, located in geographically spaced regions. Frequently, WANs are a set of separate (independently controlled) networks, possibly having different equipment for the interconnection.

WANs differ from LANs by the quantity of main routes, which are much more heterogeneous and create difficulty in predicting what kind of traffic conditions may occur in each specific situation. Another important difference is that the time constants, related to the feedback ensuring certain network conditions, vary from ten milliseconds to seconds instead of microseconds as in the LAN case. This essentially complicates adaptations of WAN applications to specific network conditions and may cause congestion, i.e. the effectiveness decrease due to the congestion in the built-in network chain between two WAN hosts. If the component has no memory in the buffer for temporary saving of data packets arriving at it, it will reject them instead of the further transfer. Therefore, reliable data transfer by WAN requires a 'complicated' transport protocol. This protocol should guarantee that in the case of loss the data packets will be transferred again, and should avoid any unnecessary repeated transfer that would intensify the congestion situation. The Internet is the well-known example of WAN.

Today Internet traffic is mainly generated by the data transfer between the clients and servers that use the TCP protocol at the transport level. The flow can be defined as a separate TCP connection which begins from the connection interfacing procedure and ends with the connection closing procedure. Each flow generates the packet sequence, which falls into the network in accordance with TCP algorithms for congestion avoidance. It is shown in References [4] and [15] that the arrival and packet processes as well as the flows demonstrate the LRD properties. In reality, the flows do not start independently but are generated by sessions. A rougher time scale of the session arrival dynamics introduces such a dependence, which can be considered the long-range dependence for typical time scales of the flows and packets.

The TCP is not the only protocol to be used in the Internet. In fact, the incremental part of the traffic is 'multicast' (frequently used for digital audio and video transfers) in which one transmitter fulfills the transfer to many recipients, which is a relatively new application that has not been studied sufficiently. Taking into consideration the wide distribution, the distinctive features related to the TCP will be analysed below in detail.

4.3.1 WAN Traffic at the Application Level

The profile of dominating Internet applications constantly varies in time. At present, the main applications in the Internet are: file transfer (FTP), the structural information search (HTTP and World Wide Web), electronic mail and network news. Just a ten years ago HTTP traffic was practically nonexistent. The 'typical mixture' of WAN applications does not exist now. The dominating WAN applications radically change from site to site. Therefore, it is necessary to be careful when assuming that the specific WAN traffic channel reflects the 'typical' traffic. The only way to solve this problem is to divide the traffic in accordance with different applications. This can be easily done for WAN traffic trace investigations because the TCP heading contains the port number, which indicates the application being used. Descriptions will focus on Telnet, FTP and HTTP [14].

FTP and HTTP are the 'group transfer' applications, the main purpose of which is the transfer of the previously known data value from one Internet host to another. While in the LAN environment the group transfer is rather simple, but becomes quite complicated because of the time overload and TCP dynamics. As a result, suitable models describing the group transfer in WAN are rare and at best they are concretized. Compared to FTP and HTTP, Telnet is the 'interactive' application. The transmitted packets from the host initiating the connection to the received host are determined by a set of push keys by the Telnet user. Actually, as these sets should be stable in the presence of widely changing network conditions, the structural modelling approach for Telnet traffic will be based on detection of 'typical' properties of these sets.

4.3.2 Some Limiting Results for Aggregated WAN Traffic

The natural approach to aggregated WAN traffic modelling is based on the 'time scale sharing' principle. This means that there are two separate processes: starting time of the session (where the session consists of one or more coupled network connections) and the packet arrival process within the session. The transmission starts at any random time moment ('session start'). The packets are transmitted (in some bursty manner) during a certain time and after that the transmission stops ('session end') until the next session begins. While the session arrivals may be generally defined unambiguously, the 'typical' peak determination (the arrival during the session) is ill-defined, just like for the host pair in the LAN environment.

The structural model investigation for the global network traffic is still at the preliminary stage. The structural modelling approach for WAN group transfer was suggested in Reference [4], which couples the WAN traffic characteristics at the macroscopic (i.e. aggregated) level with the microscopic (i.e. application) level, considering the typical properties of the group transfer such as the arrival structure, the value of transferred data or the session duration, which does not only depend on the value of transferred data but also on the network conditions during the transmission. The obtained structural models show that some of the problems occurring in practice may be avoided by covering the 'fine' details of the group transfer having a high profile for investigation, e.g. how much the network environment affects the traffic.

Some of the known approaches to model the traffic will be considered, which directly illustrate this two-stage procedure and estimate the possibility of its use for describing the experimentally observed characteristics of WAN traffic.

4.3.2.1 $M/G/\infty$ Queues

The process $M/G/\infty$ is defined in the following way [16,17]. Consider the queue $M/G/\infty$ with discrete time with some time slot Δ as the unit time interval. All Poisson arrivals during the time slot are used for servicing before the beginning of the next time slot. Let $w(S = k)$, $k = 1, 2, \ldots$, designate the PDF of the service times S in time slot units. Let $\hat{S}$ designate the time left till the end of user servicing. It is known that the queue length distribution in this system will be Poisson at the end of each slot with the mean $\lambda = \lambda_0 M[S]$, where λ_0 is the average arrival number till the time slot in the queue $M/G/\infty$. However, the queue lengths at the end of the slots following one another are correlated with the autocorrelation function $r(k) = P(\hat{S} > k)$, equal to the additional time distribution left until the service end. Therefore, if such a queue length process is used to generate the arrivals for the analysed system, the following arrival process A is obtained: the marginal distribution for A is the discrete Poisson process with intensity λ per slot and where $P(\hat{S} > k)$ plays the part of the correlation function. In practice, it would be necessary to achieve the given autocorrelation function $r(k)$, which can be used to calculate the required distribution of service time. In particular,

$$P(S > k) = P(\hat{S} = k)M[S] = [r(k) - r(k+1)]N(S) \tag{4.1}$$

Since $P(S > 0) = 1$ and $r(0) = 1$ by definition, $M[S] = 1/[1 - r(1)]$. Then for long-range dependence

$$r(k) = \alpha k^{-\beta}, \quad 0 < \beta < 1 \tag{4.2}$$

where $\alpha = r(1) = 1 - 1/M[S]$. As a result, the generated arrival process is asymptotically self-similar with the Hurst exponent $H = 1 - \beta/2$ [17].

Since the system $M/G/\infty$ describes the discrete arrival process only, the next step is generation of separate arrival times. It can be achieved by means of arrival aggregation with $K \geq 1$ slots following one another and their further redistribution over all intervals $\tau_0 = \Delta K$ seconds in size.

Let N designate the total number of arrivals over the K slots interval. As N is the Poisson process, the assignment of each arrival to the point of the interval corresponding to uniform distribution will give the exponentially distributed time moments between arrivals within the slot. (The total inter-arrival time moments distribution still remains nonexponential.) This forms the basis of comparing and creating the request process of web-pages, which is the asymptotically self-similar process with exponentially distributed increments.

The choice of using a continuous time region for a description of the arrival process is based on the fact that there is interest at the applications level in traffic generation and processing. For the same reason, the 'time slot' does not necessarily correspond to unitary transfer and can be chosen arbitrarily large. This inevitably implies that the results depend on slot sizes.

In the WAN traffic context the $M/G/\infty$ configuration presupposes that the sessions (e.g. FTP, HTTP and Telnet) arrive in accordance with the Poisson process. The packets are transmitted with constant intensity during their 'life time' or the session duration and after that the packet transmission stops. It should be noted that although the Poisson structure of the session arrival is proved, nothing is known about another stochastic characteristic such as the distribution of session lengths or its duration. It should be chosen in order to cover effectively the experimentally observed property of WAN aggregated traffic long-range dependence [4], which in its turn is linked to statistical self-similarity. The initial parameters operating in discrete time will be precisely determined.

Let X_n designate the user's number in the system at time moment n in the $M/G/\infty$ model or, equivalently, the total number of packets generated by all sessions that are active in time moment n (assuming that the packets are transmitted by groups in the time unit during the session life time). Then $(w_n)_{n\geq 1}$ is the PDF, $F(n) = \sum_{k\leq n} f(k)$ and $\overline{F} = 1 - F$ are the cumulative and complementary distribution function respectively and m is the average value of session length. It is assumed that as $n \to \infty$, the distribution function F follows the heavy tails property (2.45), i.e. $\overline{F}(n) \sim n^{-\alpha} L(n)$, as $n \to \infty$, $1 < \alpha < 2$. Under these conditions the following result was found in References [4] and [17].

Theorem 4.1
The aggregated packet process $X = (X_n : n = 0, 1, 2, \ldots)$ demonstrates long-range dependence. Having denoted the correlation function for X as $R(k)$, then

$$R(k) = m^{-1} \sum_{n=k}^{\infty} \overline{F}(n) \sim Ck^{1-\alpha} L(k), \quad \text{as}\, k \to \infty \tag{4.3}$$

for some constant $C > 0$. In addition, the degree of long-range dependence (i.e. Hurst exponent) is defined as $H = (3 - \alpha)/2$.

The main process component is the 'heavy tails' property (see Equation (2.45)) for the session duration. Instinctively, this property confirms that the 'typical' session length demonstrates high changeability, i.e. shows the fluctuations in the wide range of the time scale. This fundamental characteristic at the application level reveals itself at the network level through the (4.3) property. This property confirms that the aggregated traffic process X is an asymptotic self-similar process of the second order. In other words, if the process is observed during a large enough time scale, the statistical properties of the second order of the process X remain practically unchanged and the traffic looks like a 'similar' one in the wide range of the time scales.

In spite of its simplicity, this construction has some shortcomings, which restrict its direct application for WAN traffic modelling. First by, the Poisson nature of the session arrivals in practice is often too restricted. Secondly, and more importantly, the applications that at present form the main part of WAN traffic (e.g. HTTP), as everybody knows, transmits its packets with *variable* (not constant) intensity, but in a highly bursty manner caused by the changing network congestion and network dynamics.

4.3.3 The Statistical Analysis of WAN Traffic at the Application Level

The substantiation of structural modelling approaches for WAN traffic that are assumed in Theorem 4.1 requires checking whether the measured WAN traffic at the application level is compatible with either Poisson (or, generally, regenerative) session arrival moments or with the session life times that meet the basic property (2.45). The detailed information concerning the packet arrival structure within the session is important for determination of whether the statistically self-similar limiting process is suitable for an exact description of real WAN traffic. For this purpose it is necessary to carry out experimental studies. In conformity with WAN applications it is expedient to examine the measured traffic traces. While FTP traffic still occupies the main part of available WAN carrying capacity, the scope of HTTP traffic continues to increase and begins to displace FTP from the leading position. On the other hand, Telnet is the service qualitatively different from the two previous ones, with far less occupation of the carrying capacity but generating a large number of packets, frequently one packet to one push key.

As regards the stochastic properties of the network session arrivals, the researches show that the arrival moments of the network sessions clearly demonstrate the diurnal cycle. For example, the peak of Telnet sessions is observed in the daylight and has quite a low level early in the morning, which is almost identical by structure to the call structure observed in traditional telephony. It is shown in Reference [4] that the arrivals of both Telnet and FTP sessions are well modelled by the heterogeneous Poisson processes with intensities that are constants within one hour, but may change from one hour to another.

Using the measurement of HTTP connections (where the HTTP session is typically created by the HTTP connection set), the absolute experimental proof of the (2.45) property for the size characteristic of the HTTP session and for the life time of HTTP sessions was found [14]. Some typical measurement features and the mathematical descriptions of WAN traffic components will be considered below in the Internet example.

4.3.4 Multifractal Analysis of WAN Traffic

It was found in some researches that the scaling structure in measured WAN traffic can be divided into two categories: scaling over large time scales with self-similarity and scaling over small time scales with multifractal scaling [18–20]. The transition from multifractal scaling to self-similar scaling occurs at time scales of the order of the usual total round trip time in the network [18, 19]. Physical explanations and the consequences of these traffic properties are also considered in some publications [21].

Using the multifractals to model the network traffic is a new concept and at present only a few results are available. In particular, it is shown in Reference [22] that although self-similarity describes LAN traffic adequately, WAN traffic traces have more general, multifractal properties because the influence of network dynamics may prevail over a user's behaviour in WAN with small and medium time scales. The scaling behaviour revealed in WAN traffic can be reproduced with the help of the cascade configuration, i.e. the multiplicative process that assigns the weight over the continuously decreasing time intervals in accordance with some distribution (depending on the division level). At the limit, when the number of division levels tends to infinity, this procedure will allow multifractal generation [23,24].

4.4 Self-Similarity of Internet Traffic

At present, the best-effort service is the dominating type of service, which means that the bandwidth is divided into equal parts among all traffic flows and the transfer control is displaced to the final devices of the communication channel. On the other hand, the Internet is initially based on the client and server interconnection, which ensures a specific service to Internet users as well as applications operating in personal computers and mobile terminals requesting the service.

The network nodes and terminals share the information flow between packets to which the headings and tails are added for addressing and information control along the movement way from the application to the physical environment. In the inverted sequence the headings and tails are truncated to ensure information is available for the application. Various applications use different transfer protocols depending on their traffic requirements (e.g. TCP and UDP). These protocols use the sockets to connect the application level. The IP is situated between the transfer protocols and the Internet connection level protocols.

The communication channel measurements in the commercial Internet main during two periods (24 hours and 7 days) fulfilled in Reference [25] show that web traffic dominates in the aggregate Internet traffic, with the TCP component fraction equal to 95 % for the major part of the traffic. Web traffic is the major part of TCP traffic in the measured connections that dominate the only Internet application with a client–server component of more than half of the bytes

(65–80 %), packets (55–75 %) and flows (65–76 %). Before the Web was invented a major part of TCP traffic consisted of FTP file transfers, electronic mail and some interactive applications. After the Web, which is based on the HTTP hypertext transfer protocol at the application level and TCP at the transmit level, was introduced, web traffic began to dominate in aggregate Internet traffic [19].

Internet traffic is usually shared on the basis of the used transfer protocol (TCP and UDP) or the program (Web, Telnet, FTP or e-mail). Moreover, each of the traffic parts consists of many multiplexed flows from various connections. One user can execute one or more flow numbers simultaneously (e.g. the parallel connections in one session for acceleration or execution of several sessions from one browser). Although TCP traffic at present dominates in the Internet, there is an essential part of UDP traffic that is mainly used for the connection between the servers. The UDP is suitable for the real time service and can be used in combination with the RTP.

The analysis of IP traffic scaling will now be completed by considering the characteristics of the aggregated traffic components at the transport and applied levels and examining its influence on the aggregated traffic characteristics. It will be shown that the correlation structure of the aggregated traffic is basically determined by the components with the largest variance and correlation over the time scales.

4.4.1 Results of Experimental Studies

Figure 4.7 shows the traffic intensity in bytes per second for typical IP traffic and its main components for the measured traces. The network structure in which the measurements were fulfilled is shown in Figure 3.1.

Figure 4.7 shows the time traffic realizations for various protocols over fragmentation intervals of 100 ms. The byte-oriented traffic load is also presented. Traffic hits can be observed over the whole measurement interval and the conclusion can be made that the high hits over a short time interval can occur at any time of the day or night. Therefore, the well-known concept of the maximal load hour used in telephony cannot be applied. The data analysis demonstrates that TCP data at the transport level dominate in the traffic and the character of these data determines IP traffic characteristics.

4.4.2 Stationarity Analysis of IP Traffic

The important assumption for Internet traffic description and modelling is its stationarity, which is, however, very difficult to confirm unambiguously for the examined data. The real traffic often demonstrates local trends, load jumps, cycles, etc., over a long time interval, which is typical for the nonstationary processes. There are two main approaches to IP traffic stationarity analysis: stationary (developed, for example, in Reference [26], and nonstationary (presented in publications [4], [10], [27] and [28]). Taking into account the advantages of both types of description, they will now be considered in detail.

4.4.2.1 Stationary Approach

The obvious approach to solve the problem of stationarity is the choice of time intervals where the traffic stationarity assumption is valid (local stationarity). The simplest test to determine the

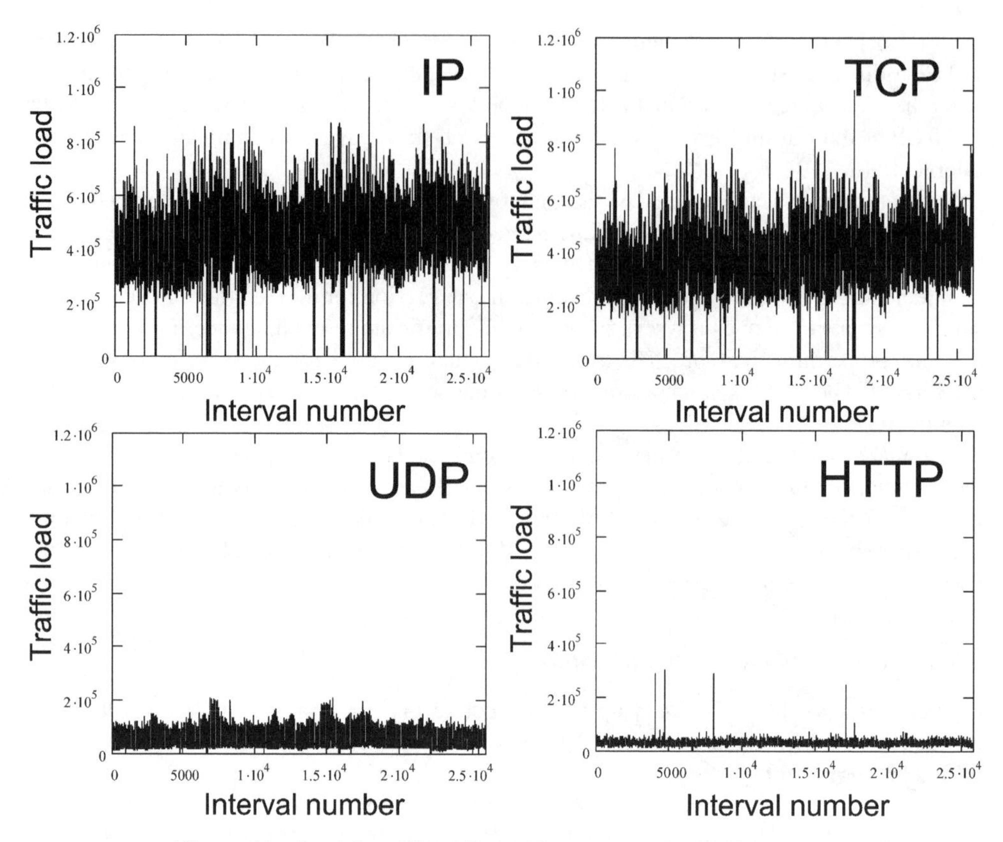

Figure 4.7 Intensity of IP traffic and some protocols of a higher level

stationarity periods in the data is to move the window through the measured data and to measure the average value variations from window to window. A diagram of such series can give information about the level shift, trends, etc. However, for such bursty data as the measured traffic this method will not give the correct result. To achieve this goal special instrumental means were used in Reference [26], based on the variation point determination method, which consists of window displacement over data and a comparison of sample distributions in the two halves of the window. If these two distributions are substantially different, the stationarity assumption for this window is rejected. The problem of executing the distribution comparison for two series can be fulfilled using the Kolmogorov–Smirnov criterion.

On the basis of the conducted tests on stationarity, several subsets can be chosen from all measured data for the analysis. These subsets for the analysis are obtained from the time intervals where IP traffic as well as each traffic component of the transport and applied level were tested for stationarity.

The possible stationarity of the analysed IP traffic characteristics and their components will now be considered. First of all, the autocorrelation function will be investigated, followed by a discussion of how the autocorrelation functions of various components affect the aggregated traffic. Then the long-range scaling of each of the IP traffic components will be studied.

There are several different components with varying contributions to the correlation structure of the aggregated IP traffic. Determination of the component characteristics that affect the correlation structure characteristics of the total IP traffic is important for an understanding of the IP traffic structure.

The superposition of several independent traffic flows, i.e. $A = \Sigma_{i=1}^{N} A_i$, will be considered. The autocorrelation function for A_i is denoted as $r_{A_i}(k)$. It can be shown that the autocorrelation of the aggregated traffic flow can be found as

$$r_A(k) = \frac{1}{\sum_{i=1}^{N} \sigma_{A_i}^2} \sum_{i=1}^{N} \sigma_{A_i}^2 r_{A_i}(k) \tag{4.4}$$

where $\sigma_{A_i}^2$ is the traffic variance in the chosen time unit in flow i.

The autocorrelations of SRD flows rapidly decay as $k \to \infty$ while the autocorrelations of LRD flows decay asymptotically as $k^{-\beta_i}$. The autocorrelation for A is determined with the help of the LRD flow which decays at the smallest rate, i.e. $r_A(k) \sim k^{-\min \beta_i}$. Thus, the LRD flow with the highest exponent H will be the main one ($\beta = 2 - 2H$) and the aggregation will have LRD with the same exponent. However, in practice, the flow variance can be examined since the variance is the weighting factor of the sum in Equation (4.4). Therefore, the situation can be found where there is flow with a faster decaying autocorrelation function but with great variance, and in this case the flow will dominate the autocorrelation function of the aggregated traffic flow over the examined time interval. In a similar manner, the small traffic part with great variance and a slowly decaying autocorrelation can determine the autocorrelation for the whole aggregated flow.

These properties will be illustrated using the measured IP traffic. Figure 4.8 shows the sample autocorrelation functions for various flows of the measured traffic. By estimating these functions (and especially analysing the correlation coefficient over large delays) it can be seen that the correlation coefficients often have small values. In these cases the confidence interval should be taken into account, which can be roughly estimated as $\pm 2/\sqrt{n}$, corresponding to 0.05. The slow decay of the IP traffic correlation was observed, which shows the possible presence of LRD. The transport level protocols (TCP and UDP) operate above the IP level and therefore IP traffic aggregates these flows. Among these components TCP traffic plays the dominating part as the type of its autocorrelation function completely defines the IP correlation structure. The TCP essentially affects the correlation structure of the aggregated IP traffic because it has the largest changeability over the analysed time scale, but not because it occupies the largest bandwidth in the aggregated IP traffic.

The traffic of HTTP, FTP, SMTP and GRE (generic encapsulation) protocols are considered as application level traffic. All these traffic flows are the components of aggregated TCP traffic. Figure 4.8 shows that the correlation coefficient of HTTP and GRE demonstrate long-range decay. The HTTP is the protocol that mainly affects the correlation structure of the aggregated TCP traffic because of its peculiar large changeability over the examined time scale. The investigation of sample variances of these components shows that HTTP traffic has the largest variance, which is the reason for TCP traffic domination when forming the aggregated TCP flow autocorrelation. It goes without saying that the correlation structures of these protocols are the result of interference with each other. For example, the TCP correlation structure is the mutual affecting ‘result’ of the protocol components (mainly the TCP) and TCP mechanism, rather than simply the ‘HTTP forming TCP correlation structure’.

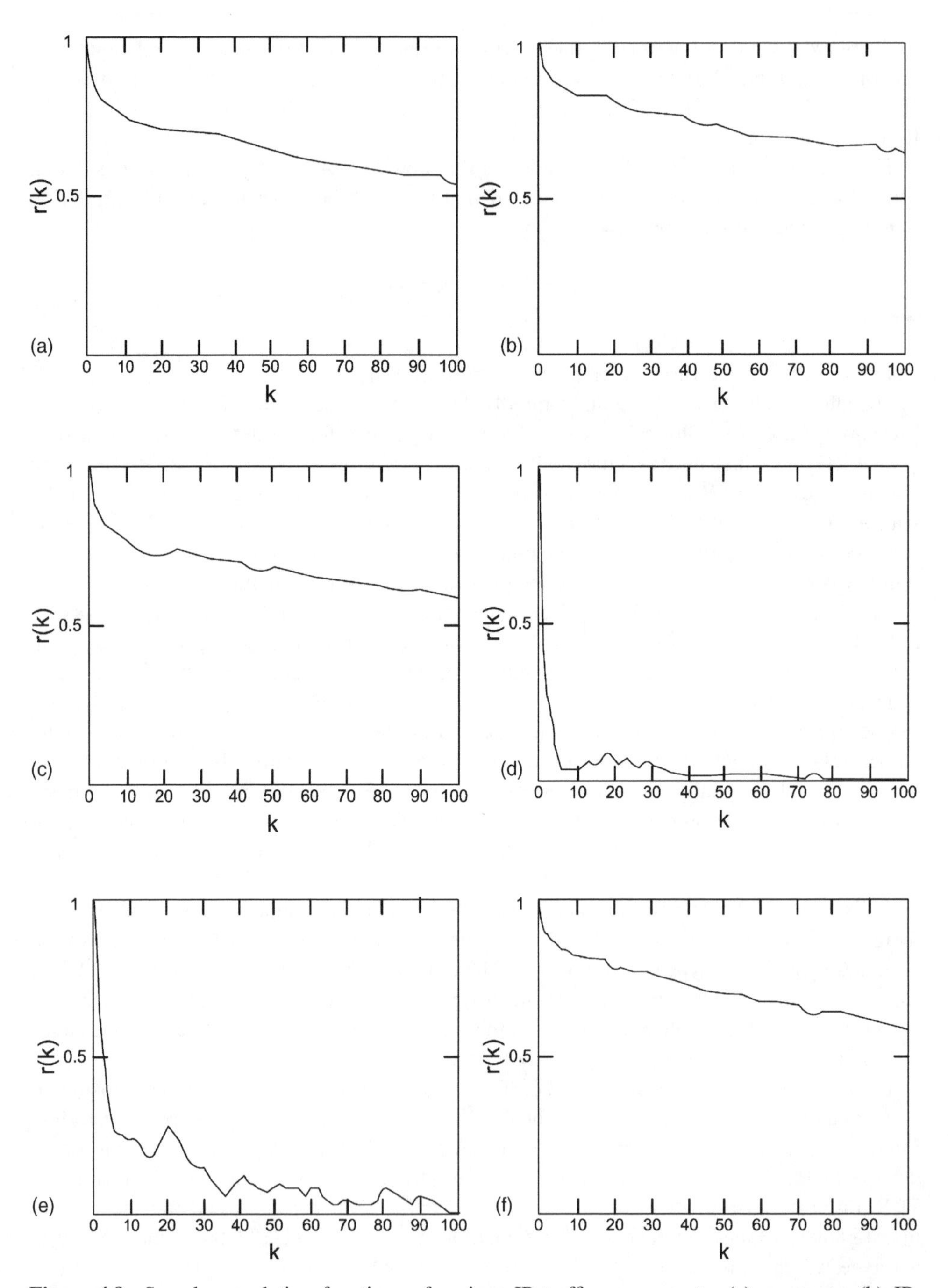

Figure 4.8 Sample correlation functions of various IP traffic components: (a) summary; (b) IP; (c) TCP; (d) UDP; (e) GRE; (f) HTTP; (g) POP3; (h) SMTP

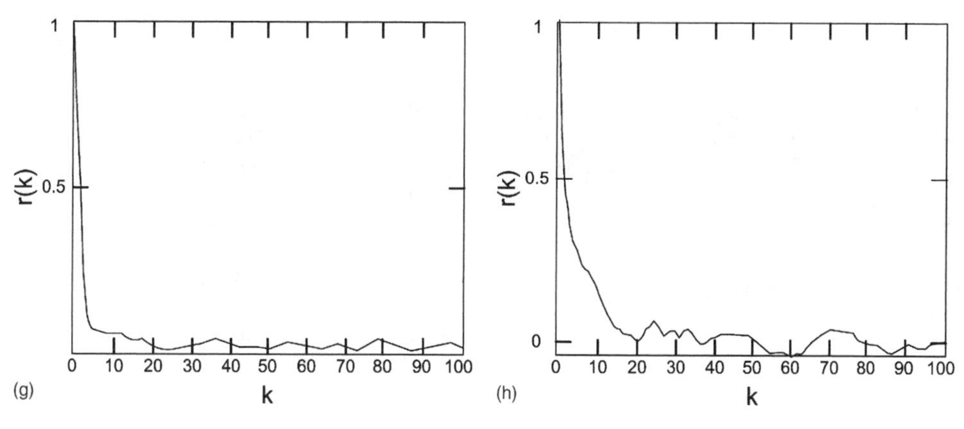

Figure 4.8 (*continued*)

4.4.2.2 Long-Range Dependence Analysis

LRD behaviour of IP traffic is defined by TCP traffic since it has the largest Hurst exponent as well as the largest variance among the transport protocols. This observation complies with the discussion in Section 4.4.2.1. It was found that UDP traffic is long-range dependent but has both a smaller variance and Hurst exponent as compared to TCP traffic.

The experimental results show the essential variations of the statistical properties of the packet sizes and the intervals between the packet arrivals for the input traffic and its considerable nonstationarity. The marginal distribution of the file sizes does not change significantly, but the Weibull distribution for inter-arrival times has essential variations tending to the exponential kind.

It is found that SMTP traffic is the process with short-range dependence. The time moments between the arrivals are nearly described by the Poisson process but this is only because the mail is transmitted from one host to a large number of hosts. The client file sizes for them as well as the full turnover times are close to independent ones. Besides TCP traffic there is a large fraction of UDP traffic, but this traffic type is smoother than TCP traffic and does not essentially affect the IP traffic structure. HTTP traffic plays the main part at the application level.

The main data part is transferred with the help of the TCP at the transport level, which covers near by 90 % of the bandwidth of the transferred data total volume. The remaining load falls mainly on the UDP. ICMP and OSPF control messages constitute only 1–2 % of the total traffic at the transport level. HTTP and FTP traffic dominate at the application level.

The traffic variables were measured after all packet headings were locked in the channel and the time mark was added to each packet. There are two variable categories: packet variables and RfT (request for transfer) variables. All traffic variables were studied depending on the protocol, as their behaviour differs for various protocols. HTTP and SMTP components of Internet traffic were considered.

The packet size and the time between packet arrivals are the packet variables. RfT variables describe the characteristics of the application requests to the TCP concerning the transmission, and also the information of the Internet environment during the request transmission. The TCP connection time was taken from the time mark of the first SYN packet. Therefore, each RfT

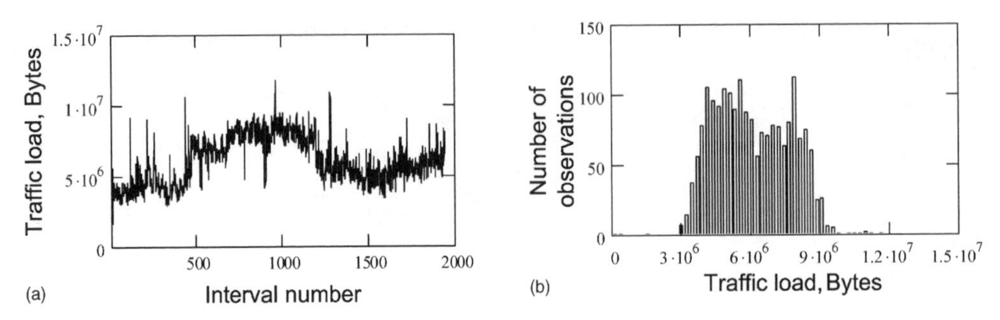

Figure 4.9 (a) IP summary traffic sample path and (b) the histogram of its distribution

variable measurement is related to the specific connection and each connection has RfT variables set by the measurements, all the time marks of which are the time marks of the connection. The following RfT variables were analysed:

- full round trip time for the server measured with the help of the time between client SYN and server SYN/ACK;
- full round trip time for the client measured with the help of the time between server SYN/ ACK and client ACK;
- client file size measured with the help of the connection serial numbers;
- server file size measured with the help of the connection serial numbers;
- inter-arrival time i.e. the time before the next connection arrival.

The sample path of the IP summary traffic and the histogram of its distribution are presented in Figure 4.9. The sample path of the Internet HTTP traffic and the histogram of its distribution, having, as a rule, the multimode character, are shown in Figure 4.10. The multimode character of the packet size distributions is mainly explained by the controlling packet mixture, such as ACK, and the data packets with different segment maximal sizes.

The packet headings data were investigated from the main database of the corporate channel to formulate the statistical models for the packet HTTP variables. The time space was divided into 1 second blocks. Such a block length is small enough to ensure the necessary stationarity inside the blocks in most cases but large enough to ensure a sufficient number of packets for

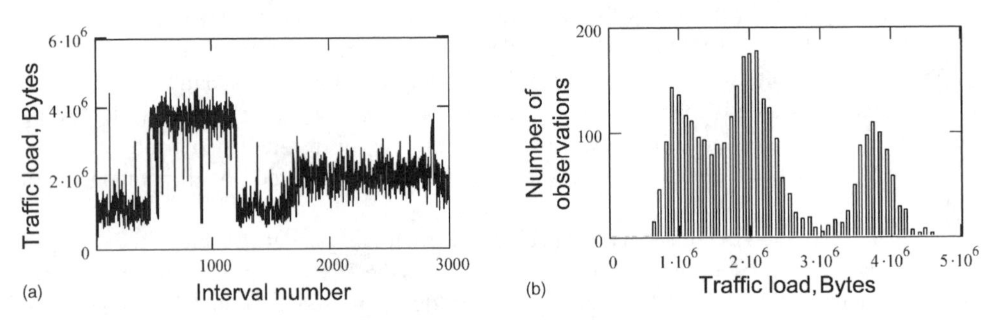

Figure 4.10 (a) HTTP Internet traffic sample path and (b) the histogram of its distribution

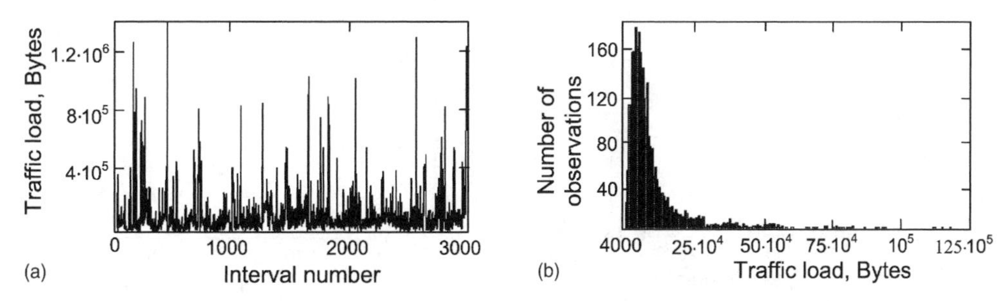

Figure 4.11 (a) SMTP Internet traffic sample path and (b) the histogram of its distribution

selection of the statistical models. The random sample with 3200 blocks was chosen for the analysis presented here.

The packets inside the channel were transmitted in two directions: inside (the packets from the server) and outside (the packets from the clients). Both directions were examined but only the results for the inside packets are presented here. The results for the outside packets are similar. The intensity of inside packets for 500 intervals varied from 1.7 packets/s to 452 packet/s. The packet size process and the time moments between the packet arrivals were analysed for each block. The marginal distribution analysis of the packet size input traffic was fulfilled for each block with the help of the quantiles plot and shows that the marginal distribution of inter-arrival time for each block is well approximated by the Weibull law. The example of the SMTP traffic realization and the appropriate histogram are presented in Figure 4.11.

The same RfT variables as for the HTTP were calculated during processing of the measurement results. In contrast to the HTTP, the connections were observed to initiate inside by the client (outgoing mail) and the connection to initiate outside of the corporate network by the client (incoming mail). Therefore, the SMTP plot can be divided into two components: incoming and outgoing. Each external client was connected for the incoming traffic to a separate host inside the corporate network, which is the reason why there was only one IP address for the SMTP server. For the outgoing component, two internal client hosts were connected outside to two hosts of the corporate network mail servers; thus there were IP addresses of only two SMTP clients and two SMTP servers. To choose between the two internal and external mail hosts, a special scheme is used for load balance and the configuration ensures is that the delay does not occur at the mail delivery to the client hosts. For the outgoing case of both the client and the server full round trip times are not calculated as the server propagation delay is too low.

RfT variables (time between the arrivals, client files, server files, client full turnover time and server full turnover time) were analysed for the obtained data. To achieve this the data were divided into blocks with 400 incoming and outgoing connections and blocks for which strong nonstationarity was observed were eliminated. The intensity of incoming connections for these blocks varied from 0.024 s/s to 0.24 s/s and the intensity of outgoing connections varied from 0.005 s/s to 0.112 s/s.

An example of the sample path of the POP3 traffic and the appropriate distribution histogram are presented in Figure 4.12. The long-range dependence estimation results for the examined traffic components with assumed stationarity are presented in Table 4.1.

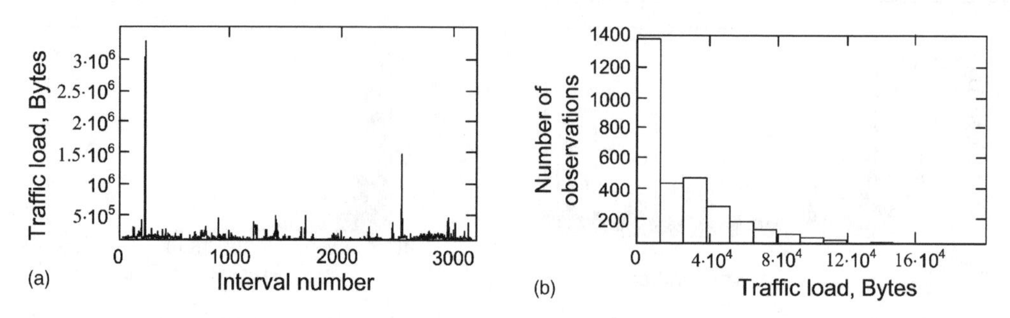

Figure 4.12 (a) POP3 Internet traffic sample path and (b) the histogram of its distribution

4.4.3 Nonstationarity of Internet Traffic

The researches [14, 16–18] show that the traffic in Internet channels not affected by the congestion displays universal nonstationarity. As the intensity ρ of TCP connections varies, both the marginal distribution and long-range dependence vary as well. Thus, as ρ grows the packets and connection arrival processes tend to Poisson, and the time sequences of the packets size, full round trip times and transmitted file sizes tend to independence. The investigations of the packet traces detect and explain the nonstationarity by the following. Firstly, as ρ varies, the complicated statistical model parameters fitted to the plot vary as well. The queueing characteristics for the packet traces also vary. The superposition effect is the reason for the nonstationarity: connection sequences mixing between separate pairs of source–recipient and packet sequences mixing from different connections. The authors of some publications [14, 16–18] suggest that nonstationarity should be one of the fundamental characteristics of Internet traffic in addition to long-range dependence and marginal distributions with heavy tails.

The times between connections and packet arrivals are long-range dependent and have a distribution that is either exponential or a tail that is longer (heavier) than exponential [10, 14, 20, 21]. A Weibull distribution approximation gives good results for Internet traffic, and these distribution parameters depend on the connection intensity ρ [27, 28].

Long-range dependence of many traffic variables can be described with the help of a simple two-parameter model: the fractal sum-differential (FSD) model and the conversion. Let $G(z) = N(0, 1)$ be the normal distribution function of the random variable with zero mean

Table 4.1 Long-range dependence estimations

Traffic type	Averaged estimate of Hurst exponent H	Correlation structure
IP	0.79	LRD
TCP	0.8	LRD
UDP	0.7	LRD
HTTP	0.8	LRD
SMTP	0.5	SRD
POP3	0.5	SRD
GRE	0.62	—

and unit variance. As earlier, let v_i be one of the variables of inter-arrival time, $F(v_i; \rho)$ be the marginal distribution function for v_i and $z_i = H(v_i)$ be the conversion of one variable of inter-arrival time v_i such that the marginal distribution for z_i is Gaussian type $N(0,1)$. Then $H(v_i) = G^{-1}(F(v_i; \rho))$. Let $\{s_i\} \in N(0,1)$ be the long-range dependent time series generated with the help of the fractal ARIMA model:

$$(I - B)^d s_i = \varepsilon_i + \varepsilon_{i-1}$$

where ε_i is Gausian white noise with zero mean value and the variance

$$\sigma_\varepsilon^2(\rho) = \frac{(1-d)\Gamma^2(1-d)}{2\Gamma(1-2d)}$$

B is the backward shift operator $Bs_i = s_{i-1}$ and $0 \le d \le 0.5$. Let $n_i - N(0,1)$ be the white noise series independent of ε_i. Then the FSD model can be defined as

$$z_i = \sqrt{1-\theta(\rho)}\, s_i + \sqrt{\theta(\rho)}\, n_i \tag{4.5}$$

where $0 \le \theta(\rho) \le 1$.

The term 'sum-differential' is used because the fractionally differenced z_i is the sum of the output from applying a summation operator to a white noise series and the output from applying a difference operator to a white noise series independent of the first process:

$$(I - B)^d z_i = \sqrt{1-\theta(\rho)}(I + B)\varepsilon_i + \sqrt{\theta(\rho)}(I - B)^d n_i$$

The autocorrelation function z_i from Reference [10] is determined as

$$R_z(k) = [1-\theta(\rho)]\frac{2k^2(1-d) - (1-d)^2\Gamma(1-d)\Gamma(k+d)}{[k^2 - (1-d)^2]\Gamma(d)\Gamma(k+1-d)}$$

for $k = 1, 2, \ldots$. The power spectrum is determined as

$$S_z(f) = [1-\theta(\rho)]\frac{(1-d)\Gamma^2(1-d)|1+\mathrm{e}^{2\pi if}|^2}{2\Gamma(1-2d)|1-\mathrm{e}^{2\pi if}|^{2d}} + \theta(\rho)$$

The component $\sqrt{1-\theta(\rho)}s_i$ variance is $1-\theta(\rho)$ and the component $\sqrt{\theta(\rho)}\, n_i$ variance is $\theta(\rho)$. It can be seen that parameter θ depends on ρ, but does not depend on d. For $\theta(\rho) \to 1$, z_j tends to the white noise.

In order to select the FSD model with conversion to the measurements, $F(v_i; \rho)$ is assumed to be the experimental marginal data distribution. Its simplicity is the attractive feature of this model since it is described by two parameters, d and θ only. Another advantage of this model consists in the fact that it shows that use of the Hurst exponent value $H = d + 0.5$ is not enough to describe the long-range dependence. It was found for each variable that H is not significantly changed as ρ varies, but θ increases with ρ growth, sometimes very considerably. In other words, long-range dependence decreases significantly but H remains permanent. Therefore both d and H are necessary to describe long-range dependence.

A comparative analysis of the two presented approaches to Internet traffic shows the following. The stationary models have been studied in some depth but they do not reflect in full measure the real process complexity because they do not take into account many factors connected with Internet technologies.

The nonstationary models are complicated in practical applications since they require the execution of a large number of measurements and the estimation of many auxiliary parameters. The obtained results show that due to their concrete character they are not able to be extrapolated in full measure to other networks and technologies.

4.4.4 Scaling Analysis

The multifractal and monofractal (e.g. self-similar) scaling structures of Internet traffic components will now be considered. According to the above-introduced terminology, the time series $\{X_i, i = 1, 2, \ldots, n\}$ is referred to as multifractal if the logarithms of $S^m(q)$ fragmentation functions (or, equivalently, the absolute moments) linearly depend on the logarithm of the aggregation level m, i.e.

$$S^m(q) = \tau(q)\log(m) + c_1(q)$$

where $S^m(q) = \Sigma_{k=1}^{n/m}|Z_k^{(m)}|^q$ for $Z_k^{(m)} = \Sigma_{i=1}^{m} X_{(k-1)m+i}$ and $c_1(q) = \text{constant}$.

The scaling behaviour can also be tested with the help of the methods based on wavelet presentation [29, 30]. The discrete wavelet conversion presents X series of n size at the j scaling level with the help of the set of wavelet coefficients $d_X(j,k), k = 1, 2, \ldots, n_j$, where $n_j = 2^{-jn}$. The q-th order of the logarithmic diagram (q-LD) can be defined as the logarithmically linear plot of the estimated qth moment $\mu_j(q) = (1/n_j)\Sigma_{k=1}^{n_j}|d_X(j,k)|^q$ of octave j. The linearity of the logarithmic diagrams for the order of various moments illustrates the series scaling property, i.e. $\log_2 \mu_j(q) = j\alpha(q) + c_2(q)$, where $\alpha(q)$ is the scaling exponent and $c_2(q) = \text{constant}$. The $\alpha(q)$ function shows the scaling type.

Both the method of the partition function and the method based on wavelet conversion were used in the present analysis.

The multifractal analysis results for the most typical components of Internet traffic are presented in Figures 4.13, 4.14, 4.15 and 4.16. Figure 4.13 shows the logarithmic diagrams of wavelet coefficients with the subtracted mean value for various q orders with a 0.5 step. In Figure 4.14 the multifractal Legendre spectra is given for $q \in [0.5; 4]$ with a 0.5 step for the partition interval range varying from 1 to 1000 meanings per interval. The logarithms of the partition functions $S^m(q)$ and τ(q) for various traffic types are presented in Figures 4.15 and 4.16 respectively. The estimation of the scaling function $\tau(q)$ carried out in Figure 4.16, based on the partition functions presented in Figure 4.15, also confirms the assumptions concerning multifractal scaling.

The scaling analysis in the course of data series studies for transport level protocols shows that the traffic of the transport level protocols has the LRD property. In the case of the TCP it is found that its scaling structure is similar to the IP traffic scaling structure and the estimated $\tau(q)$ functions shown in Figure 4.16 resemble those obtained for IP traffic. This leads to the conclusion that TCP traffic also demonstrates multifractal scaling. Similar conclusions are valid for the UDP.

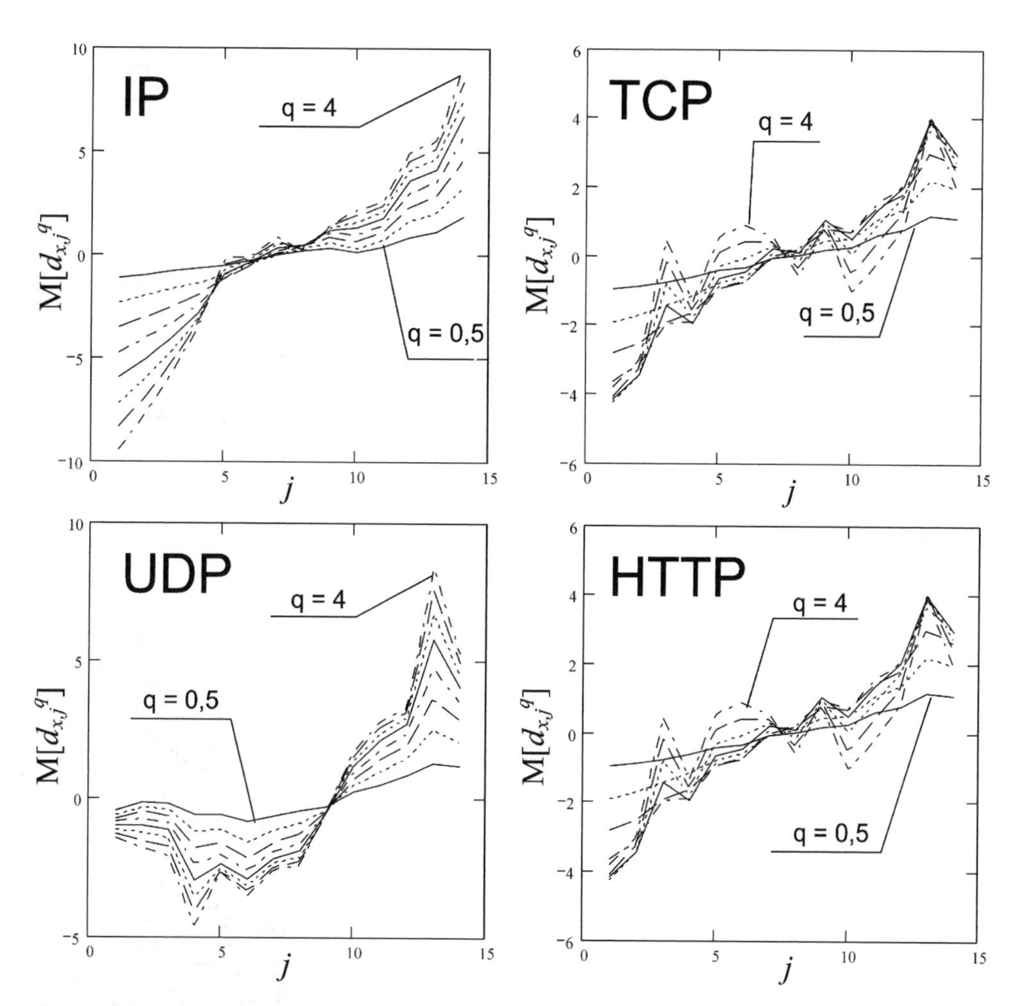

Figure 4.13 The logarithmic diagrams of wavelet coefficients with the subtracted mean value for various orders q with a 0.5 step

The scaling structure analysis for the application level protocols, in particular the HTTP, shows that the scaling function $\tau(q)$ is a convex curve, which proves the presence of multifractality. The estimated $\tau(q)$ is similar to that for IP traffic as well as for UDP traffic. It should be noted that SMTP traffic does not demonstrate the scaling structure.

The presented results confirm the conclusions made in References [6], [7] and [9], demonstrating that WAN traffic is LRD. Obtained results also demonstrate that WAN has a complicated multifractal structure, not only on small but also on large time scales. Moreover, the analysis shows that the aggregation consists of components with strongly varying scaling behaviour (without scaling, multifractal scaling and monofractal scaling).

At the transport level both the TCP and the UDP demonstrate LRD and have a multifractal structure. It is found that other protocols at this level correspond to SRD without scaling properties. Thus a conclusion may be made that the analysed IP traffic is LRD, containing

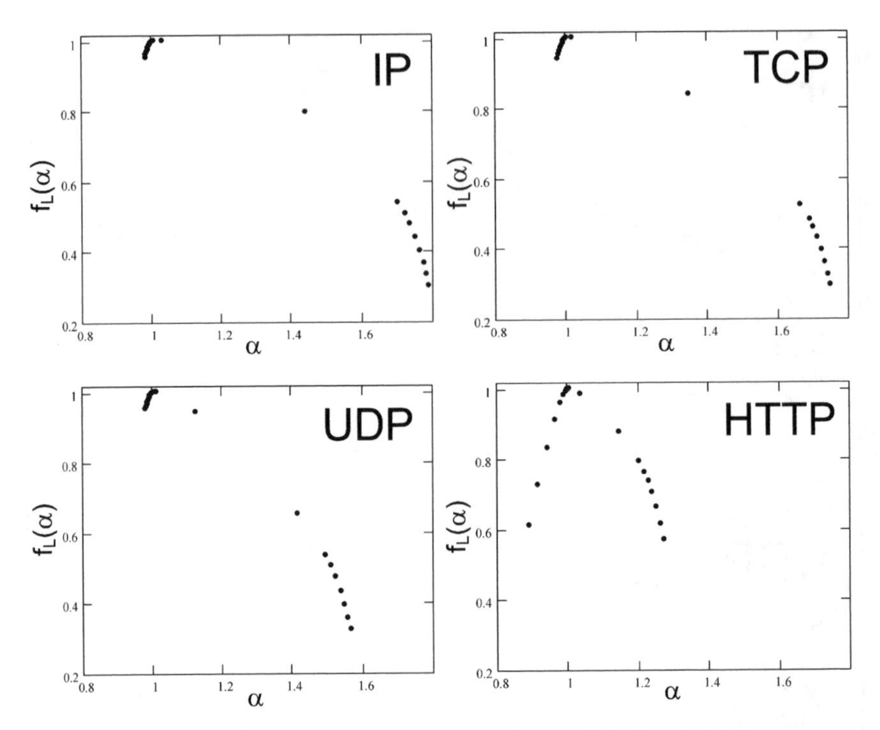

Figure 4.14 Multifractal Legendre spectra for $q \in [0.5; 4]$ with a 0.5 step

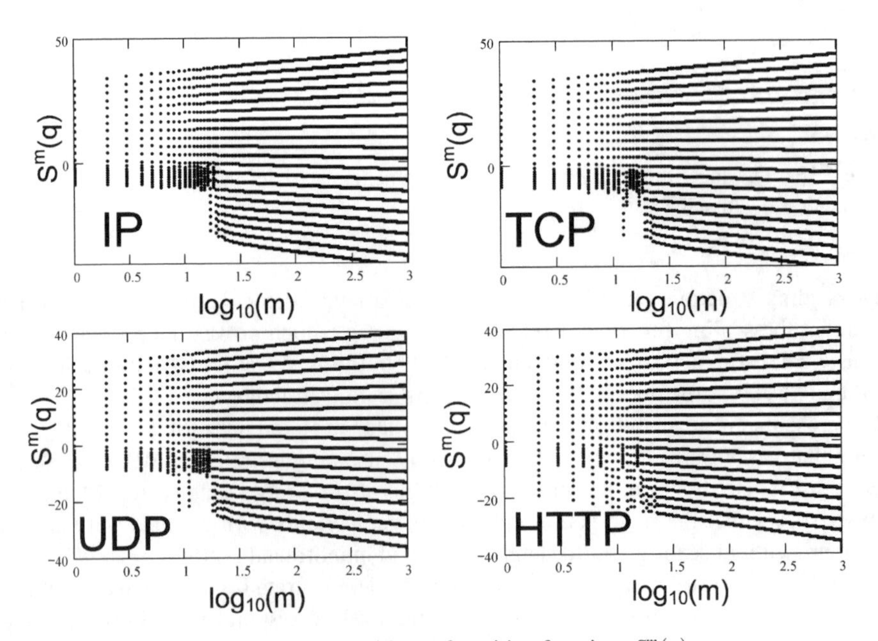

Figure 4.15 Logarithms of partition functions $S^m(q)$

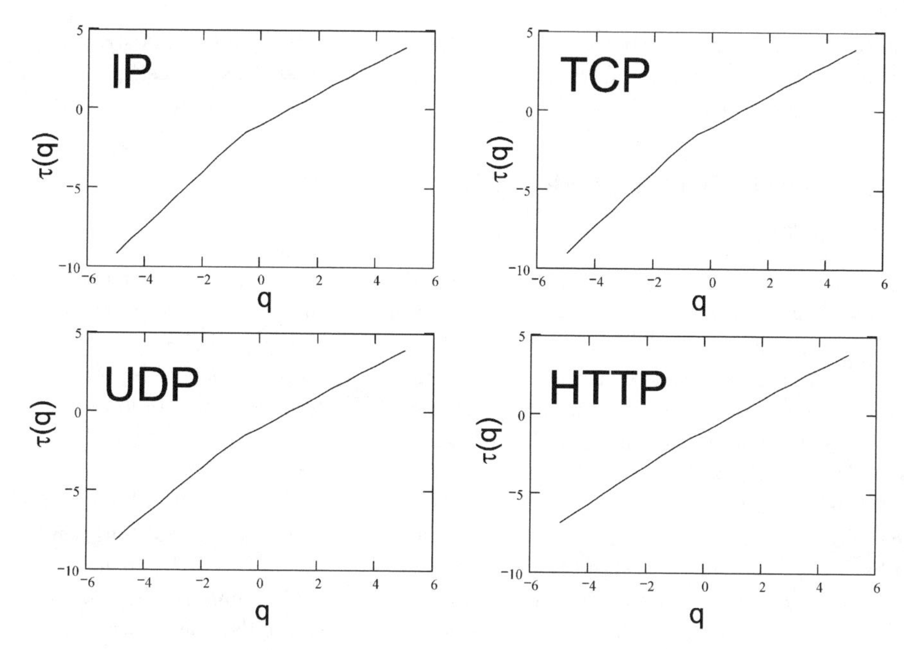

Figure 4.16 Function $\tau(q)$ for various traffic types

components with various scaling properties, which leads to the complicated multifractal structure of the aggregated WAN traffic, even on large time scales.

Wavelet analysis demonstrates that Internet traffic is monofractal on the large time scales (5–10 minutes or more), which is mainly caused by the slow decay of the file size distribution 'tail' transmitted through the Internet [7, 8, 31, 32]. However, the traffic behaviour on small time scales is much more complicated and is multifractal [33–36], which is first of all caused by protocol dynamics, such as TCP flows inspection, the network congestions, the packet losses and the packet retransmission. Taqqu *et al.* [37] explain the monofractal behaviour on large time scales by the totality of the large number of independent ON/OFF flows. The duration of ON and OFF intervals in their models has the slowly decaying tail distribution corresponding to the total one-file transmission time and to the user waiting (cogitation) time. They have proved that the traffic totality converges asymptotically to the well-known fractional Brownian motion [38, 39] when the flow number tends to infinity. They also found the simple relations between the distribution shape parameter with a slowly decaying tail and the Hurst exponent describing the self-similarity. However, the one-level ON/OFF model [7,13,40–44] is unable to explain the multifractal behaviour found on small time scales when the constant rate in ON intervals is assumed. On the other hand, Riedi offered the multifractal wavelet model to collect information on the second-order statistical behaviour on all time scales [45, 46]. As for other cascade models [21, 35, 45], these cascade models are not able to explain the observed behaviour by the simple mechanisms of network operation. Moreover, these cascade models usually require many parameters to match the statistical behaviour of real Internet traffic. Therefore, it is important to ensure a more accurate model which will be able to fix the traffic behaviour on all time scales, to explain more satisfactorly the relation between the observed and factual traffic statistical properties and to understand the mechanisms of simple network operation.

Internet traffic multifractal behaviour on small time scales mostly affects the mechanism of TCP flow control. The appropriate communication parameters, such as round-trip time (RTT), TCP session duration and pulse active time, play an important part in traffic behaviour.

4.5 Multilevel ON/OFF Model of Internet Traffic

4.5.1 Problem Statement

A large number of protocols has been developed and applied in Internet, but the traffic controlled by the TCP (the protocol of the transmission control) has dominated for decades. The TCP has the well-known control mechanism for reliable communication and congestion avoidance. To avoid the overload of a 'narrow' router, the hit size (the packet size) is limited by a certain size of the congestion control window. The TCP determines the window size in accordance with its current state and the occurring packet losses. The algorithm of the window size variations depends on TCP versions (e.g. Reno and Tahoe). In particular, the window size for TCP/Reno has the small initial value (1 MSS is the maximal segment size) and increases its size by one MSS after receiving confirmation from the recipient. This stage is referred to as 'slow start'. The TCP finishes the slow start stage and transfers to the overload avoidance stage if TCP counters (the packet losses or the window sizes) exceed the value of the slow start threshold parameter, which is called 'ssthresh'. The standard meaning for ssthresh is set for 64 kbyte and its value changes to the half-minimum of two values: the current overload window size or the recipient window size.

The TCP considers the packet losses as an indicator that the network is overloaded. At the congestion avoidance stage the TCP slowly increases its window size by one packet during each round-trip time and decreases its window size to half when a new packet loss is detected.

The typical web traffic transmitted through the Internet will be analysed. Since most of the objects in the usual web-page are small graphical or textual files, the appropriate TCP connection usually transmits the information during most of its life time at the slow start stage, and the packet arrival structure is similar in many respects to the ON/OFF process. The TCP transmits the series of packets in the ON period. The OFF period is roughly equal to the full round-trip time (RTT) in the network.

The multilevel ON/OFF model (Figure 4.17) imitating the operation character of a typical Internet connection is offered in Reference [47] to model the traffic on small time scales for one TCP connection. The higher level is ON/OFF process modelling of the life time of the TCP session (T_{11}) and the user waiting time (T_{10}). To cover TCP mechanism behaviour there is another ON/OFF process inside the ON period T_{11} for the ON/OFF process at a higher level.

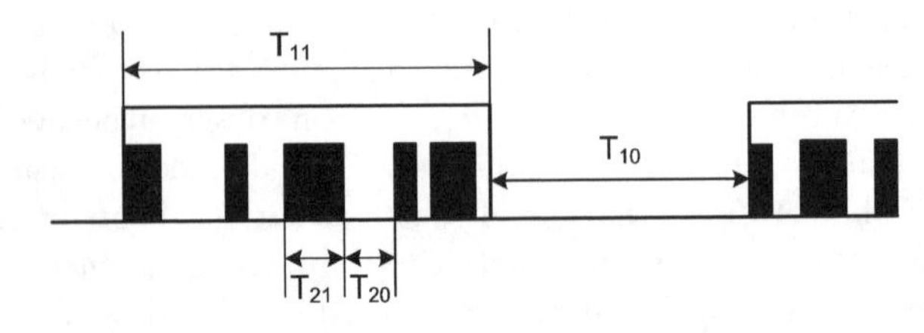

Figure 4.17 Traffic model for the single TCP session

This approach imitates the hit arrival structure by the activity time (the hit duration is T_{21}) and by the absence of the activity time (T_{20}) within the limits of the same TCP connection. The packet intensity B [bytes/Δ] in T_{21} is assumed to be permanent.

The variables T_{11} and T_{21} are described by Pareto distribution 'Pareto I' with parameters (K_{11}, a_{11}) and (K_{21}, a_{21}) respectively: $T_{11} \in$ Pareto (K_{11}, a_{11}) and $T_{21} \in$ Pareto (K_{21}, a_{21}). The complementary Pareto distribution function has the following form:

$$P[T > t] = \begin{cases} (K/\tau)^a, & \text{if } t \geq K \\ 1, & \text{if } t<K \end{cases}$$

The intervals of the user waiting time T_{10} and the absence activity time T_{20} are chosen as the exponentially distributed random variables $T_{10} \in$ constant $\exp(1/\lambda_{10})$ and $T_{20} \in$ constant $\exp(1/\lambda_{20})$ with the mean values $1/\lambda_{10}$ and $1/\lambda_{20}$ respectively. All these random variables are statistically independent of each other. The artificial traffic is generated by the aggregation of N independent multilevel ON/OFF processes with the transmission rate B in bytes/s.

4.5.2 Estimation of Parameters and Model Parameterization

To estimate the offered traffic model parameters the following information is necessary in order to describe the format of real Internet traces observed in the 'bottleneck' router:

- time mark: packet arrival time;
- packet size: the packet length;
- source address: IP address of the source host;
- destination address: IP address of the destination host;
- source port: the number of TCP ports of the source;
- destination port: the number of TCP ports of the destination;
- TCP flag: shows SYN, SYN-ACK and FIN packets.

Each TCP connection can be chosen using the source–recipient pair (the source IP address and the recipient IP address). The aggregated traffic X_i is formed with the help of the time mark and the respective packet size corresponding to the time interval i. The time interval between SYN and FYN packets of some TCP sessions is defined as the session life time. Similarly, the time between SYN and SYN-ACK packets is defined as the round-trip time value. Proceeding from these observations, the following statistical characteristics can be estimated:

- average traffic intensity (MX_i);
- autocorrelation function $R(k)$;
- logarithmic diagram L_j;
- average round-trip time (MT_{20});
- average session time (MT_{11}).

Having an real Internet trace, the appropriate model parameters can be estimated, the artificial traffic trace can be generated with its help and it can be shown that the artificial traffic possesses similar statistical properties and queueing character of the real traffic trace. In accordance with the multilevel model definition, the aggregated traffic can be written as the sum of the independent and identically distributed indicated functions. The total number of bytes in the multilevel ON/OFF process can be determined by the expression

$$U(t) = B \int_0^t \sum_{k=1}^{n} U_k(u) V_k(u)\, du \tag{4.6}$$

where $U_k(t)$ and $V_k(t)$ are the indicated functions defined as

$U_k(t) = 1$ {kth connection is at the ON state at moment t}
$V_k(t) = 1$ {kth connection is in the 'active' state at moment t}

The separate ON/OFF process can be considered as the specific case of the multilevel ON/OFF process with the set $B = 1$ and $V_k(t) = 1$ for all t. Consequently, the total number or bytes in the one-level ON/OFF process on the $[0, Tt)$ interval is determined as

$$\hat{Y}(Tt) = \int_0^{Tt} \sum_{k=1}^{N} U_k(u)\, du \tag{4.7}$$

As shown in Theorem 2.1 (see Chapter 2), for large N and T the aggregated cumulative process $\{\hat{Y}(Tt), t \geq 0\}$ behaves statistically (on account of the above designations) as

$$TN \frac{MT_{11}}{MT_{11} + MT_{10}} t + T^H \sqrt{N}\, \sigma_{\lim} B_H(t) \tag{4.8}$$

where $H = (3 - a_{11})/2$ is the Hurst exponent and $B_H(t)$ is the standard fractional Brownian motion. The expression (4.8) illustrates the relation between the Hurst exponent H (the self-similarity parameter) and the shape parameter a_{11} (the Pareto distribution parameter). It also shows that Internet traffic self-similarity is mainly caused by the heavy tail of the file size distribution, which is usually transmitted through the Internet network.

Since the ON/OFF process of the lower level exists only during the ON period of the higher level process, under the condition $\max(MT_{21}, MT_{20}) \ll ET_{11}$ the following relation can be obtained between $Y(t)$ and $\hat{Y}(t)$:

$$\lim_{t \to \infty} \frac{Y(t)}{\hat{Y}(t)} = M[V]B \tag{4.9}$$

As $t \to \infty$ the aggregated cumulative traffic of the multilevel ON/OFF process $Y(t)$ behaves statistically as FBM.

To coordinate the second-order statistical properties of the model and the real traffic, the model parameters will be estimated on the basis of the real route. The increment X_i process for the analysed traffic will be defined instead of using the cumulative $Y(t)$ process as

$$X_i = Y[(i+1)\Delta] - Y(i\Delta), \quad i = 0, 1, 2, 3, \ldots \tag{4.10}$$

where Δ is the minimal time resolution of interest. The increment X_i process is considered for the total byte load arriving at the $[i\Delta, (i+1)\Delta)$ interval. In the case when $t \to \infty$, the aggregated integral traffic of the multilevel ON/OFF process $Y(t)$ is similar statistically to fractional Brownian motion. The logarithmic diagram of the X_i process over large time intervals has the slope $\alpha = 2H - 1 = 2 - a_{11}$.

The asymptotically objective and effective method of estimating the logarithmic diagram slope within some region [47] will be applied to calculate the model parameter a_{11} by way of the α slope estimation over large time intervals. Analogous reasoning can be applied when considering the traffic X_i over small time intervals. If it is assumed that $T_{11} \gg T_{21}$ and T_{20}, then the lower level $V(t)$ of the ON/OFF process dominates over the small time intervals.

As the higher level $U_k(t)$ of the ON/OFF process behaves as a constant value over the small time intervals, then

$$\begin{aligned} \lim_{t\to 0} Y(t) &= \lim_{t\to 0} B \int_0^t \sum_{k=1}^{N} U_k(u)V_k(u)\,du \\ &\approx B \sum_{k=1}^{N} U_k(0) \lim_{t\to 0} \int_0^t V_k(u)\,du \end{aligned} \tag{4.11}$$

where the increment of the $Y(t)$ process can be determined as

$$\begin{aligned} X_i &= Y[(i+1)\Delta] - Y(i\Delta) \\ &= B \int_{i\Delta}^{(i+1)\Delta} \sum_{k=1}^{N} U_k(u)V_k(u)\,du, \quad i = 0, 1, 2, 3, \ldots \end{aligned} \tag{4.12}$$

The ON/OFF process will be the process of another (lower) level over the small time intervals. Due to the multilevel ON/OFF model structure, another linear region can be observed in the logarithmic diagram in the small time interval area. It can be explained by ON/OFF processes of the lower level. The method of the logarithmic diagram slope estimation can also be used within a certain region to estimate the model parameter a_{21} in the small time interval area.

Parameter K_{11} can be estimated from the first moment of the session time duration:

$$\hat{K}_{11} = \frac{a_{11} - 1}{a_{11}} MT_{11} \tag{4.13}$$

In contrast to estimation of K_{11} using the method of the average session duration T_{11} fitting, real traffic records have no controlling packet indicating the start and the end of each active period T_{21}. Therefore, it can be assumed that at the process observation over the small time intervals, the ON/OFF process of a higher level always keeps its state and parameter K_{21} can be estimated using the autocorrelation function $R_n(t)$ of X_i. It was found that parameter K_{21} can be estimated with the help of the following procedure:

$$\hat{k}_{21} = \Delta(a_{21}\widehat{R}_n(\Delta))^{1/(a_{21}-1)} \quad \text{as } \Delta \to 0$$

where

$$\hat{R}_n(\Delta) := \frac{\sum_i X_i X_{i+1}}{\sum_i X_i^2}$$

To estimate parameter $1/\lambda_{20}$ or (which is equivalent) the average period of the activity absence, the round-trip time in the network out of the experimental trace is to be measured. It can be found from the real trace on the basis of the time interval between SYN and SYN-ACK packets following the beginning of each TCP session. In the considered model the round-trip time is equal to the OFF period T_{20} of the lower level model:

$$\frac{1}{\lambda_{20}} = MT_{20} \tag{4.14}$$

Assuming that the connections are independent,

$$R_1 = \frac{MT_{11}}{MT_{11} + MT_{10}} \quad \text{and} \quad R_2 = \frac{MT_{21}}{MT_{21} + M_{20}}$$

As a result,

$$MX_i = N\Delta BR_1R_2 \tag{4.15}$$

$$MX_i^2 = N(\Delta B)^2 R_1 R_2 \tag{4.16}$$

Thus, parameter B, which estimates the constant rate of data transmission during the activity period T_{21}, can be expressed by the formula $B = MX_i^2/(\Delta MX_i)$, and its estimate has the form $\hat{B} = (\sum_i X_i^2)/(\Delta \sum_i X_i)$.

Expression (4.15) implies that there is one freedom degree to choose N and R_1 (or similarly MT_{10}). To satisfy the Theorem 2.1 assumption (see Chapter 2) the large integer for N should be chosen so that the average OFF states time $1/\lambda_{10}$ an be determined with the help of R_1 in (4.15). Since T_{10} is the exponential random variable and $ET_{10} \gg ET_{11}$, the starting time of each TCP session can also be approximated by Poisson process as $N \to \infty$.

The described model can be a useful tool for estimating the network parameters that influence the TCP connection operation.

4.5.3 Parallel Buffer Structure for Active Queue Control

For a small queuing delay the buffer size in the router is rather small. However, due to the essential rippling of Internet traffic, a router with a small buffer size usually has a high loss coefficient. In the case of packet loss, the TCP considerably decreases the flow rate during the congestion avoidance stage. Therefore, after the case of buffer overflow in a drop-tail queue, all connections recognize the fact of packet loss and together decrease the transmission rate. To avoid the presence of this global synchronization and to increase use of the connection, many schemes of active queuing were offered, such as random early detection (RED) [46].

The main idea of RED is to keep the queue length within the given region by means of the omission of random packets among various connections before the buffer is overflowed. The loss probability is an increasing function with respect to the queue length. The connection with a higher rate has a higher risk of packet loss and of transmission rate decrease. Since the queue length is controlled and saved within the desired region, the carrying network capacity is fully used and the packets ensure a smaller average delay and smaller delay variations.

At the same time, RED operation is very sensitive to established parameters. A RED adaptive scheme dynamically renews the maximal losses probability in accordance with the value of the exponentially weighted moving average (EWMA) queue length, and ensures protection against the congestion level.

The strategy of the adaptive RED provides a good rate control for TCP connections operating at the congestion avoidance stage [48]. However, the larger part of Internet traffic is Web traffic and UDP traffic. Since most Web connections lead to several small file transmissions, these connections have a short life time and operate mainly at the TCP slow start stage with a small congestion window.

The truncation of Web packets at this stage is not an effective way to provide traffic rate control and facilitation of router congestion. Moreover, from the Web user's point of view, the loss of single or several packets at the slow start stage would lead to additional delay for recurring transmission or even to TCP time-out. This would also make the TCP enter the congestion avoidance stage with the small congestion window ahead of time and would result in a low carrying capacity. The delay and the low carrying capacity would lead to strong deceleration in the short message transmission, such as web-pages, and Web browsers will experience long waiting times, even at the high connection rate. On the other hand, the adaptive RED will not be able to keep the queue length within the desired region due to the bursty character of Web traffic.

In other words, any scheme with random truncation/marking, such as RED, cannot effectively control the congestion without taking into account the influence of short-life TCP (and UDP) traffic. Moreover, the loss of a single or several packets at the slow start stage leads not only to a very low carrying capacity and additional delay but also to the high probability of communication interruption.

To solve these problems, in Reference [24] the virtual buffer structure is offered for the queue active control. In this structure the real time traffic (Web and UDP) and the nonreal time traffic (FTP) are divided into two different virtual buffers, which distribute the same physical buffer memory.

The first virtual buffer operates with short-term bursty traffic in real time (Web, and UDP). Since the omission of these packets cannot facilitate the congestion level and essentially increases the transmission delay, it would be good to save them in the buffer until the whole buffer (used mutually with another buffer) is overflowed. The drop-tail policy is used in the first virtual buffer for servicing real time applications. To obtain a small average delay, the service rate of this drop-tail buffer is defined dynamically by the queue length of the virtual buffer. In order to decrease the transmission delay for a Web reviewer and UDP connections, the service rate $C_1(t)$ is changed dynamically depending on the queue length $q_1(t)$ of the virtual buffer.

The second virtual buffer maintains long-term TCP communication sessions, like the FTP, with large file sizes when the adaptive RED applies. Although the available service rate of this queue is determined as $C_2(t) = C - C_1(t)$, it is expected that the adaptive RED scheme will keep the queue length $q_2(t)$ of the virtual buffer in the desired region due to the following reason. When there is a strong load in the drop-tail buffer, $C_2(t)$ rapidly decreases. FTP receivers feel the small packet arrival rate and send back the acknowledgements (ACK) more slowly. Without the loss probability increase in the adaptive RED buffer, the slow rates of ACK arrivals make the FTP sender decrease the transmission rate automatically without a decrease in the congestion window. On the other hand, when the congestion level is reduced, the adaptive RED buffers obtain a larger carrying capacity. Since the congestion window sizes for FTP servers are still large, the FTP carrying capacity is rapidly restored because of the higher rates of ACK packet arrivals from the receivers.

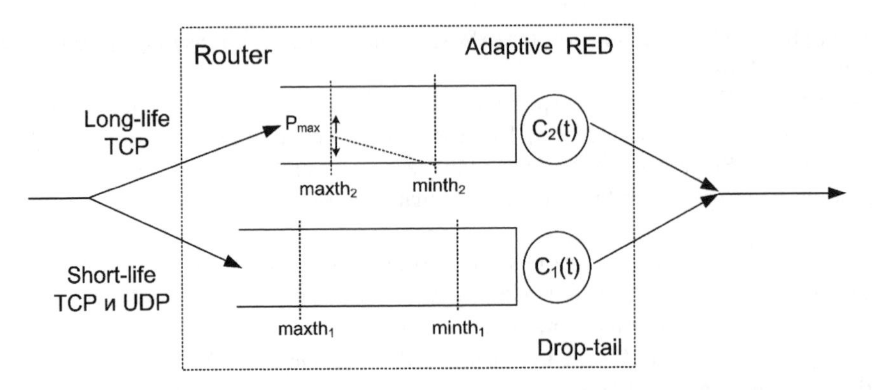

Figure 4.18 The parallel virtual buffer structure for queue active control

This parallel virtual buffer structure (which is described in Reference [24] as the RED+TAIL policy) enables the advantages of the adaptive RED, such as high (100 %) communication usage, to be used. Moreover, the loss rate for the short-life packets of TCP and UDP connections is essentially reduced by the drop-tail policy and the shared buffer. The packet loss rate of the long-term TCP traffic is also reduced due to a decrease in the carrying capacity, big thresholds (longer RTT) and a more stable average length of the virtual buffer for the adaptive RED buffer.

Figure 4.18 illustrates the parallel structure of the RED+TAIL buffer in the router. The variables $C_1(t)$ and $C_2(t)$ designate in t time the service rates in the drop-tail buffer and the adaptive RED buffer respectively. The maximal $\mathrm{maxth_i}$ and minimal $\mathrm{minth_i}$ thresholds $(i = 1, 2)$ will be defined for the dynamic distribution of the carrying capacity of both buffers and the desired region estimation for the queue length of the adaptive RED buffer. The service rates $C_1(t)$ and $C_2(t)$ are given by the following algorithm:

If $q_1 = 0$, then $C_1(t) := 0$
If $0 < q_1 < \mathrm{minth}_1$, then $C_1(t) := C_{1\,\mathrm{min}}$
If $\mathrm{min}_1 \leq q_1$, then $C_1(t) := \min\left(C\frac{q_1}{\max \mathrm{th}_1}, C_{1\,\mathrm{max}}\right)$
$C_2(t) := C - C_1(t)$

where C is the carrying capacity of the connection. The variable q_1 designates the queue length of the drop-tail buffer. The constant $C_{1\mathrm{max}}$ represents the minimal available service rate $C - C_{1\,\mathrm{max}}$ for the RED buffer to avoid the time-out for FTP connections. The simple identification algorithm is used to divide the short-life and long-life TCP connections in different virtual buffers.

The main advantage of the this technology is that it keeps the average packet loss rate within the range of very small values, so that the average size of the TCP window at the sender has a high profile at the congestion avoidance stage. Therefore, any congestions in the router cannot be the cause of other congestions on the side of the TCP sender.

The packet loss coefficient and the average delay can be essentially decreased by the carrying capacity dynamic distribution and active control of the buffer with the parallel structure. This scheme combines the advantages of drop-tail and the adaptive RED policy. The modelling results illustrate that this scheme achieves a less average delay for real time applications and keeps the high carrying capacity for best-effort connections; it also essentially decreases the packet loss coefficient in each of the buffers.

The structure with parallel buffers also ensures the large freedom for router control since it applies various policies of the carrying capacity distribution and the dynamic thresholds for the adaptive RED. In the considered case the carrying capacity distribution policy is a simple function of the virtual buffer current length. The parallel scheme of queue control unequally processes the bursty and nonbursty traffic (in real time) to increase the router carrying capacity with strong loading.

The present forecasting approach can help the network manager to determine which carrying capacities should be distributed among all buffers so that the requirements for the quality of service are respectively met under the conditions of two-level TCP traffic.

References

[1] W.E. Leland, M.S. Taqqu, W. Willinger and D.V. Wilson, 'On the self-similar nature of Ethernet traffic', *Computer Communications Review*, 1993 **23**, 183–193. Proceedings of the ACM/SIGCOMM'93, San Francisco, September 1993. Reprinted in '*Trends in Networking – Internet*', the Conference Book of the Spring 1995 Conference of the National Unix User Group of the Netherlands (NLUUG). Also reprinted in *Computer Communication Review*, **25** (1), 1995, 202–212, a special anniversary issue devoted to '*Highlights from 25 years of the Computer Communications Review*'.

[2] W.E. Leland, M.S. Taqqu, W. Willinger and D.V. Wilson, 'On the self-similar nature of Ethernet traffic' (extended version), *IEEE/ACM Transactions on Networking*, **2**, 1994, 1–15.

[3] V. Paxon and S. Floyd, 'Wide-area traffic: the failure of Poisson modelling', in Proceedings of the ACM Sigcomm '94, London, 1994, pp. 257–268.

[4] V. Paxson and S. Floyd, 'Wide-area traffic: the failure of Poisson modelling', *IEEE/ACM Transactions on Networking*, **3**, 1995, 226–244.

[5] R. Addie, M. Zukerman and T. Neame, 'Fractal traffic: measurements, modelling and performance evaluation', in Proceedings of IEEE INFOCOM '95, Boston, MA, 1995. pp. 977–984.

[6] I. Norros, 'A storage model with self-similar input', *Queueing Systems*, **16**, 1994, 387–396.

[7] K. Park, G. Kirn and M. Crovella, 'On the relationship between file sizes, transport protocols, and self-similar network traffic', in Proceedings of IEEE International Conference on '*Network Protocols*', Hyatt Regency Hotel, Columbus, Ohio, 1996, pp. 171–180.

[8] M. Crovella and A. Bestavros, 'Self-similarity in world wide web traffic: evidence and possible causes', in Proceedings of the ACM SIGMETRICS '96, Philadelphia, PA, 1996, pp. 160–169.

[9] Kihong Park, Gitae Kim and Mark E. Crovella, 'The protocol stack and its modulating effect on self-similar traffic', in '*Self-Similar Network Traffic Analysis and Performance Evaluation*' (eds K. Park and W. Willinger), Wiley–Interscience, New York, 1999.

[10] Jin Cao, William S. Cleveland, Dong Lin, Don X. Sun, 'On the nonstationarity of Internet traffic', *ACM SIGMETRICS*, **29**, (1), 2001, 102–112.

[11] O.I. Sheluhin, A.M. Tenyakshev and A.V. Osin, '*Fractal Processes in Telecommunications*' (in Russian), Radiotekhnika, Moscow, 2003.

[12] H.J. Fowler and W.E. Leland, 'Local area network traffic characteristics with implications for broadband network congestion management', *IEEE Journal on Selected Areas in Communications*, **9**, 1991, 1139–1149.

[13] W. Willinger, M.S. Taqqu, R. Sherraan and D.V. Wilson, 'Self-similarity through high-variability: statistical analysis of Ethernet LAN traffic at the source level' (extended version), *IEEE/ACM Transactions on Networking*, **5**, 1997, 71–86.

[14] W. Willinger, V. Paxson and M.S. Taqqu, 'Self-similarity and heavy tails: structural modeling of network traffic', in '*A Practical Guide to Heavy Tails: Statistical Techniques and Applications* (eds J. Adler, R. E. Feldman and M. S. Taqqu), Birkhauser, Boston, MA, 1998.

[15] A. Feldman, 'Characteristics of TCP connection arrivals', in '*Self-Similar Network Traffic Analysis and Performance Evaluation*' (eds K. Park and W. Willinger), Wiley–Interscience, New York, 1999.

[16] M. Parulekar and A. Makovski, '*M/G/1* input processes: a versatile class of models for network traffic', in Proceedings of IEEE Infocom 97, Kobe, Japan, April 1997, pp. 419–426.
[17] D.R. Cox, 'Long-range dependence: a review', in '*Statistics: An Appraisal*' (eds H.A. David and H.T. David), Iowa State University Press, 1984, pp. 55–74.
[18] A. Feldmann, A.C. Gilbert and W. Willinger, 'Data networks as cascades: investigating the multifractal nature of internet WAN traffic', *ACM Computer Communication Review*, **28**, September 1998, 42–55.
[19] A. Feldmann, A.C. Gilbert, W. Willinger and T.G. Kurtz, 'The changing nature of network traffic: scaling phenomena', *ACM Computer Communication Review*, **28**, April 1998, 5–29.
[20] A.C. Gilbert, W. Willinger and A. Feldmann, 'Scaling analysis of conservative cascades, with applications to network traffic', *IEEE Transactions on Information Theory*, **45**(3), April 1999, 971–991.
[21] R.H. Riedi and W. Willinger, 'Toward an improved understanding of network traffic dynamics', in *Self-Similar Network Traffic Analysis and Performance Evaluation* (eds K. Park and W. Willinger), Wiley–Interscience, New York, 1999.
[22] R. Holley and E.C. Waymire, 'Multifractal dimensions and scaling exponents for strongly bounded random cascades', *Annals of Applied Probability*, **2**, 1992, 819–845.
[23] A. Feldmann, A.C. Gilbert and W. Willinger, 'Data networks as cascades: investigating the multifractal nature of Internet WAN traffic', in Proceedings of the 1998 ACM SIGCOMM, Vancouver, Canada, 1998, pp.42–55.
[24] Jia-Shiang Jou, 'Multifractal Internet traffic model and active queue management', PhD Dissertation, Faculty of the Graduate School of the University of Maryland, 2003.
[25] K. Thompson, G.J. Miller and R. Wilder, 'Wide-area Internet traffic patterns and characteristics', *IEEE Network*, November/December 1997.
[26] T. Elteto and S. Molnar, 'On the distribution of round-trip delays in TCP/IP networks', in the 24th Annual Conference on '*Local Computer Networks*' (LCN'99), Lowell, Boston, MA, October 1999.
[27] W.S. Cleveland, D. Lin and D.X.Sun, 'IP packet generation: statistical models for TCP start times based on connection-rate superposition', in proceedings of ACM SIGMETRICS, Santa Clara, CA, 2000, pp. 166–177.
[28] Jin Cao, William S. Cleveland, Dong Lin and Don X. Sun, 'Internet traffic tends toward Poisson and independent as the load increases', in '*Nonlinear Estimation and Classification*', (eds C. Holmes, D. Denison, M. Hansen, B. Yu and B. Mallick), Springer, New York, 2002, pp. 83–109.
[29] R. Riedi, 'An improved multifractal formalism and self-similar measures,' *Math. Anal. Appl.*, **189**, 1995, 462–490.
[30] V.J. Ribeiro, R.H. Riedi, M.S. Crouse and R.G. Baraniuk, 'Simulation of non-Gaussian long-range-dependent traffic using wavelets', in Proceedings of the ACM SIGMETRICS Conference, atlanta, GA, 1999, pp. 1–12.
[31] M. Grossglauser and J. Bolot, 'On the relevance of long-range dependence in network traffic', *IEEA/ACM Transactions on Networking*, **7**(5), October 1999, 629–640.
[32] M.M. Krunz and A.M. Makowski, 'Modeling video traffic using *M/G/*∞ input processes: a compromise between Markovian and LRD models', *IEEE Journal on Selected Areas in Communications*, **16**(5), 1998, 733–748.
[33] R. Riedi and J.L. V'ehel, 'Multifractal properties of TCP traffic: a numerical study', Technical Report 3129, INRIA, February 1997.
[34] A. Feldmann, A.C. Gilbert, P. Huang and W. Willinger, 'Dynamics of IP traffic: a study of the role of variability and the impact of control', in Proceedings of SIGCOMM, Cambridge, MA, 1999, pp. 301–313.
[35] D. Veitch, J. Backar, J. Wall, J. Yates and M. Roughan, 'On-line generation of fractal and multi-fractal traffic', in PAM2000 Workshop on '*Passive and Active Networking*', Hamilton, New Zealand, 2000.
[36] J. Gao and I. Rubin, 'Multiplicative multifractal modeling of long-range dependent traffic', in Proceedings of ICC, 1999.
[37] M. Taqqu, W. Willinger and R. Sherman, 'Proof of a fundamental result in self-similar traffic modeling', *Computer Communication Review*, **27**(2), 1997, 5–23.
[38] I. Norros, 'On the use of fractional Brownian motion in the theory of connectionless networks', *IEEE Journal of Selected Areas in Communications*, **13**(6), 1995, 953–962.
[39] F. Kelly, 'Notes on effective bandwidths', in '*Stochastic Networks: Theory and Applications*', Oxford University Press, 1996, pp. 141–168.
[40] P. Barford and M. Crovella, 'Generating representative web workloads for network and server performance evaluation', in Proceedings of the ACM SIGMETRICS Conference, Madison, WI, July 1998, pp. 151–160.
[41] K. Park, 'On the effect and control of self-similar network traffic: a simulation perspective', in Winter Simulation Conference, atlanta, GA, 1997, pp. 989–996.
[42] A.M. Makowski, 'Bounding on–off sources – variability ordering and majorization to the rescue', Technical Report ISR TR2001-13, ISR University of Maryland, 2001.

[43] S. Resnick, 'Heavy tail modeling and teletraffic data', *Annals of Statistics*, **25**, 1997, 1805–1869.

[44] D. Heath, S. Resnick and G. Samorodnitsky, 'Patterns of buffer overflow in a class of queues with long memory in the input stream', *Annals of Applied Probability*, **7**(4), November 1997, 1021–1057.

[45] K. Daoudi, A.B. Frakt and A.S. Willsky, 'Multiscale autoregressive models and wavelets', *IEEE Transactions on Information Theory*, **45**(3), April 1999, 828–845.

[46] S. Floyd and V. Jacobson, 'Random early detection gateways for congestion avoidance', *IEEE/ACM Transactions on Networking*, **1**(4), 1993, 397–493.

[47] D. Veitch and P. Abry, 'A wavelet based joint estimator for the parameters of long-range dependence', *IEEE Transactions on Information Theory*, **45**(3), 1999, 878–897.

[48] S. Floyd, R. Gummadi and S. Shenker, 'Adaptive RED: an algorithm for increasing the robustness of RED', available at http://www.icir.org/floyd/papers/adaptiveRed.pdf, August 2001.

5

Queuing and Performance Evaluation of Telecommunication Networks under Traffic Self-Similarity Conditions

5.1 Traffic Fractality Influence Estimate on Telecommunication Network Queuing

The results of many investigations show that measurements of the characteristics of queuing systems with fractal traffic may essentially differ from those predicted by the corresponding systems with traditional traffic models. In this context, the tail behaviour of Q in the queue length distribution $P\{Q > B\}$ in the stable state will be investigated for one server with the queue infinite capacity. For Markovian traffic processed in such a queue, the tail distributions are approximately exponential [1,2], i.e.

$$P\{Q > B\} \sim e^{-\eta B}, \quad \text{as } B \to \infty \tag{5.1}$$

where $\eta > 0$ is the asymptotic decay degree.

The function (5.1) is assumed to be the basis for the effective capacity concept, where the access control or the distributed capacity of the service channel is based of the probability distribution of the random variable tail choice. In contrast to (5.1), the traffic flows with long-range dependence (in particular, the models based on the fractional Brownian motion) lead to the Weibull type asymptotic of tails probability distribution, i.e.

$$P\{Q > B\} \sim \mathrm{e}^{-\gamma B^{\beta}}, \quad \text{as } B \to \infty \tag{5.2}$$

where γ is a constant and $\beta = 2 - 2H \in (0, 1]$[3,4].

Formulas (5.1) and (5.2) are essentially different. The first allows more optimistic forecasts compared to the second. The question of whether other traffic models lead to correct forecasts of the network productivity compared to the experimental data is still kept open.

Self-Similar Processes in Telecommunications O. I. Sheluhin, S. M. Smolskiy and A.V. Osin
© 2007 John Wiley & Sons, Ltd

At present there are no general analytic results for queuing, the influence of self-similarity and LRD on the quality of service (QoS). Only certain specific analytical results are known. Several best-known specific cases are considered below. At the same time, the imitation modelling methods are evidently the most effective ones for an estimate of the operation efficiency of telecommunication networks. It is exactly from this position that the problems of the efficiency of the traffic self-similarity degree of influence on telecommunication systems are considered.

5.1.1 Monofractal Traffic

When designing any telecommunication network there will be the possibility of channel capacity restrictions. Under these conditions, the effective capacity estimate becomes one of the key problems. The calculations on the basis of the classical approaches of queue theory oriented at the noncorrelated request flow give overoptimistic results under conditions of self-similar traffic. The queuing analysis for fractal input traffic becomes problematic within the limits of classical queuing theory after detecting the fractal structure in the network traffic. To date some important results have been published [5–9].

Studying the influence of fractality on queuing is a very important problem. Some network designing applications, such as the buffer size assignment and traffic control, are related to this problem, which makes it an important issue.

5.1.1.1 Queuing Model with the Traffic Described by Fractional Brownian Motion

A simple queuing model will be considered, which is the separate server queue. It can be considered in continuous time and the service discipline is given as FIFO. It is assumed that the queue has an infinite buffer and constant service intensity r. Let the total traffic arriving in the queue from time moment $-t$ in the past to the present time moment $t = 0$ be designated as $A(t)$. The so-called traffic process $Q(t)$ is the total traffic, saved in the buffer over the interval $(-t; 0)$. The current buffer length of the queue is defined as $Q(t, r)$, which is the queue length in the equilibrium state, when the system functions during a long time interval and the initial queue length does not have any influence. If this system state exists, i.e. the assumptions of the traffic process stationarity and ergodicity is valid, and the system stable state can be achieved, then

$$Q(t, r) = \sup_{0 \leq s \leq t} [A(t) - A(s) - r(t - s)] \tag{5.3}$$

Here $A(t) - A(s)$ is the traffic value arriving for processing during the time interval $[s, t]$ and $r(t - s)$ is the traffic value which was processed during the same time interval.

The input arrival process $A(t)$ is considered as the fractal process of the form

$$A(t) = \lambda t + \sqrt{a\lambda}\, Z(t), \quad t \in (-\infty; \infty) \tag{5.4}$$

where $Z(t)$ is the normalized fractional Brownian motion. Here $\lambda > 0$ is the averaged input intensity, $a > 0$ is the variation coefficient, $H \in [\frac{1}{2}; 1)$ is the Hurst exponent for the $Z(t)$ process and $r > \lambda$ is the service intensity.

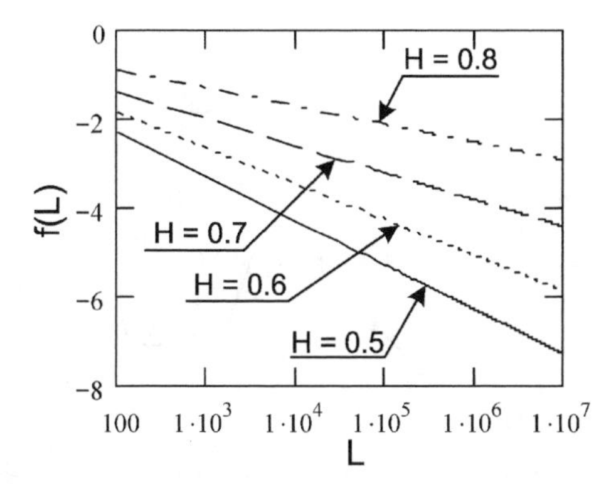

Figure 5.1 The queue tail approximation versus the queue length L for $r = 1$ and fixed H

It can be seen that the equation system of (5.3) and (5.4) is fully determined by four parameters: λ, a, H and r. Z_t process self-similarity allows a more accurate relation between the network parameters to be obtained from Equation (5.4): the buffer size L, the channel capacity C and the traffic parameters r, a and H for the boundary values.

The analysis of a single queue construction with FBM at the input was presented for the first time in Reference [8], where it was shown that the queue length distribution can be approximated by Weibull distribution. In Reference [8] it was particularly found that the queue distribution tail in the FBM case at the input satisfies the equality

$$\log(P[Q > L]) \approx -\tfrac{1}{2}L^{2(1-H)}r^{2H}(1-H)^{-2(1-H)}H^{-2H} \tag{5.5}$$

for large L. Figures 5.1 and 5.2 show the functions of the queue tail approximation versus the queue size in log–log scale for fixed H and r:

$$\log(P[Q > L]) = f(L) = -\log\left[\tfrac{1}{2}L^{2(1-H)}\,r^{2H}(1-H)^{-2(1-H)}H^{-2H}\right]$$

The observed plot linearity illustrates the probability decay in accordance with the Weibull law.

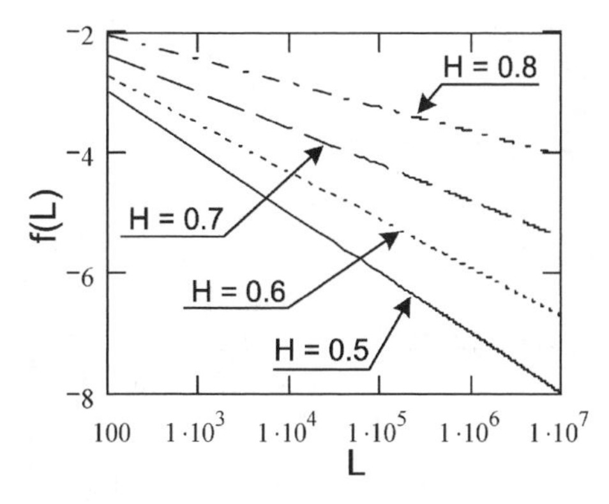

Figure 5.2 The queue tail approximation versus the queue length L for $r = 5$ and fixed H

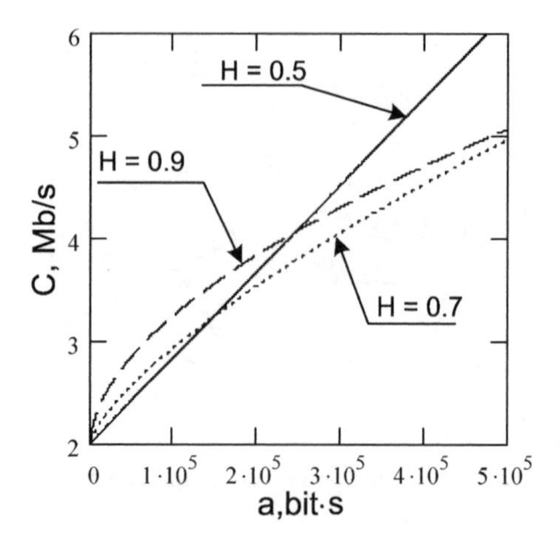

Figure 5.3 The channel capacity as a function of a for $r = 2\,\text{Mb/s}$ and $H = 0.5$, 0.7 and 0.9 for $L = 100\,\text{kb}$

Assuming the probability $P(Q > L) = \varepsilon$ and $\rho = r/C$, it is possible to solve (5.5) with respect to C and to find that the QoS is achieved approximately when

$$C = r + \{k(H)\sqrt{-2\ln\varepsilon}\}^{1/H} a^{1/(2H)} L^{-(1-H)/H} r^{1/(2H)} \tag{5.6}$$

where $k(H) = H^H(1-H)^{1-H}$. For a practical application of (5.6) as the formula giving the channel size, it is of interest to examine its sensitivity to a and H. Figures 5.3 and 5.4 show the channel characteristics with different values of a and H for $r = 2\,\text{Mb/s}$, $\varepsilon = 10^{-3}$ and for two buffer sizes $L = 100\,\text{kb}$ and $1\,\text{Mb}$. Of course, similar reservations were shown in the previous figure and should be made for strict a and H variation independence. In any case, it can be seen that when the buffer is small, the requirements to the channel are less dependent

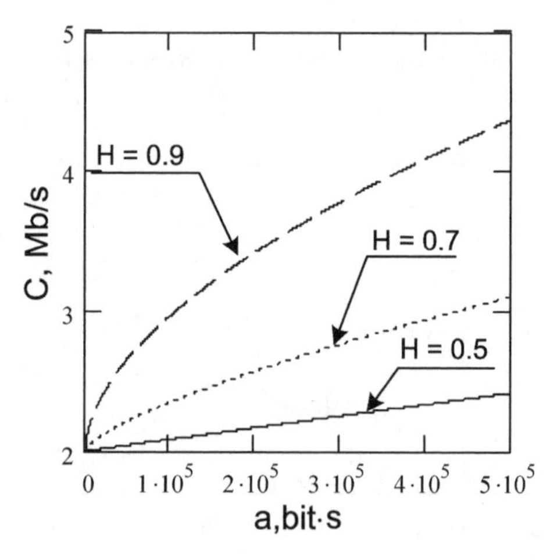

Figure 5.4 The channel capacity as a function of a for $r = 2\,\text{Mb/s}$ and $H = 0.5$, 0.7 and 0.9 for $L = 1\,\text{Mb}$

on H than when the buffer is large. The observed result illustrates the well-known fact that it is very difficult to fill a large buffer by short-range dependent traffic.

The obtained results show that the queue distribution with FBM at the input has much less decay than in the exponential case. However, this approach is based on the Gaussian property of the input process and cannot be extended to other processes with scaling properties. There are only a few analytical results for queuing in the cases where the traffic has a more complex scaling behaviour. For example, the result is known when the input traffic is asymptotically self-similar and is described by Pareto distribution and when the Levy distribution is used for a traffic description. Both cases will be considered in more detail.

5.1.2 Communication System Model and the Packet Loss Probability Estimate for the Asymptotic Self-Similar Traffic Described by Pareto Distribution

Consider the communication system in which the input packet flow $Y = (\ldots, Y_{-1}, Y_0, Y_1, \ldots.)$ is the superposition of the packets generated by different sources:

$$Y_t = \sum_{s \in \mathbb{Z}} \theta_s(t - \omega_s + 1), \quad t \in \mathbb{Z} \tag{5.7}$$

Here Y_t is the total number of packets generated by the sources active at moment t. The numbers of the generated sources are designated as $s \in \mathbb{Z} \stackrel{\Delta}{=} \{\ldots, \mathbf{-1}, \mathbf{0}, \mathbf{1}, \ldots\}$. The source s begins the generation of its packets at the moment $\omega_s (\omega_s \leq \omega_{s+1})$ and generates $\theta_s(i) \in \mathbb{Z}^+$ packets at the moment $\omega_s + i - 1$ in the interval $\omega_s, \ldots, \omega_s + \tau_s - 1, i \in \{1, \ldots, \tau_s\}$. The sequence $\{\theta_s(1), \ldots, \theta_s(\tau_s)\}$ is referred to as the active period of source s, where τ_s is its length. Before moment ω_s and after moment $\omega_s + \tau_s - 1$, source s does not generate the packets; therefore $\theta_s(i) = 0$ for $i \leq 0$ and $i \leq \tau_s + 1$. Thus, $\theta_s(t - \omega_s + 1), t \in \mathbb{Z}$, is the sequence of the packet numbers generated by source s at successive time moments.

For example, the specific cases of the active periods may be:

1. The constant $\theta_s(i) = R \in N, 1 \leq i \leq \tau_s$, where R is the source rate (this traffic is considered below).
2. The random constant $\theta_s(i) = R$, where $R = R(\tau_s)$.
3. The independent and identically distributed variables $\theta_s(i)$, possessing the values 0 and 1 with probabilities p_0 and p_1, respectively.
4. The independent and identically distributed variables $\theta_s(i)$, possessing the values from the set $\{0, 1, \ldots, k\}$ with the binomial distribution or from $\mathbb{Z}_+$ with the geometrical, Poisson or any other given distribution.
5. Markovian, semi-Markovian, stationary or other well-known sequences $\theta_s(i)$, etc.

The moment t may be the moment ω_s for several sources simultaneously. Let ξ_t be designated as the source number for which $\omega_s = t$. The necessary and sufficient conditions for Y, the exactly second-order asymptotic self-similar (H-sssa) (see Definition 1.5), are presented in the following theorem using $\mu^{(l)} = M\theta(t)$ and $B^{(l)}(k) = M\theta(t)\theta(t + k), k \in Z$, where $(\ldots,\theta(-1),\theta(0)\ \theta(1),\ldots)$ is the random stationary process depending on the lth distribution where the length l coincides with the conditional distribution of the source active period at $\tau = l$ (here and later, τ is the random variable having similar distribution as τ_s).

Theorem 5.1[10]
The process Y is the H-sssa with $0 < \beta < 1$ if, when $P\{\tau = l\}, \mu(l)$ and $B^{(l)}(k)$ lead to $P\{\tau = l\}B^{(l)}(l) \sim L(l)l^{-(\beta+2)}, l \to \infty, \sum_{l=1}^{\infty} l\, P\{\tau = l\}B^{(l)}(0) < \infty$ and $\sum_{l=1}^{\infty} l\, P\{\tau = l\}\mu^{(l)} < \infty$, where $L(x)$ is any function slowly varying to infinity.

In particular, in accordance with this theorem, the Y process becomes asymptotically self-similar in the wide sense with $H = (3 - \alpha)/2$ if $\mu(l)$ does not depend on l, $B^{(l)}(k)$, does not depend on l and k, and the length distribution of the source active periods $P\{\tau = l\}$ will have the form of Pareto distribution

$$P\{\tau = l\} = c_0 l^{-\alpha-1}, \quad 1 < \alpha < 2, l \in \mathbb{N} \tag{5.8}$$

where

$$c_0 = \left(\sum_{l=1} l^{-\alpha-1} \right)^{-1}$$

The communication system with discrete time $t \in \{\ldots, -1,0,1,\ldots\} \stackrel{\Delta}{=} \mathbb{Z}$ consists of the finite buffer memory and the channel. The packet number is denoted as Y_t, arriving at moment t. To make this more definite, Y is the traffic having the sources generating the packets with the constant velocity $R (R \in \mathbb{N})$. In this case Y_t/R has Poisson distribution such that $MY_t = \lambda R M_\tau$.

The finite buffer size means that it can not save more than h packets in any time moment t. In each window t (the window is the time interval $[t, t+1)$, containing only one time moment t of the discrete axis $\mathbb{Z}$) the channel can transmit (supply) not more than C packets, which can be taken either from the buffer or from new Y_t packets. The packet from the buffer, which is transmitted in window t, departs from the channel and from the analysed system as a whole at moment $t + 1$. The value $C \in \mathbb{N}$ is referred to as the channel carrying capacity.

The analysed queuing system (QS) is denoted as $Y/D/C/h/d$, where Y means that the input traffic is Y, D is the determined service time equal to unity, C is the number of service devices, h marks the buffer size and d shows that the discipline d, acting in the system, is taken into account. It is assumed that in the QS at each moment t the discipline decides which of the following alternatives should be applied to each packet lying in the system: (a) to begin the packet transmission (service) at moment t; (b) to save the packet in the buffer until moment $t + 1$; (c) to clear the packet (loose) at moment t.

The $D_C(h)$ class is the most interesting class of d disciplines satisfying the following conditions:

1. If $Y_t + Z_t > 0$ (where Y_t is the new packet number arriving at moment t, Z_t is the packet number in the system at moment t before the new packet arrival), then min $\{Y_t + Z_t, C\}$ packets are used for the transmission (service) at moment t.
2. If $Y_t + Z_t \leq h + C$, none of the packets is lost at moment t. Which packets are lost and which are transmitted depends on the specific discipline $d \in D_C(h)$.

The probability of the buffer overflow P_{over} will be determined. The event $\{Y_t + Z_t - h - C > 0\}$ is referred to as the buffer overfilling at moment t. The time moment t is called the overflow moment if at least one packet is lost at this moment.

5.1.2.1 The Upper Limit for Overflow Probability

In Reference [11] it is proved that the following asymptotic upper limit for the probability of buffer overflow is valid for the $Y/D/C/h$ system, where Y is the traffic with τ, distributed by the Pareto type law (5.8), and $d \in D_C(h)$:

$$P_{\text{over}} \le c_1 h^{(-a+1)k}, \qquad k = 1 + \left[\left(\frac{C}{R} \right) - \lambda M\tau \right], \qquad h \to \infty, \qquad C > \lambda RM\tau \qquad (5.9a)$$

where $[\cdot]$ is the integer part of a number, the inequality $f(x) \le g(x)$ as $x \to \infty$ is understood as lim sup $f(x)/g(x) \le 1$, and c_1 is some function of the packet source rate R, the channel carrying capacity C, the sources arrival intensity λ, a parameter Pareto type distribution (5.8) and the average length of the source active period $M_\tau = c_0 \sum_{l=1}^{\infty} l^{-\alpha}$ as $x \to \infty$. It should be noted that c_1 does not depend on h. For exactly the same system $Y/D/C/h$, for which the boundary (5.9a) is found, the lower asymptotic limit for the probability of buffer overflow is determined in Reference [12] in the form

$$P_{\text{over}} \ge c_2 h^{(-a+1)k}, \qquad h \to \infty, \qquad C \ge \lambda RM_\tau \qquad (5.9b)$$

where k is the same as in (5.9a), the inequality $f(x) \ge g(x)$ is understood as lim inf $f(x)/g(x) \ge 1$ and c_2 is some function of R, C, λ, α that does not depend on h. Therefore the limits (5.9a) and (5.9b) give the true decay rate of the buffer overflow probability for the buffer size growth h.

It is noteworthy that the P_{over} probability does not decay in accordance with the exponential law, which is usual in teletraffic theory, but according to the power law of h. Therefore, the exponent is proportional to the C–λRM_τ value and to the C carrying capacity exceeding the input flow rate $MY_t = \lambda RM_\tau$. The latter means that the buffer overflow probability nevertheless decays (for large h) in accordance with the exponential law, and depends not on the buffer size but on $\tilde{N}$ exceeding MY_t. This shows that it is easier to overcome buffer overflow by means of increasing the carrying capacity, rather than increasing the buffer size.

To illustrate this conclusion, the system is considered in which $C/R > 1, C/R$ is integer and $\lambda M_\tau < 1$. In this system $k = C/R$. Increasing the C carrying capacity by factor $b > 1$ (for simplicity it is assumed that C/R is the integer number) decreases the $h^{(-\alpha+1)k}$ function by the same amount, which takes place for the buffer size growth by h^{b-1} times. If, for example, $h = 10^4$, a twofold increase in carrying capacity is equivalent to a buffer size increase by 10^4 times (to size 10^8). Note that a carrying capacity increase is accompanied by a delay decrease in the number of packets transmitted through the channel and a buffer size increase by a delay increase.

5.1.3 Queuing Model with Fractional Levy Motion

An analysis analogous to the above-mentioned one will be performed, but for a more general case, where the network traffic is self-similar and stable instead of Gaussian. It is well-known [13–15] that α-stable models can be related to the most developed real self-similar traffic models, from the point of view of queuing theory. Displaced fractional stable noise, the adequacy of which as a model was proved for http and also for the classical record of Bellcore traffic, allows expressions to be obtained for the queue with the appropriate fractal input flow in an explicit form.

The separate server queue with the constant service intensity $r > 0$ and an infinite buffer will be examined, where the stable self-similar FLM process is an input process. The distribution density character, completely displaced to the positive axis, is a fortunate property of FLM, allowing the given process to be used for network traffic modelling. In analogy with the model (5.4), the traffic volume arriving at the channel during the interval $[0, t)$ is equal to

$$A(t) = mt + (\bar{\sigma}m)^{1/\alpha} L_{\alpha,H}(t)$$

Correspondingly, the traffic model is given by four parameters: $m > 0$ is the average input intensity; $\alpha \in (1;2]$ is the stable distribution exponent; $\bar{\sigma} > 0$ is the scaling parameter defining the traffic values dispersion around the mean intensity value; and $H \in [1/\alpha, 1)$ is the Hurst exponent. The fractional Levy process

$$L_{\alpha,H}(t) = \frac{1}{\Gamma(H + 1/2)} \int_0^t \mathrm{d}L_\alpha(\tau)(t - \tau)^{H-1/2}$$

where $L_\alpha(t)$ is the ordinary symmetric α-stable Levy motion (OLM), $\rho = m/r$ is the queue usage coefficient and $r > m$ for stability ensurance.

The process $Q(t, r)$ of buffer occupation at time moment t (the queue size or the queue length) can be described similarly to Equation (5.3):

$$Q(t, r) = \sup_{0 \le s \le t} [A(t) - A(s) - r(t - s)]$$

It is obvious that $Q(t, r)$ is really the stationary fractional stable process, which is the consequence of the stationarity, self-similarity and stability of the FLM increment process. The equation

$$\varepsilon = P[Q(0, r) > L] = P\left[\sup_{\tau \ge 0}[A(\tau) - r\tau] > L\right]$$

can be considered as a requirement for the quality of service (QoS), defining the requirement to the buffer capacity $L > 0$ and related to its overflow probability. It was shown in Reference [14] that the QoS requirement is equivalent to the formula of carrying capacity distribution:

$$r = m + q^{-1}(1, \varepsilon)\sigma^{-1/\alpha(H-1/2+1/\alpha)} L^{-(3/2-H-1/\alpha)/(H-1/2+1/\alpha)} m^{1/[\alpha(H-1/2)+1]} \tag{5.10}$$

and to the formula of the buffer size assignment

$$\frac{1 - \rho}{\rho^{1/\alpha(H-1/2+1/\alpha)}} L^{(3/2-1/\alpha-H)/(H-1/2+1/)} r^{(H-1/2)/(H-1/2+1/\alpha)} = \bar{\sigma}^{1/\alpha(H-1/2+1/\alpha)} q^{-1}(1, \varepsilon) \tag{5.11}$$

Here $q(1,\varepsilon)$ was found from $q(L,\varepsilon) = P\{\sup_{\tau\geq 0}[L_{\alpha,H}(\tau) - \varepsilon\tau] > L\}$ for $L = 1$. Substituting $\rho = m/r$ into Equation (5.11), it is possible to obtain the buffer size assignment formula.

The above-mentioned formulas will be applied to various types of input traffic:

1. The input process is modelled by ordinary Brownian motion, i.e. $H = \frac{1}{2}$ and $\alpha = 2$. In this case Equation (5.11) is reduced to

$$L = L(\rho) = \text{constant} \times \rho(1-\rho)^{-1} \tag{5.12}$$

2. The input process is modelled by the ordinary Levy motion, i.e. $H = \frac{1}{2}$ and $0 < \alpha \leq 2$. In this case,

$$L = L(\rho) = \text{constant} \times \rho^{1/(\alpha-1)}(1-\rho)^{-1/(\alpha-1)} \tag{5.13}$$

As for Brownian motion, the service intensity r disappears from Equation (5.13).

3. In the case of FBM, for $\alpha = 2$ and $H > \frac{1}{2}$ the situation is different. From formula (5.11), the buffer size assignment, fixing the service intensity r and solving the equation for the buffer size requirement L as the ρ function gives

$$L = L(\rho) = \text{constant} \times \rho^{1/[2(1-H)]}(1-\rho)^{-H/(1-H)} \tag{5.14}$$

which coincides with the result found in Reference [8].

From the carrying capacity distribution formula (5.10), fixing L and solving with respect to r gives

$$r = r(\rho) = \text{constant} \times \rho^{1/(2H-1)}(1-\rho)^{-H/(H-0.5)} \tag{5.15}$$

i.e. it coincides with the results of Reference [8].

$$L = L(\rho) = \text{constant} \times \rho^{[1/\alpha(3/2-1/\alpha-H)]}(1-\rho)^{(-H-1/2+1/\alpha)/(3/2-1/\alpha-H)} \tag{5.16}$$

4. The fractional Levy motion is used as the input process, which is a more general case. Again from Equation (5.9) the requirement for the buffer size L is expressed a function of a ρ usage. As a result, the function of the required service intensity in the QS of the FLM/D/1 type with respect to the traffic ρ can be written in the form

$$r = r(\rho) = \text{constant} \times \rho^{-1/[\alpha(H-1/2)]}(1-\rho)^{-(H-1/2+1/\alpha)/(H-1/2)} \tag{5.17}$$

which can be considered as a generalization of the well-known Norros relations [8] for FGN (Gaussian case $\alpha = 2$). Therefore, the value of the constant coefficient takes into account the buffer volume and the quality of the service parameters. The function (5.17) plot is shown in Figure 5.5.

It can be seen from Figure 5.5 that for big channel loading non-Gaussian traffic a much higher channel carrying capacity is required with the same intensity and keeping the QoS requirements.

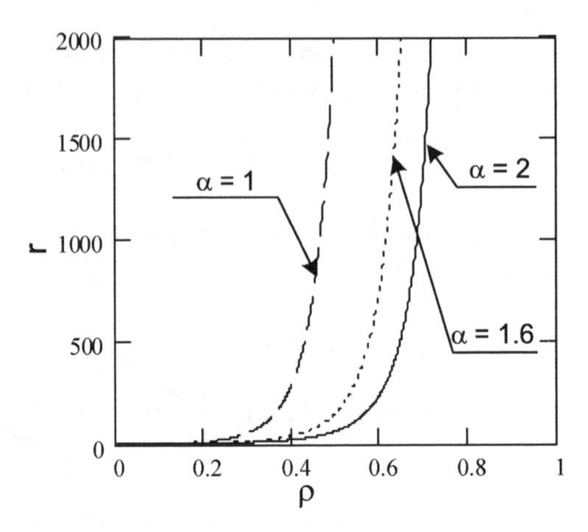

Figure 5.5 Service intensity function with respect to the usage coefficient for $\alpha =$ constant

5.1.3.1 Lower Asymptotic Limit for the Buffer Overflow Probability

The lower asymptotic limit for the queue length distribution in QS will be determined, having a constant service time and service intensity, that meets QoS requirements. It is known from Reference [14] that the lower asymptotic limit for the buffer overflow probability is determined with the help of

$$\varepsilon = P(Q(0, r) > L) \geq \Delta_\alpha L^{-\alpha(3/2-H-1/\alpha)}, \quad L \to \infty \tag{5.18}$$

where

$$\Delta_\alpha = M_\alpha(\overline{\sigma} m)\left[\frac{(3/2-H)\alpha - 1}{\alpha}\right]^\alpha \left\{\frac{\alpha(H-1/2)+1}{[(3/2-H)\alpha - 1]\,(r-m)}\right\}^{\alpha(H-1/2)+1}$$

and

$$M_\alpha = \frac{\overline{\sigma}}{\alpha\pi}\Gamma(\alpha+1)\sin\frac{\pi\alpha}{2}$$

The required service intensity r in the QS of the FLM/D/1 type satisfying QoS criterion can be found from the (5.18) solution with respect to r, the result of which leads to the relation

$$r = m + (M_{\alpha/\varepsilon})^{1/[\alpha(H-1/2)+1]}\overline{\sigma}^{1/[\alpha(H-1/2)+1]} m^{1/[\alpha(H-1/2)+1]} b^{(H-3/2+1/\alpha)/(H-1/2+1/\alpha)} \tag{5.19}$$

Comparing the above-mentioned approximated requirement with the exact one, obtained with the help of the carrying capacity distribution formula (5.10), it can be seen that they differ in the $(M_\alpha/\varepsilon)^{1/[\alpha(H-1/2)+1]}$ coefficient only. For the Brownian case, i.e. for $H = \frac{1}{2}$ and

$\alpha = 2$, expression (5.18) is reduced to the well-known asymptotic form, obtained for the exponential distribution.

The important conclusion from the relations for queues and for α-stable traffic models is that in the general case it is impossible to describe the behaviour of the QS with the input fractal flow without going beyond the intensity and Hurst exponent. When studying the velocity limitation typical algorithm effect used in data transmission equipment ('leaky bucket') on α-stable traffic, remarkable results were obtained concerning parameter changing [15]. The rate limitation procedure leads to any initial α-stable traffic in the 'limited' channel being reduced to Gaussian ($\alpha = 2$), while fully keeping the fractal properties: the Hurst exponent H value does not practically change. Undoubtedly, the applications experience an essential delay due to buffering or truncation of the packet.

5.1.4 Estimate of the effect of traffic multifractality on queuing

The queuing formulas in the case of the Gaussian input process lead to queuing results that conform well with Gaussian process theory. The new practical approach to estimate the queuing productivity is introduced for generalized multifractal traffic.

5.1.4.1 Approximation for Queue Tail Probabilities

It was shown in Reference [16] that the probabilities of the queue distribution tail asymptotic for the one-server queuing model with a generalized multifractal input process are well approximated by the expression

$$\log(P[Q > L]) \approx \min_{q>0} \log\left\{ c(q) \frac{\left[\frac{L\tau_0(q)}{r(q-\tau_0(q))}\right]^{\tau_0(q)}}{\left[\frac{Lq}{q-\tau_0(q)}\right]^{q}} \right\} \tag{5.20}$$

for large L, where $\tau_0(q) = \tau(q) + 1$. It was noted above that the scaling functions $\tau(q)$ and $c(q)$ are the functions that fully define the multifractal input process.

Examining (5.20) it can be observed that it has the exact form of the scaling function $\tau_0(q)$ and the moment coefficient can give the final result. The reason for this lies in the definition of the multifractal process class, which does not apply limitations on functions $c(q)$ and $\tau_0(q)$ (except for $\tau_0(q)$ being convex). The queuing system analysis with the generalized multifractal input process shows that it could give certain generalized results similar to the case of the multifractal input processes. This implies that there is no general queuing character for the systems similar to having the decay according to Weibull law in the case of Gaussian self-similar processes [8]. The real mutifractal model will define, for example, the queue length probability for the system.

The Gaussian process with the scaling property is monofractal, having the following parameters:

$$\begin{aligned} \tau(q) &= \frac{q}{2}[\tau(2) + 1] - 1 \\ c(q) &= \frac{[2c(2)]^{q/2}}{\sqrt{\pi}} \Gamma\left(\frac{q+1}{2}\right) \end{aligned} \tag{5.21}$$

where $\Gamma(z) = \int_0^{+\infty} x^{z-1} \exp^{-x} dx, z > 0$, is the Gamma function. For FBM at $q = 2$ it is found that $c(2) = 1$ and $\tau(2) = 2H - 1$. Thus

$$\begin{aligned} \tau(q) &= qH - 1 \\ c(q) &= \frac{2^{q/2}}{\sqrt{\pi}} \Gamma\left(\frac{q+1}{2}\right) \end{aligned} \tag{5.22}$$

In the Gaussian case, $\tau_0(q) = \tau(q) + 1 = qH$.

Substituting the obtained relations in (5.20) the following expression is found after some transforms:

$$\log(P[Q > L]) \approx -\tfrac{1}{2} L^{2(1-H)} r^{2H} (1 - H)^{-2(1-H)} H^{-2H} \tag{5.23}$$

For large L the obtained relation coincides with the expression found in Reference [8], which confirms the statement that FBM is the specific case of the multifractal processes.

5.1.4.2 Queuing in the Case of Multifractal Input Traffic

It is assumed that the input process demonstrates the multifractal scaling properties, and the scaling function $\tau(q)$ and the function $c(q)$ can be estimated from the experimental data for several possible $q > 0$ parameters. It is necessary to pay special attention to the $c(q)$ function as well as to the quantitative coefficient of the multifractal process. The scaling function $\tau(q)$ defines the multiscaling quality only, so the function alone cannot describe the multifractal model, nor, consequently, analyse the queuing models using the multifractal input processes.

There are two methods used to estimate queue length:

1. Using (5.20), the approximation for $\log P[Q > L]$ can be calculated for each L value, taking into account the service intensity r and two sets $\{c(q)\}$ and $\{\tau(q)\}$. This approach is not very simple but it is more useful from the point of view of network planning and size assignment, and only several probability values at the distribution tail are of interest.
2. The input process corresponds to the multifractal model. Two measured sets $c(q)$ and $\tau(q)$ are selected with the help of approximations $\tilde{c}(q)$ and $\tilde{\tau}(q)$, as illustrated in Chapter 2. As a result, the analysis of (5.20) using these functions can lead to an analytic form of the probabilities of the queue distribution tail part. This approach is used for studies of the queue tail behaviour for the multifractal model.

At the same time, the characteristic functions of the multiscaling processes are often complex. Therefore in these cases it is difficult to find the final solution for queuing. The approximation (5.20) of the queue distribution tail part is considered in the form

$$\log P[Q > L] \approx \min_{q>0} \left\{ \log c(q) + \tau_0(q) \log \frac{L\tau_0(q)}{r[q - \tau_0(q)]} - q \log \frac{Lq}{q - \tau_0(q)} \right\} \tag{5.24}$$

where

$$\tau_0(q) = \log_2 \frac{\Gamma(\alpha)\Gamma(2\alpha + q)}{\Gamma(\alpha + q)\Gamma(2\alpha)} \tag{5.25}$$

$$c(q) = \mathrm{e}^{mq+\sigma^2 q^2/2} 2^{N(q-\log_2\{[\Gamma(\alpha)\Gamma(2\alpha+q)]/[\Gamma(\alpha+q)\Gamma(2\alpha)]\})} \tag{5.26}$$

The function (5.25) is presented in Figure 2.41 while Figure 2.42 shows the function

$$\ln c(q) = \left(mq + \frac{\sigma^2 q^2}{2}\right) + \left\{q - \log_2\left[\frac{\Gamma(\alpha)\cdot\Gamma(2\alpha+q)}{\Gamma(\alpha+q)\cdot\Gamma(2\alpha)}\right]\right\} N \ln(2)$$

$$\text{for } m = 0.6; \sigma = 0.2; N = 20; q = 0\ldots 15$$

Using approximation (5.24), the logarithm of the buffer size excess probability on the distribution tail can be found for the appropriate queue size value L by means of the numerical $\log P[Q > L]$ minimization with the estimated sets $\{c(q)\}$ and $\{\tau_0(q)\}$ with the help of the auxiliary functions presented in Figures 2.41 and 2.42.

5.1.4.3 Multifractal and Monofractal

Consider the multiplexed multifractal process with the symmetrically distributed beta(α,α) multiplier. For this multifractal the characteristic functions can be calculated exactly at the specific cascade level.

It is assumed that there is a (mono) fractal process with exactly the same moment coefficient $c(q)$ as the multifractal, but having a single-scaled fractal structure $\tau_0(q) = qH$. Then approximation (5.24) is transformed to the form

$$\log P[Q > L] = \ln c(q) + \tau_0(q) \ln\left[\frac{L\tau_0(q)}{r[q - \tau_0(q)]}\right] - q \ln\left(\frac{Lq}{q - \tau_0(q)}\right) \tag{5.27}$$

Correspondingly for FBM,

$$\log P_G[Q > \mathrm{L}] = \log\left[\frac{2^{q/2}}{\sqrt{\pi}}\Gamma\left(\frac{q+1}{2}\right)\frac{\left[\frac{LH}{r(1-H)}\right]^{qH}}{\left(\frac{L}{1-H}\right)^q}\right]$$

Knowing the characteristic functions of these scaling processes, it is possible to calculate the estimate of the probability on the distribution tail of the queuing system for large queue sizes using the numerical approach. Calculation results from formula (5.24) are presented in Figures 5.6 and 5.7 [17]. (The service intensity is set equal to $r = 2.0$ in the numerical example.) It is observed that the approximated probabilities on the queue distribution tail in the multifractal case are much higher than in the monofractal case. Figure 5. 8 shows the function

$$\log P[Q > L] = f(q) = c(q) + \tau_0(q) \ln\left\{\frac{L\tau_0(q)}{r[q - \tau_0(q)]}\right\} - q \ln\left[\frac{Lq}{q - \tau_0(q)}\right]$$

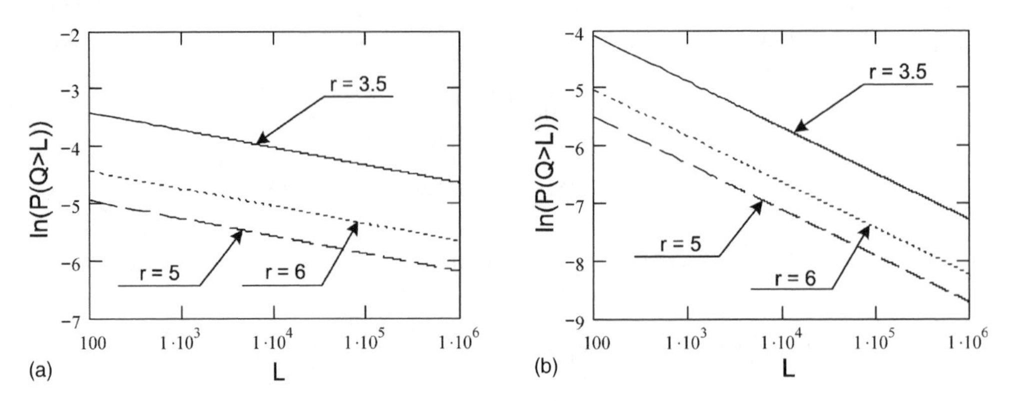

Figure 5.6 Function log $P[Q > L]$ of the queue size value L for given r values: (a) $\alpha = 15$; (b) $\alpha = 5$

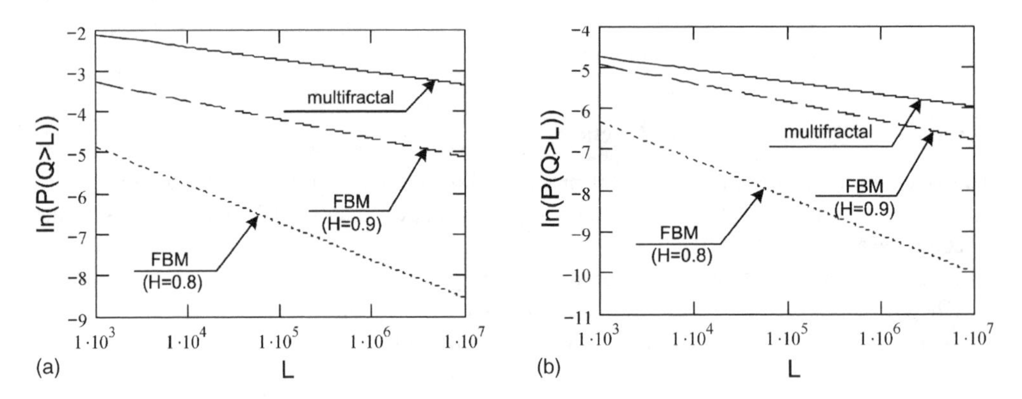

Figure 5.7 Functions log $P[Q > L]$ of the queue size value L: (a) $r = 2$; (b) $r = 5$

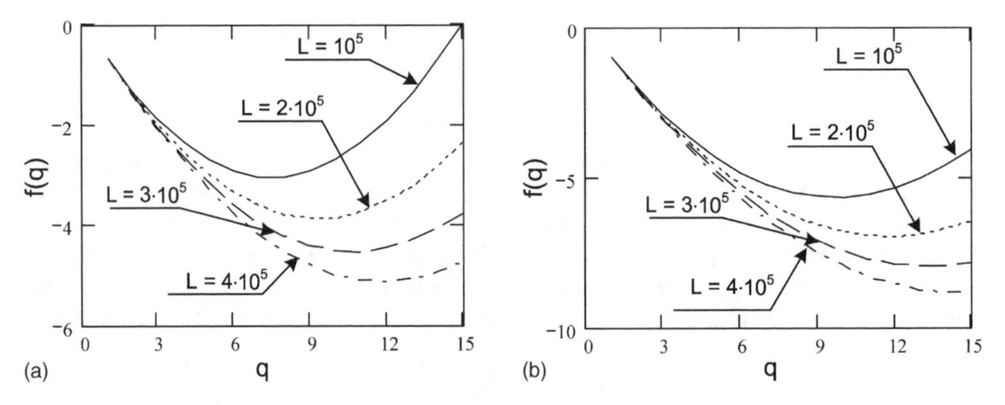

Figure 5.8 Function log $P[Q > L]$ of q at constant L: (a) $r = 3.5$; (b) $r = 5$

for $L=$ constant, where

$$c(q)=\left(mq+\frac{\sigma^2q^2}{2}\right)+[q-\tau_0(q)]N\ \ln(2)$$

illustrating the minimum presence depending on the q parameter value.

The theoretical probability of the queue distribution tail part for each value L of the queue size is the minimum of the log $P[Q>L]$ function. Besides, there is no necessity to plot the log $P[Q>L]$ function for each q value in order to find a minimum. This procedure can be realized for all appropriate L values with the help of an easy computer program.

5.1.4.4 Conlusion

Researchers of the queuing performance for the separate server with an infinite buffer capacity showed that the asymptotic approximation of the queue length distribution probabilities in the stable state may be obtained for the constant service intensity, when the generalized multifractal process arrives at the buffer. It is shown that the approximation leads to the familiar tail of the queue distribution in accordance with the Weibull law, when the fractional Brownian motion is chosen as the input process. Some multifractality consequences are analysed and presented. It is shown that the applied formulas give correct results in the course of the analysis of both multifractal and monofractal traffic.

5.2 Estimate of Voice Traffic Self-Similarity Effects on IP Network Input Parameter Optimization

5.2.1 Problem Statement

In telecommunication network (TN) design and operation it is often required to ensure a certain quality of service (QoS) for the user or to define the network input parameter range, within the limits of which the required quality of service level will be sustained in the network parameter metrics. The problem of TN input parameter optimization in particular with regard to network traffic self-similarity is illustrated by the example of voice services.

The use of the offered approach in TN designing contributes to the introduction of the new services, the functioning reliability of which can be guaranteed in advance. Taking into account the essential computational complexity, such problem solution, as a rule, is fulfilled by the methods of imitation mathematical modelling.

5.2.2 Simulation Structure

Consider the TN, the structural diagram of which in the ns2 environment is shown in Figure 3.26. The experimental investigations of voice traffic for two types of codecs, G.723 and G.729B, using voice activity detectors became the basis of imitation modelling. The voice flows were subject to processing and the duration of the ON and OFF periods was found for them from the packet flows. The average values of ON/OFF periods depend on VAD settings. During these measurements the standard VAD settings of Cisco equipment were used. The experimental distribution functions of the ON/OFF period duration for different codecs have practically the same shape and almost coincide with each other. This implies that the voice codecs slightly affect the main features of ON/OFF periods in the packet sources.

The statistical analysis of the measurement results showed [18] that the popular models of the summed VoIP traffic based on exponential distributions of the ON/OFF period duration of the voice sources cannot be used. These distributions are obviously nonexponential and in

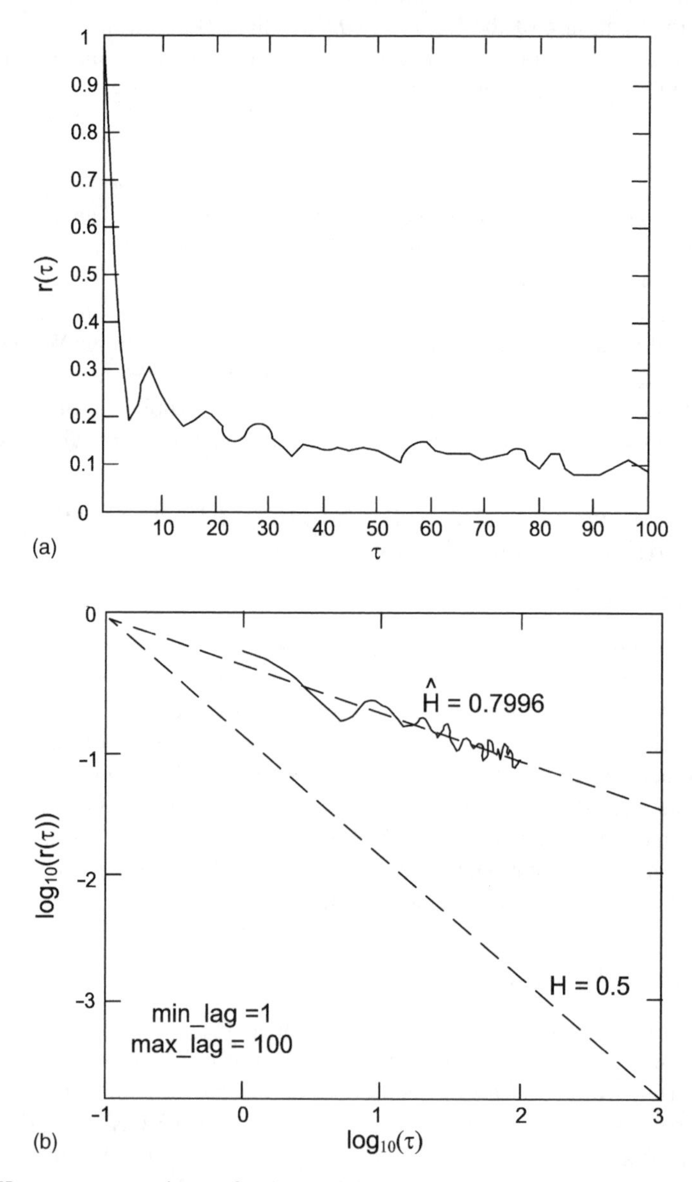

Figure 5.9 Hurst exponent estimates for the multiplexed traffic for $\alpha = 1.4$: (a) correlation coefficient; (b) correlation coefficient in log-log scale; (c) variance–time plot; (d) plot of R/S statistics

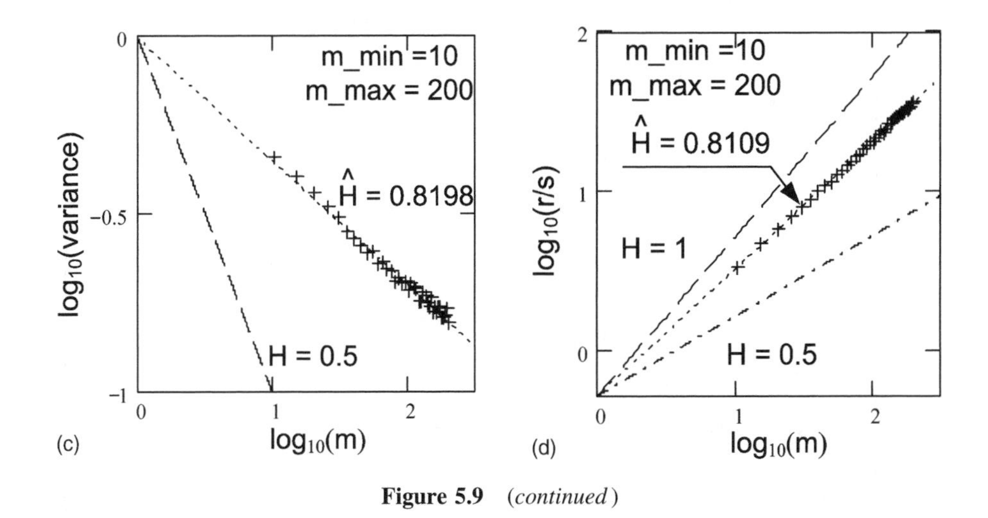

Figure 5.9 (*continued*)

this case the distributions with heavy tails can be used as a model. The Pareto distribution $w(x) = ab^a/x^{a+1}$, for $x \geq b$, where a is the shape parameter of Pareto distribution and b is a scaling parameter, shows the best results.

It was shown earlier that the aggregation of a large number of ON/OFF sources with heavy tails for ON and/or OFF periods leads to self-similarity, and self-similar models are the most acceptable approximation for aggregation of VoIP traffic. The Pareto traffic generator, implemented in the ns2 system, was chosen to describe operation of the voice sources. The generator parameterization was fulfilled on the basis of the experimental data analysis: the average activity interval is equal to 500 ms and the average silence interval to 1500 ms, and the Pareto distribution parameter α was changed for different experiments depending on the multiplexed flow fractal properties.

As an example, Figure 5.9 shows the multiplexed flow self-similarity estimate found as a result of the imitation modelling for a Pareto distribution parameter value equal to $\alpha = 1.4$. It can be seen that the traffic obtained as a result of the imitation modelling possesses fractal properties and can be used for an analysis of the self-similarity influence on the system productivity parameters using the properties of the imitation modelling environment ns2.

5.2.3 Estimate of the Traffic Self-Similarity Influence on QoS

Special software was developed to estimate the influence of traffic self-similarity on the QoS, and the imitation modelling was fulfilled to estimate the main QoS TN characteristics (the percentage of lost packets for each source, the average delay per IP packet for each source, the jitter average value per IP packet and the standard deviation of jitter per IP packet) under conditions of voice traffic self-similarity [19].

The results of the system perfromance studies on the influence of multiplexed traffic fractal properties are shown in Figure 5.10. It was found that, on the whole, voice traffic self-similarity worsens the quality of service characteristics.

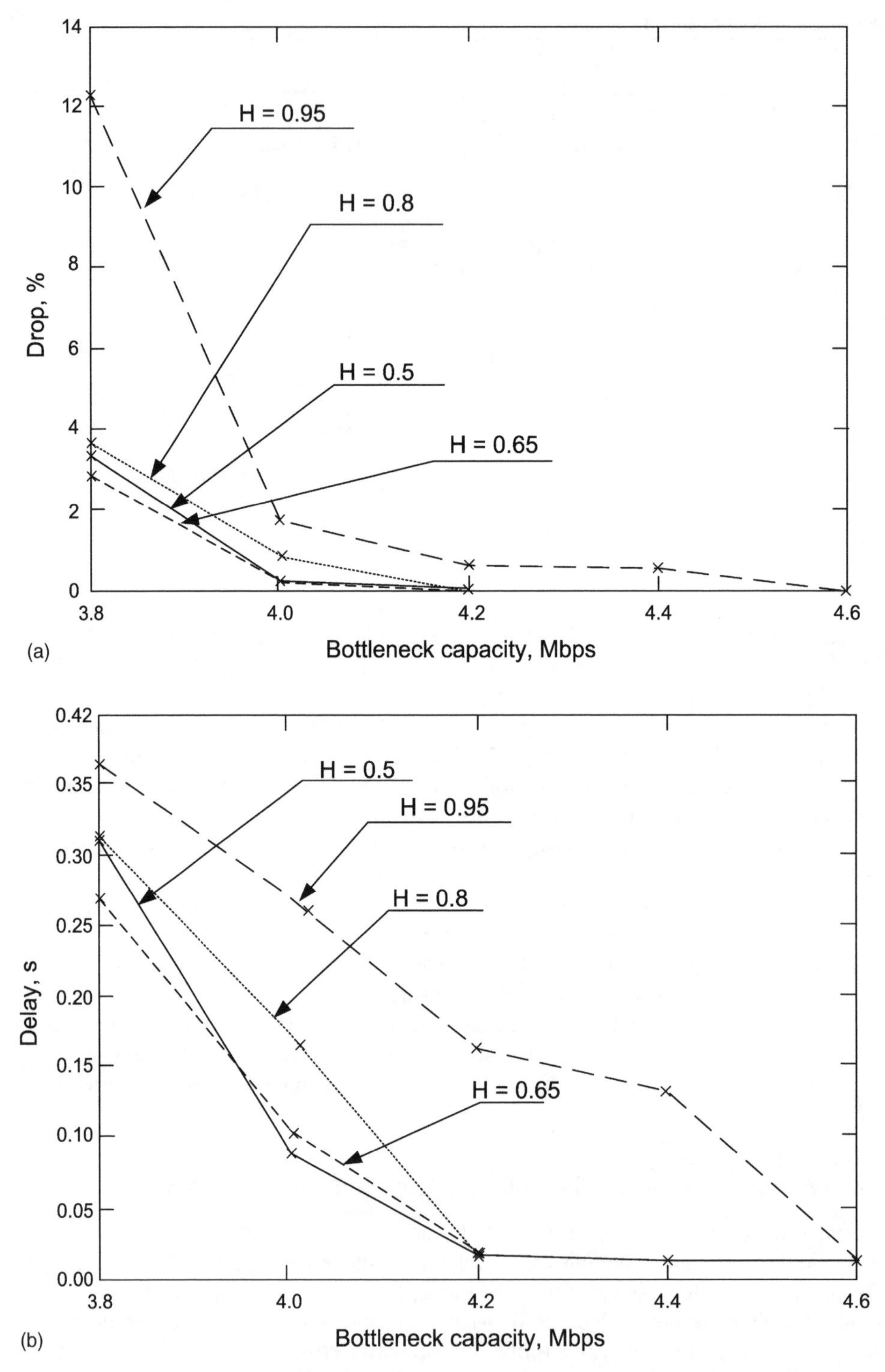

Figure 5.10 Estimate of the multiplexed flow Hurst exponent effect on the quality of service characteristics: (a) the percentage of lost packets; (b) the average delay; (c) standard deviation of the jitter; (d) the system utilization factor

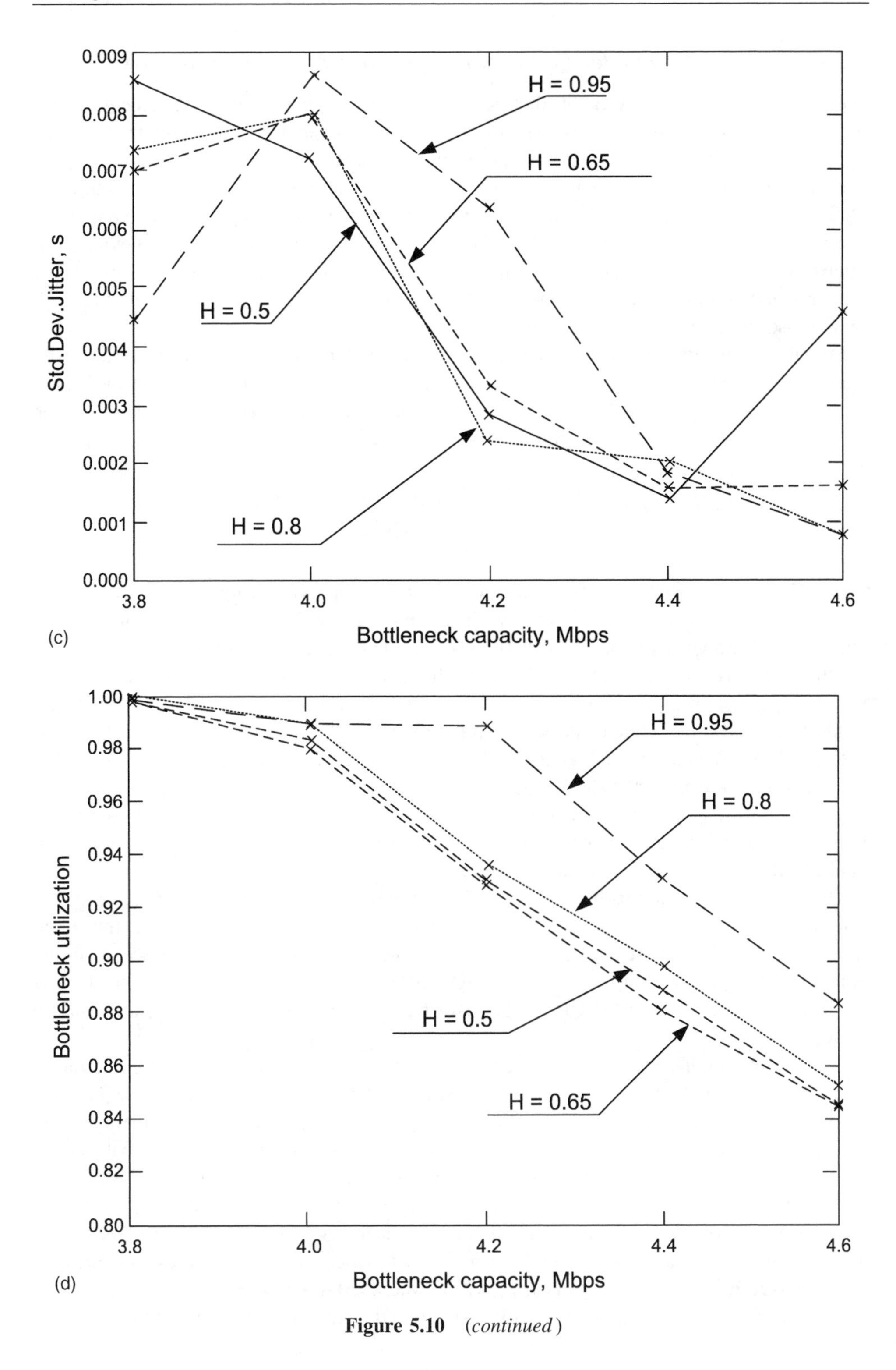

Figure 5.10 *(continued)*

5.2.4 TN Input Parameter Optimization for Given QoS Characteristics

Having finished the analysis of the degree of traffic self-similarity influence on TN productivity parameters, optimization of the analysed system input parameters will be carried out in order to ensure given performance characteristics. To make this clearer, consider the three-dimensional required vector of quality of service characteristics $\mathbf{QoS}_0 = (drop_0, delay_0, jitter_0)^{\mathrm{T}}$, with the given characteristics $(drop_0)$, the percentage of lost packets in %; $(delay_0)$, the VoIP packet delay in s; $(jitter_0)$, the standard deviation of jitter in s; $\mathbf{X} = (C, L, N)^{\mathrm{T}}$, the system input parameter vector; C, the carrying capacity value for the worst communication channel; L, the FIFO buffer capacity value; and N, the VoIP service user number connected behind the appropriate router.

For the given $\mathbf{QoS}_0$ vector, the required TN parameters are found, ensuring the given quality of service. It can be shown that the search for the $\mathbf{X}^*$ optimum consists of the search for the $\mathbf{X}^* = \arg\min_{x \in X} f(\mathbf{X}_n)$ functional extremum. Here $f(\mathbf{X})$ is understood to be the functional discrepancy value for the quality of service parameters:

$$f(\mathbf{X}_n) = \text{constant} \left\| \boldsymbol{w} \left(\frac{\mathbf{QoS}_n(\mathbf{X}_n)}{\mathbf{QoS}_0} - \boldsymbol{e} \right) \right\|_2 \tag{5.28}$$

where $\mathbf{QoS}_n$ is the vector of the quality of service characteristics found at the nth minimization step, constant is the definitely chosen constant; $\boldsymbol{e}$ is the unit vector having the dimension of the **QoS** vector and $\boldsymbol{w}$ is the vector reflecting the weights of each component of the QoS vector (in the case when all quality characteristics are equilibrium, $\boldsymbol{w} = (w_{\text{drop}}; w_{\text{delay}}; w_{\text{jitter}})^{\mathrm{T}}$ vector is equal to the unit vector $\boldsymbol{e}$, where w_{drop} defines the weight of the lost packets percentage and varies in the range (0;1], w_{dealy} defines the weight of the average delay per packet and varies in the range (0;1], w_{jitter} defines the weight of the standard deviation of jitter of the delayed packets and varies in the range (0;1]). The subjective character of the offered optimization criterion described by the weight coefficient vector $\boldsymbol{w}$ is conditioned by the subjective character of the voice quality estimate (e.g. intellegibility or mean opinion scores (MOS)).

The computational algorithm consists of the iteration sequence around the basis point. The input parameter vector at the nth iteration step can be written in the form $\mathbf{X}_n = (C_n, L_n, N_n)^{\mathrm{T}}$, where C_n is the botleneck channel carrying capacity value at the nth step of minimization, L_n is the FIFO buffer capacity value at the nth step of minimization and N_n is the VoIP services user number, connected behind the appropriate router.

The appropriate increments of the basis point coordinates are interpreted as the iteration step. The vectors included in $f(\mathbf{X})$ are presented in the form $\mathbf{QoS}_n = (drop_n, delay_n, jitter_n)^{\mathrm{T}}$ and $\mathbf{QoS}_0 = (drop_0, delay_0, jitter_0)^{\mathrm{T}}$, where $drop_n$, $delay_n$ and $jitter_n$ are respectively the lost packets percentage, the average delay and the standard deviation of jitter that should be obtained as a result of optimization. During the optimization the required values of the quality of service characteristics are given with some inaccuracy as $\Delta\mathbf{QoS}$. To estimate the expediency of the fulfilled optimization it is necessary to compare the results with the optimization inaccuracy which can be estimated by the expression error $=$ constant$(\| \Delta\mathbf{QoS}/\mathbf{QoS}_0 \|_2 + \| \mathbf{error}_{\text{model}} \|_2)$, where $\Delta\mathbf{QoS}$ is the vector of the assignment inaccuracy of the output network characteristics (the lost packets percentage, the average delay per packet and the standard deviation of jitter) and $\mathbf{error}_{\text{model}}$ is the vector of the output characteristics errors introduced by the model.

Figure 5.11 shows the TN input parameter function of the Hurst exponent for various initial conditions. The obtained results, on the whole, indicate that with Hurst exponent growth the point in the input parameter space leading to the required QoS values moves towards a quantitative parameter increase for optimized parameters. Figure 5.12 shows the plots of output QoS characteristics (the lost packets percentage, the average delay per packets and the standard deviation jitter per packet) versus Hurst exponent values of the aggregated traffic.

The investigation of the influence of the multiplexed flow Hurst exponent on the fulfilled optimization integrated quality for the selected objective function $f(\mathbf{X}_n)$ was the main optimization goal. The plots of integral QoS parameters and of the quality of service parameter vector obtained in the optimal parameter points versus the Hurst exponent (Figure 5.13) are based on the results of separate QoS characteristics for various Hurst

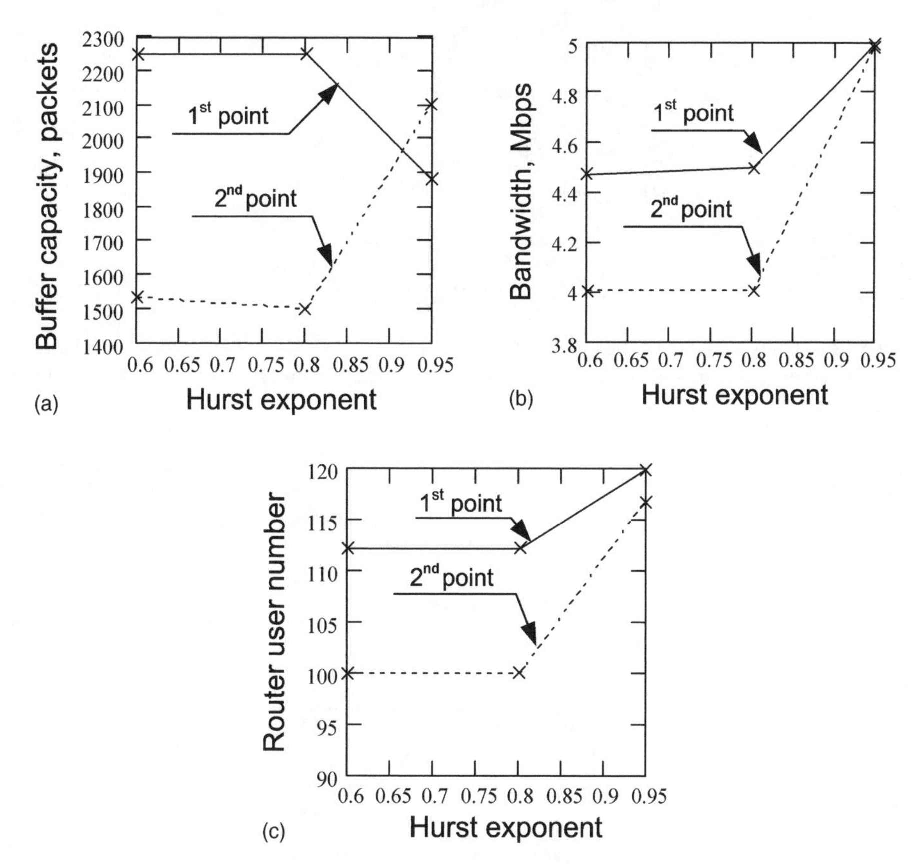

Figure 5.11 TN input parameters versus the Hurst exponent for found optimal points when choosing various initial conditions: (a) the buffer size function versus H; (b) the bottleneck capacity function of H; (c) the number of router users versus H

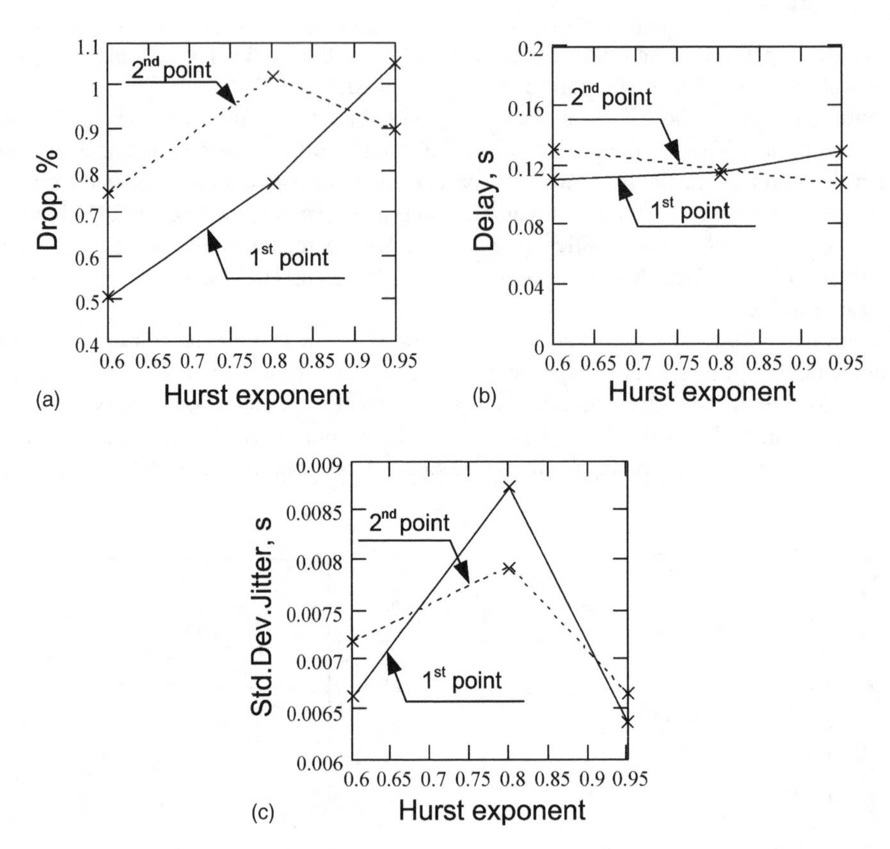

Figure 5.12 Output characteristics versus the Hurst exponent for the found optimal points: (a) the lost packet percentage versus H; (b) the average delay per packet versus H; (c) standard deviation of the jitter per packet for each source versus H

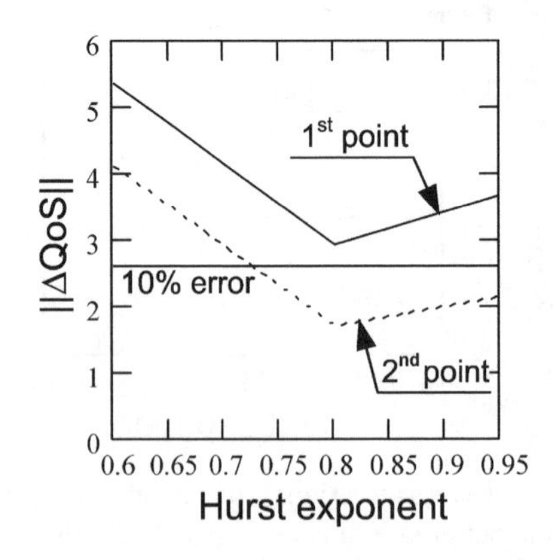

Figure 5.13 Discrepancy values versus the Hurst exponent in the optimal points for the different modelling scenarios

exponents. The value corresponding to the 10 % optimization error is shown here as well. It can be seen that for Hurst exponent growth the obtained discrepancy decreases in the optimal point, but for large H values the discrepancy again increases. Nevertheless, note that for the second modelling scenario at some Hurst exponent values <0.8 the optimization results keep within the limits of 10 % error, which is a good illustration of a successfully executed optimization. For the second modelling scenario a similar tendency is observed for the full range of analysed Hurst exponent values, but the errors in this case exceed the 10 % threshold for all experiments. On the basis of the numerical results found in the course of imitation modelling of the functional minimization of (5.28), it can be concluded that using the functional minimization algorithm of (5.28) allows TN input parameter vector optimization, providing a given quality of service with acceptable errors.

During the experiments various initial points of the minimization algorithm were used. On the basis of the numerical results the conclusion was drawn that the farther the initial point is from the position obtained due to optimization, the larger the number of iterations required and the less accurate the optimization results. This implies that at TN modelling it is necessary to have *a priori* information concerning the network functioning process, which can be used for optimization execution.

The observed instability of the solutions found that depend on the selected optimization initial value demonstrates the ill-posed character of the optimization problem mathematical statement; the numerical errors increase and require further perfection of the used optimization methods. Tikhonov's regularization approach [20] is one of the most known optimization methods that can be offered in the present case.

5.2.5 Conclusions

It is found that, on the whole, voice traffic self-similarity worsens QoS quality of service characteristics. The optimization algorithm for the TN input parameter vector is offered, which ensures a given quality of service and uses the integral discrepancy functional for the quality of service characteristics as the minimization criteria. On the basis of the results obtained in the course of imitation modelling, the conclusion was drawn that use of the developed optimization algorithm allows the optimization of the TN input parameter vector, ensuring a given quality of service for acceptable errors.

The functions obtained as a result of the optimization of the input parameter variations with respect of the minimization step number allow the conclusion to be made that to ensure the result convergence the iteration number should not exceed 50. It is found that the optimization results depend on the telecommunication traffic self-similarity degree. The higher the Hurst exponent the higher is the fulfilled optimization accuracy, but for $H \to 1$ the accuracy begins to decrease.

5.3 Telecomminication Network Parameter Optimization Using the Tikhonov Regularization Approach

5.3.1 Problem statement

One of the problems that is most often met in telecommunication engineering is the problem of determining the system input parameters satisfying the given output characteristics

[18, 19]. The TN model is presented as previously defined and is perfectly realizable in the analytical or computational form. In the general form the system description can be presented as a reflection of the definite parameter set in a certain characteristic space. Thus, the problem of the system parameter choice can be formulated as a solution search for the following operational equation:

$$\mathbf{Ap} = \mathbf{f} \tag{5.29}$$

where $\boldsymbol{A}$: $\boldsymbol{D}(\boldsymbol{A}) \rightarrow \boldsymbol{F}$ is the operator describing the system operation, $\boldsymbol{p} = \{\boldsymbol{p}_1, \boldsymbol{p}_2, \ldots, \boldsymbol{p}_m\} \in \boldsymbol{M}_p \subset \boldsymbol{U}$ are parameters, $\boldsymbol{f} = \{\boldsymbol{f}_1, \boldsymbol{f}_2, \ldots, \boldsymbol{f}_n\} \in \boldsymbol{F}$ are characteristics and $\boldsymbol{U}$ and $\boldsymbol{F}$ are Hilbert spaces.

The user capacity, their node element performance, their coupling, etc., can be considered as TN optimization parameters $\boldsymbol{p}$. These parameters are given for each node (element) of network i ($i \leq n$) and thus $\boldsymbol{p}$ usually represents the multidimensional value array.

The network characteristics $\boldsymbol{f}$ are the network carrying capacity for various traffic types, the number of connected users, the percentage of lost packets or the error coefficient in the channels, the delay time of the signal transmission and its variation, etc. Usually the numeric and quality values of these characteristics are included in TN contracts with the users or into the service standard agreement.

Such problems occur at different stages of the TN life cycle and actually require the step-by-step solutions of a whole series of similar problems for qualified or changing initial data (as the operator modelling the system or the required characteristics). In practice, all successive solutions of such problems should match each other in order not to lead to 'reclining from side to side', i.e. they should be stable.

A correct approach to the problem solution is badly needed and can be observed on the example of TN development planning. The gist of the problem in this case can be reduced to the network structure formation and its parameter choice. Therefore the TN should have a set of characteristics that allow the users' requirements to be met.

It is shown in Reference [21] that the telecommunication system functioning optimization by quality characteristics is an ill-posed problem. To obtain optimization the regularization methods should be used where Tikhonov's functional minimization allows the regularization parameter α to be obtained, which keeps model errors within admissible limits and within required QoS characteristics deviations. By applying the regularization methods of the found solutions, the network parameter set should be found, with the probability of keeping a given quality of service characteristics values with in certain tolerances.

Before applying the regularization methods to found solutions, the direct optimization problem is considered by means of the discrepancy functional minimization of QoS characteristics. This approach is chosen to illustrate the advantages of the method based on the regularization using Tikhonov's functional over the classical approach to optimization.

Consider the TN input parameter optimization problem for TN quality of service parameters with the help of simulation in the ns2 environment for the condition of the ill-posed problem. To imitate TN operation the imitation complex ns2 was chosen as the modelling environment. The developed software allows the execution of one-way and grouped routing algorithms, transport and session protocols (including reliable as well as unreliable broadcasting protocols), based on IP service reservation and integration, and the application level protocols of the HTTP type. The structural diagram of imitation modelling is shown in Figure 3.26.

The general TN model for ns2 modelling, including video services, voice services, Internet services, the main TN routers with MAC addresses, stack switches and also N services, which are used by the users of Internet services, was developed on the basis of a statistical characteristic analysis for experimentally obtained traffic data. The developed software allows TN simulation modelling to be provided in order to study the choice method of its parameters on the basis of accurate and approximated initial data.

To analyse QoS characteristics of voice traffic, multiplexing of 500 ON/OFF sources, each of which is considered as one of the telephone conversation parties, is carried out. The traffic is received for a time resolution of 0.1 s, which ensures a high level of averaging and allows reliable statistical estimates to be received. The effect of separate source property variations on the multiplexed flow properties is analysed.

5.3.2 Telecommunication Network Parameter Optimization on the Basis of the Minimization of the Discrepancy Functional of QoS Characteristics

5.3.2.1 Hooke–Jeeves Algorithm of Discrepancy Functional Minimization

The Hooke–Jeeves algorithm of discrepancy functional minimization consists of the step sequence around the basis point, which is illustrated by the following example. The basic point is understood as the point with three coordinates (the buffer size, the channel rate and the number of users in the system). The step is described as appropriate increments of the basic point coordinates. The function $f(x)$ is the discrepancy functional value of the system of QoS characteristics. The procedure description is presented below.

Step 1.

The initial basic point $\mathbf{b}_1$ and the step with length h_j are chosen for each variable $x_j, j = 1, 2, \ldots, n$. In developed software step h is used for each variable.

Step 2.

Calculate the function $f(x)$ in the basic point $\mathbf{b}_1$ to obtain the information about the function $f(x)$ local behaviour. This information will be used to find the proper search direction, with the help of which the achievement of a larger decrease in the function value can be expected. The function $f(x)$ in the basic point $\mathbf{b}_1$ can be found as follows.

Each variable in turn varies by the step length rise. Thus, the function value $f(\mathbf{b}_1 + h_1\boldsymbol{e}_1)$ can be calculated, where $\boldsymbol{e}_1$ is the unit vector in the x_1 axis direction. If this leads to the function value decrease, $\mathbf{b}_1$ is changed to $\mathbf{b}_1 + h_1\boldsymbol{e}_1$. Otherwise, the function value $f(\boldsymbol{b}_1 - h_1\boldsymbol{e}_1)$ is calculated and, if its value decreases, $\mathbf{b}_1$ is changed to $\mathbf{b}_1 + h_1\boldsymbol{e}_1$. If none of the recommended steps leads to a function value decrease, the point $\mathbf{b}_1$ stays unchanged and the variations in the x_2 axis direction are considered, i.e. the function value $f(\boldsymbol{b}_1 + h_2\boldsymbol{e}_2)$ is calculated, etc. When all n variables are considered, there will be a new basic point $\mathbf{b}_2$.

If $\mathbf{b}_2 = \mathbf{b}_1$, i.e. a function decrease is not achieved, the analysis is repeated around the same basic point $\mathbf{b}_1$, but with a shortened step length. In practice, a decrease in the step (steps) by ten times from the initial length is acceptable.

If $\mathbf{b}_2 \neq \mathbf{b}_1$, the search has been executed.

Step 3.

When searching for an example, the information obtained during the analysis is used and the function minimization is completed with a search in the direction given by the specimen. This procedure is fulfilled as follows:

(a) It is quite reasonable to move from the basic point $\mathbf{b}_2$ in the direction of $\mathbf{b}_2 - \mathbf{b}_1$, since the search in this direction has already led to a function value decrease. Therefore, the function in the specimen point can be calculated as

$$P_1 = \mathbf{b}_1 + 2(\mathbf{b}_2 - \mathbf{b}_1) \tag{5.30}$$

In the general case,

$$P_i = \mathbf{b}_i + 2(\mathbf{b}_{i+1} - \mathbf{b}_i) \tag{5.31}$$

(b) The investigation should then continue around the point $P_1(P_i)$.

(c) If the least value in step 3 point (b) is smaller than the value in the basic point $\mathbf{b}_2$ (in the general case $\mathbf{b}_{i+1}$), a new basic point $\mathbf{b}_3(\mathbf{b}_{i+2})$ is obtained, after which it is necessary to repeat step 3 point (a). Otherwise, the search from the point $\mathbf{b}_2(\mathbf{b}_{i+1})$ should not be executed, but the analysis in the point $\mathbf{b}_2(\mathbf{b}_{i+1})$ is to be continued.

Step 4.

Finish this process when the step (steps) length decreases to a given small value.

5.3.3 Optimization Results

TN operation optimization will be executed on the example of the VoIP transmission, using the principle of discrepancy functional minimization of the system functionality quality. Using the above-mentioned minimization algorithm gives the discrepancy functional values.

To fulfil the optimisation the following basic QoS value points were chosen (the assignment error for the model output parameters was taken to be equal to 10 %):

- average packet delay at system passing (τ): 100 ± 10 ms;
- average percentage of the lost packets (%): $1 \pm 0.1\%$;
- standard deviation of jitter for all sources (γ): 8 ± 0.8 ms.

In accordance with the given error values equal to 10 % of the initial values, the solution is carried out with respect to the fulfilled optimization adequacy on the basis of the error discrepancy norm, which was assumed to be equal to 5 % at regularization with the reference values.

For the minimization algorithm it was necessary to assign the initial point and the initial steps of the appropriate coordinate variations. For this aim the following values were chosen:

- buffer size at the input of the bottleneck channel (L): 2000 packets;
- initial step of the buffer size: 500 packets;
- bottleneck channel rate (C): 4.0 Mb/s;

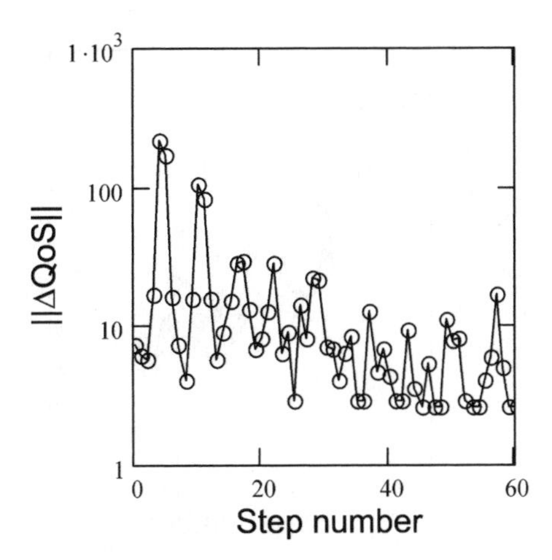

Figure 5.14 VoIP system optimization results (QoS discrepancy functional versus the step number of the functional minimization for the case of the three-parameter optimization) under Ist initial point choice

- initial step of the narrow channel rate: 1 Mb/s;
- number of the system users (*N*): 100×5 users;
- initial step of the number of users: 20 users.

The system optimization was carried out for three input parameters: the bottleneck channel rate, the buffer capacity and the number of users in the system. As seen from Figure 5.14, the optimal point obtained at step 4 has the coordinates: $C = 4.007\,\text{Mb/s}$, $L = 1.125$ packets and $N = 100$ users. The following values of QoS characteristics were obtained at this point in the modelling:

- average packet delay at passing the system (τ) $\sim$87.4 ms;
- average percentage of the lost packets (%) $\sim$0.87 %;
- standard deviation of jitter for all sources (γ) $\sim$6.5 ms.

From the found values it can be seen that the QoS characteristics are quite similar to the required values, but the examined functional can have several local minima, and there is the probability that the functional global minimum has never been found during the modelling process. Applying the present approach, the local minimum of the discrepancy functional can be found using the quality of service parameters, but a more effective procedure of the functional minimization can be used, which would reduce the minimization step number and ensure a reliable search of the functional global minimum.

To estimate the stability of the obtained results, another modelling initial point was chosen and the optimization process was repeated. To fulfil the optimization the following reference points of QoS values were chosen with the errors:

- average packet delays at passing the system (τ): $100 \pm 10\,\text{ms}$;
- average percentage of lost packets (%): $1 \pm 0.1\%$;
- standard deviation of jitter for all sources (γ): $8 \pm 0.8\,\text{ms}$.

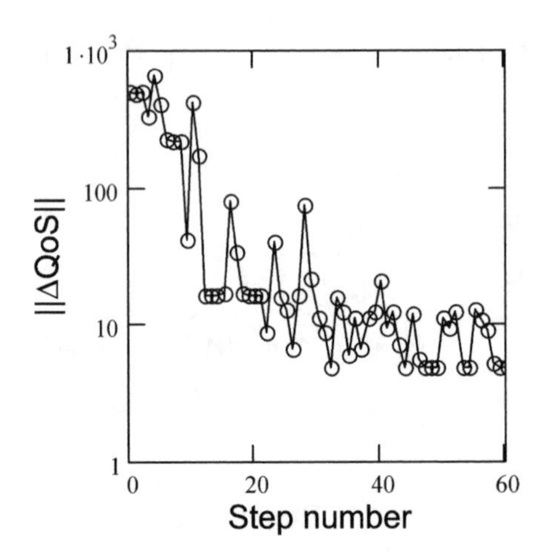

Figure 5.15 Results of VoIP system optimization (QoS discrepancy functional versus the step number of the functional minimization for the case of the three-parameter optimization) under 2nd initial point choice

The following values were chosen as the initial point and the initial steps of appropriate coordinate variations:

- buffer size at the input of the narrow channel (L): 2000 packets;
- initial step of the buffer size: 500 packets;
- narrow channel rate (C): 3.0 Mb/s;
- initial step of the narrow channel rate: 1 Mb/s;
- number of users in the system (N): 150×5 users;
- initial step of the number of users: 20 users.

The optimization results are presented in Figure 5.15. Comparing them with the results of the analogous optimization shown in Figure 5.14 it can be seen that the discrepancy functional value has increased almost twice (4.8 instead of 2.6). Such behaviour indicates model incorrectness as a whole, and the regularization approaches should be applied to find stable solutions for similar modelling. Moreover, it is obvious that the fulfilled optimization does not meet the given problem conditions as it exceeds the given error values.

The observed instability of the obtained solutions indicates incorrectness of the assigned optimization problem with increasing numerical errors and requires further modernization of the used optimization methods. The Tikhonov regularization approach is one of the best-known optimization methods, and can be suggested in this case.

5.3.4 TN Parameter Optimization on the Basis of Tikhonov Functional Minimization

5.3.4.1 Algorithm of Tikhonov Functional Minimization [21]

The models are, in fact, a combination of the operator and the parameter set, part of which is varied to find the optimal characteristics in accordance with some criteria usually used to solve the problem of TN configuration parameter choice. However, if the problem assignment is analysed (5.29), it is not evident that it is formulated correctly, i.e. that the solution exists, is unique and continuous if the initial data changes. Therefore, to obtain solution stable behaviour, it makes sense to apply the regularization methods, e.g. Tikhonov's regularization method, in which the initial assignment (5.29) is substituted for the problem of the minimum search of the following functional [20]:

$$\Phi_\alpha(p) = \| Ap - f \|_F^2 + \alpha\Omega^2(p) \tag{5.32}$$

where $\alpha > 0$ is the regularization parameter and $\Omega(p)$ is the normalizing functional.

The expression $\| p - p_0 \|_U$ is often used as the normalizing functional $\Omega(p)$. There are other methods that may be used to solve similar problems (optimal discrepancy and quasi-solutions), but they require the *a priori* definition of the set $M_0 \subseteq M_p \subset U$, in which the solution search is executed. Factually, this leads either to trivial solutions (like the guaranteed result) or to the considerable complication of the applied algorithms.

The operator describing TN operation is usually a computational model on which various solution types are analysed and where any model corresponds to the real network behaviour with a definite error. In other words, it is necessary to use the approximated operator $\tilde{A}$ instead of the real operator A. Therefore the required TN characteristics should also be assigned approximately (usually the minimal, maximal and expected values are assigned). The problem statement should be changed to

$$\tilde{A}p = \tilde{f} \tag{5.33}$$

where $\| Ap - \tilde{A}p \|_F \leq h \| p \|_U$ or $h\Omega(p), p \in M_p \subset U, \| f - \tilde{f} \|_F \leq \delta, f, \tilde{f} \in F$ and the Tikhonov functional appropriate to the given problem is

$$\tilde{\Phi}_\alpha(p) = \| \tilde{A}p - \tilde{f} \|_F^2 + \alpha\Omega^2(p) \tag{5.34}$$

It should be noted that in this case the problem (5.33) does not need to be solved, which simplifies the TN model formation.

Besides, it was not assumed when stating the problem that operator A is linear. The choice of the regularization parameter α in the case of the nonlinear operator assigned with an error can be effected according to the principle of the smoothing functional:

$$\tilde{\Phi}_\alpha(p_\alpha) = \Delta^2 \tag{5.35}$$

for discrepancy

$$\| \tilde{A}p_\alpha - \tilde{f} \|_F = \Delta \tag{5.36}$$

or (rarely, due to the above-mentioned reason) quasi-solutions:

$$\Omega(p_\alpha) = \Delta$$

The value Δ can be determined as

$$\Delta = hR + \delta \tag{5.37}$$

where the numerical parameter $R \geq \Omega(p_\alpha)$. The solution p_α, minimizing Tikhonov's functional (5.34), converges to the problem (5.29) solution as the errors of the initial data assignment (and the model) decrease.

5.3.4.2 Initial Data

The following parameters appear as the initial data for the analysed system optimization by the classical or normalizing approaches:

1. *Buffer size at the bottleneck channel input.* With the help of this parameter the initial buffer size value is assigned from which the functional minimization algorithm is 'launched'. The buffer size is assigned in packet number units.
2. *Optimization step on the buffer size.* With the help of this parameter the initial step value at optimization of the buffer size, with which the functional minimization algorithm starts, is given. The buffer size step is also assigned in packet number units.
3. *Bottleneck channel rate.* With the help of this parameter the appropriate characteristic of the bottleneck channel is assigned, from which the system optimization of the channel rate starts. The channel rate at modelling is assigned in Mb/s.
4. *Optimization step on the channel rate.* This parameter defines the initial step value at optimization by the channel rate. The rate step is also assigned in Mb/s.
5. *Number of system users.* This assigns the initial number of user hosts functioning in the system.
6. *Optimization step on users.* This parameter defines the initial step of the number of system users at functional minimization.
7. *Hurst exponent.* This defines the self-similarity exponent value at the router input. The given value is re-counted based on HTD shape indexes for ON/OFF periods of the separate traffic source.
8. *Average number of truncated packets.* This defines the appropriate characteristic value with respect of which the optimization is fulfilled on the input model parameters. This characteristic is assigned in percent.
9. *Average delay per packet.* This defines the appropriate characteristic value with respect of which the optimization is fulfilled on the input model parameters. This characteristic is assigned in ms.
10. *Standard deviation of jitter.* This defines the appropriate characteristic value with respect of which the optimization is fulfilled on the input model parameters. This characteristic is assigned in ms.
11. *Step of the regularization parameter.* This parameter defines some step for the calculation of the discrepancy functional of the output parameters versus the regularization parameter.
12. *Modelling time.* This defines the time interval which will be imitated in the model. This parameter is measured in seconds.

5.3.5 Regularization Results

To perform the regularization the initial point of the minimization algorithm was chosen which corresponds to the knowingly worst conditions of the network functioning. The following initial point and initial steps of the appropriate coordinate variation were assigned for the minimization algorithm:

- buffer size at the bottleneck channel input(L): 2000 packets;
- initial step for buffer size: 500 packets;
- bottleneck channel rate (C): 3.0 Mb/s;
- initial step of the bottleneck channel rate: 1 Mb/s;
- number of system users (N): 150×5 users;
- initial step of the number of users: 20 users;
- Hurst exponent: 0.8.

The quality of service characteristics, their errors and also the model error remained the same as at optimization on the discrepancy functional for quality of service characteristics, namely:

- average packet delay at passing the system (τ): 100 ± 10 ms;
- average percentage of the lost packets (%): $1 \pm 0.1\%$;
- standard deviation of jitter for all sources (γ): 8 ± 0.8 ms.

Tikhonov's functional was chosen as the normalizing functional, which was estimated at various minimization steps and also for different values of the regularization parameter α. The values of the minimal discrepancy functional were plotted for all α values, beginning from 17 to zero. The value $\alpha = 0$ corresponds to the case of usual optimization considered earlier.

As a result of running the developed regularization algorithm and the realization of its software, the following conclusions were obtained. Figure 5.16 (a) shows the plots of the Tikhonov functional minimization process for $\alpha = 14$ in the semi-logarithmic scale. It can be seen that the normalizing functional essentially affects the Tikhonov functional minimization procedure, introducing the considerable inertia to the minimization procedure for $\alpha > 14$ (Figure 5.16(a)). For $\alpha = 14$ the discrepancy functional value equal to $\sim$2.39 was obtained, which is near the given error boundary (2.6) and ensures the most stable solution under the given conditions.

The plots of the Tikhonov functional minimization process for $\alpha = 6$ in the semi-logarithmic scale are shown in Figure 5.16(b). It is obvious that the regularizing functional affects the Tikhonov functional minimization procedure, providing less minimization procedure inertia for $\alpha = 6$ compared to the case of $\alpha = 11$. The minimization procedure for $6 < \alpha < 14$ gives the same results; therefore, only the extreme point is given. The discrepancy functional value equal to $\sim$2.41 was obtained for $\alpha = 6$, which is situated almost on the boundary given for the error (2.6) and ensures the most stable solution under given conditions. It should be noted that the discrepancy functional value obtained at this optimization stage is closer to the error value and consequently it ensures a greater stability of the found solution.

The plots of the Tikhonov functional minimization process for $\alpha = 3$ are shown in Figure 5.16(c) in the semi-logarithmic scale. It can be seen that the regularizing functional affects the Tikhonov functional minimization procedure, reducing the inertia of the minimization procedure for $\alpha = 3$ further as compared to the case of $\alpha = 6$. The discrepancy functional value equal to $\sim$4.21 was obtained for $\alpha = 3$, which is more than the given error (2.6) and does not satisfy the given conditions. It can be assumed that a further decrease in

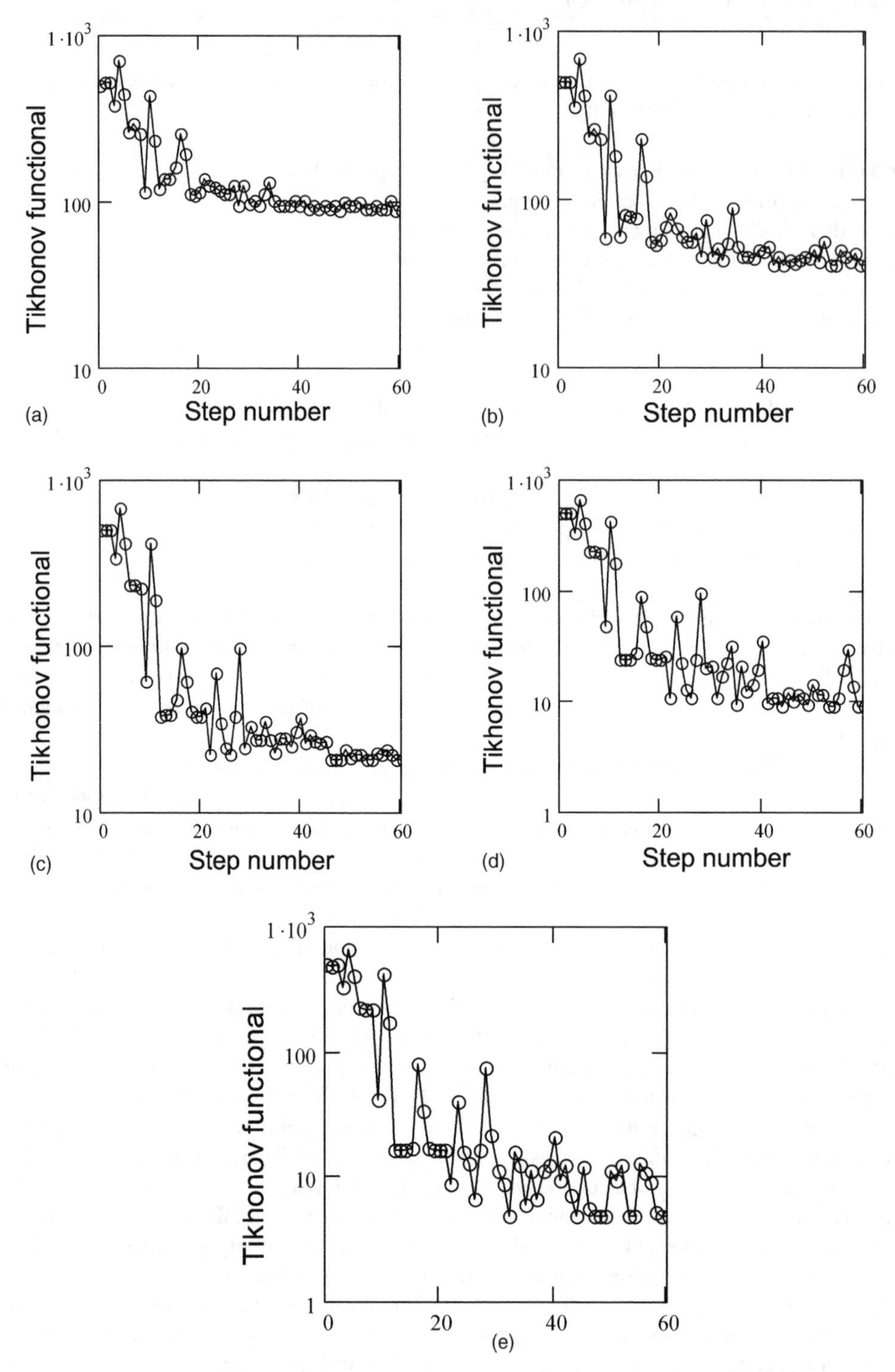

Figure 5.16 The optimization procedure on the basis of Tikhovov functional minimum criteria for various values of the regularization parameter: (a) $\alpha = 14$; (b) $\alpha = 6$; (c) $\alpha = 3$; (d) $\alpha = 1$; (e) $\alpha = 0$

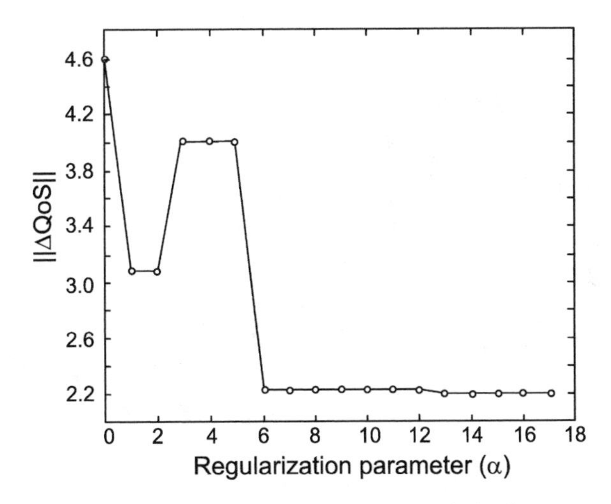

Figure 5.17 Discrepancy functional versus regularization parameter

the regularization parameter, which leads to reduction of the inertia, introduced in the functional minimisation procedure leads to a decrease in the result stability.

Figure 5.16(d) shows the plots of the Tikhonov functional minimization process for $\alpha = 1$ on the semi-logarithmic scale. For $\alpha = 1$ the discrepancy functional value equal to $\sim$3.27 was obtained, which is more than the given error (2.6) and does not satisfy the given conditions.

The quality of service discrepancy versus the regularization parameter is plotted in Figure 5.17. It can be seen from this figure that for small values of the regularization parameter, when the inertia introduced by the regularization functional is inessential, the unstable behaviour of the found results is observed. On the contrary, for $\alpha < 6$ the convergence indicates that the stable optimization results were obtained due to the value increase of the Tikhonov normalizing functional. Together with the obtained solution stabilization, the reduction of the discrepancy functional values is observed below the required threshold, and therefore they can be chosen as the stable solution of the optimization problem. The value for $\alpha = 6$ was chosen as the optimal point.

The plots in Figure 5.18 are given in one coordinate array and illustrate the discrepancy functional minimization procedure for different values of $\alpha = 0$ and $\alpha = 14$. Note that Tikhonov functional values, which are the real optimization criteria, are not marked on the ordinate axis, but the discrepancy functional values with respect to QoS are.

Figure 5.18 shows the channel carrying capacity variation, the buffer size and the number of users during minimization for different values of the regularization parameter. The convergence of the indicated values to several optimal values is observed, which differ by the discrepancy functional and the Tikhonov functional for the optimal case. This confirms once again that due to the regularization approach it is possible to seek more stable solutions than in the case of optimization using the discrepancy functional. For large α it should be possible with the help of the regularization algorithm to improve all input parameters compared with the optimization results on the basis of the functional of the quality of service characteristics discrepancy.

The effectiveness of the Tikhonov optimization approach is compared to the previous case considered in Section 5.2. Using Tikhonov regularization approaches, the discrepancy values

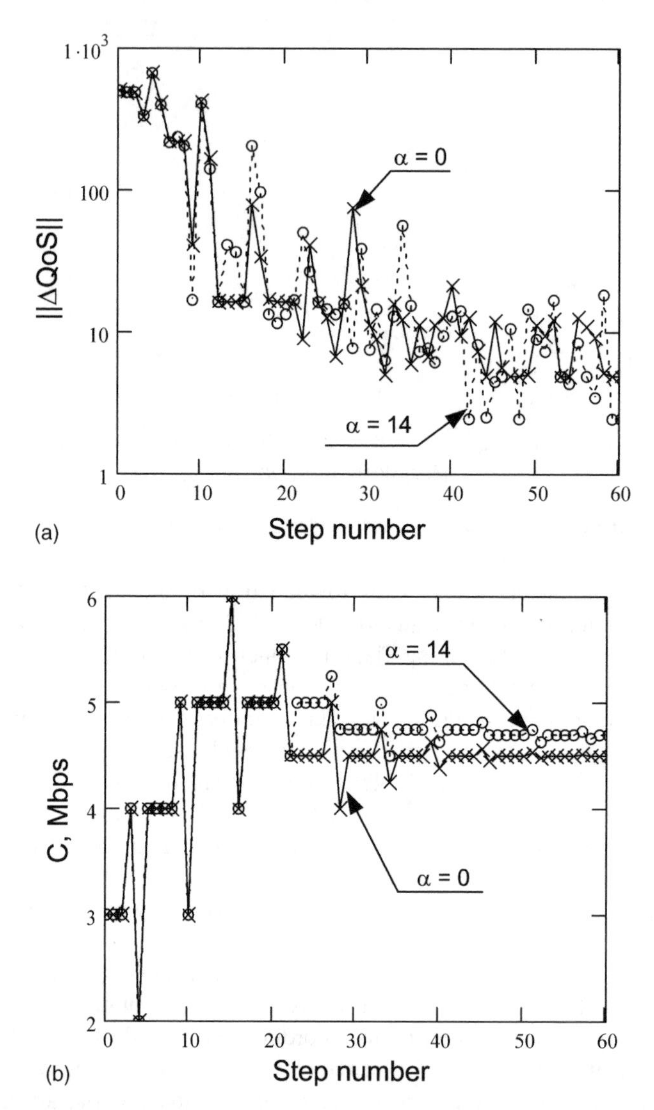

Figure 5.18 Tikhonov functional minimization results for various normalization parameter values versus the minimization step: (a) minimization of the functional of QoS discrepancy; (b) the bottleneck capacity variation during optimization; (c) the buffer capacity variation at the bottleneck channel input; (d) the user number variation on single router

of the chosen objective function $f(\mathbf{X}_n)$ (5.28) versus the Hurst exponent were found in the optimal points for various modelling scenarios (Figure 5.19).

It can be seen that Tikhonov regularization approach applications give similar results (first and second regularization points), which are slightly dependent on the chosen scenario. Therefore, application of the Tikhonov regularization approach permits the possible ambiguity of the multicriteria optimization problem to be eliminated for TN input parameters in order to ensure the given integral **QoS** characteristics.

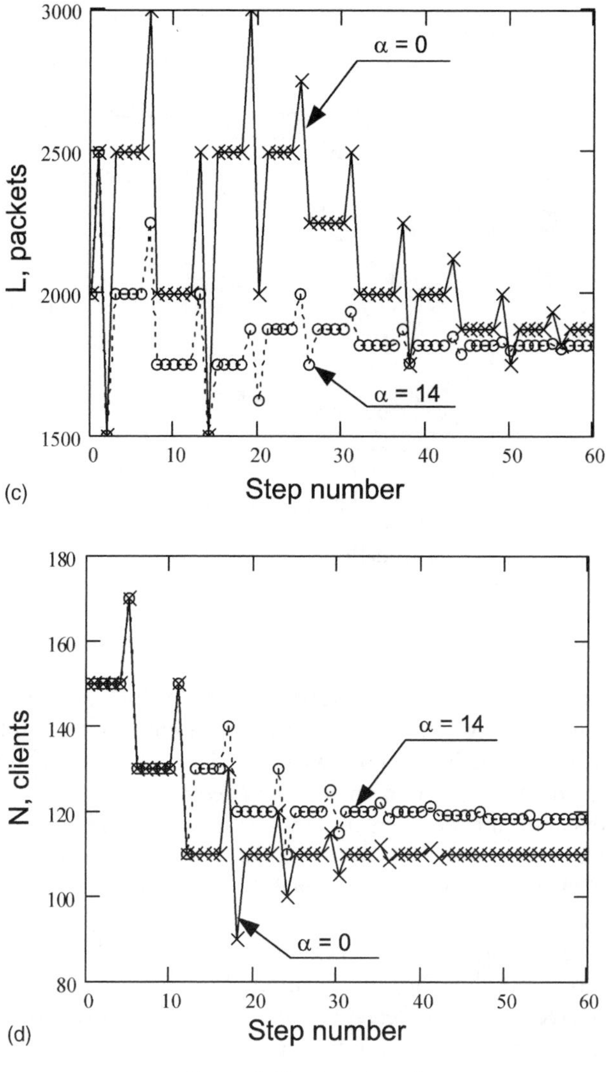

Figure 5.18 (*continued*)

5.3.6 Conclusions

The strategy and algorithms of TN input characteristics optimization for the given QoS parameters at the network output are offered. The numerical results obtained by the modelling confirm the efficiency of the given algorithm.

The observed instability of the obtained solutions when using the classical optimization methods demonstrates the incorrectness of the optimization problem mathematical statement as well as the increase in numerical errors and requires further improvements of the applied methods. The Tikhonov regularization approach is one of the well-known optimization methods, and can be used in this case.

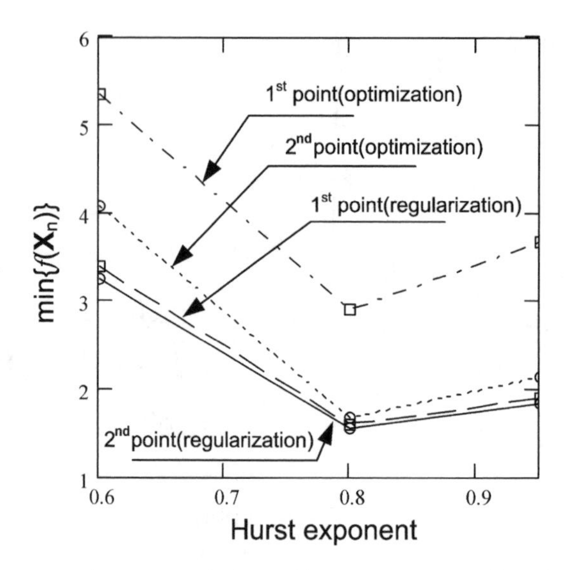

Figure 5.19 Discrepancy values versus the Hurst Exponent for two arbitrarily chosen initial values for the usual optimization types (curves 1 and 2) and for application of Tikhonov regularization approaches (curves 3 and 4)

It is shown that for the large regularization parameter ($\alpha > 6$) the discrepancy functional value slightly varies, which proves the optimization result stability. For $\alpha < 6$ the obtained solution stability falls abruptly, which is caused by a reduction in the regularization functional effect on the minimization procedure and by the growth of the TN model unstable dynamic influence on the optimization procedure. As a result, the considerable oscillation of the discrepancy functional value is observed to depend on the α value as well as QoS discrepancy functional values, which exceed the requirements assigned by errors.

As the regularization parameter α grows, the model optimal input parameters chosen by the regularization algorithm vary. Application of the regularization approaches proves to be very advantageous when designing new telecommunication services, as well as examining the existing services for the given quality of service characteristics and for errors.

5.4 Estimation of the Voice Traffic Self-Similarity Influence on QoS with Frame Relay Networks

The traffic self-similarity influence on the example of the frame relay (FR) will be considered. The frame relay network node designing oriented on the voice message transmission requires the solution of two main tasks:

- ensuring the required quality of service for users;
- parameter choice of the packets transmission system, which will guarantee the first problem solution with the least economic loss, most often with the least contractual carrying capacity requested from the frame relay network.

For a small error probability during the transmission through the network main line, the quality of service for users, i.e. the perception quality of the transmitted voice signal, depends primarily on:

- voice signal distortions related to its encoding and decoding algorithms;
- interfering effects caused by total delay during the network transmission, which should be small enough;
- interfering effects related to the interrupted transmission, caused by the random character of the frame arrival from end to end, by possible frame loss due to buffer overflow, by late packets truncating, etc.

In the general case the network designers should control not only the delay time mathematical expectation but also the variance.

5.4.1 Packet Delay at Transmission through the Frame Relay Network

The total transparent packet delay can be calculated as

$$D(t) = V + h + d(t) + B \tag{5.38}$$

where V is the delay in the analogue-to-digital converters, the essential part of which is defined by the algorithmic delay and is different for the possible coding methods (for the analysed coding method LD-CELP (codec G.728) it is equal to 5 ms only); h is the packetization time, which is determined by the packet length and by the rate of its generation; $d(t)$ is the delay inside the network at moment t, determined by the packet arrival time before the beginning of its transmission (waiting time in the queue) and the packet transmission time proper; and B is the packet processing time on the receiving side, which is often determined by the buffer capacity, connected in front of the decoder and serving, in turn, for delay time fluctuation smoothing and reducing its dispersion.

The delay $D(t)$ should not exceed 200–250 ms from the point of view of voice signal perception. Using the LD-CELP algorithm, the considerable delay part can be attributed to the buffer size growth.

5.4.2 Frame Relay Router Modelling

The bridge router, which consists of the necessary number of connection boards for the user or connection lines, is the main part of the equipment providing the user access to the services of VoFR (voice over frame relay) technologies. The voice signal coding methods and appropriate packet formats are defined by the FRF.11 document. It is important to emphasize that this standard ensures the possibility of changing the length of the voice packet informative part by the network operator, which is the tool used to increase the effectiveness of the main line carrying capacity for the frame relay network. A node model, such as the queuing model, is often used as the optimal parameter choice of the nodes for packet voice transmission.

Consider the router operation, which has m ports for the voice signal source connection and one output port to connect the frame relay network. It is assumed that the output port is used for voice packet transmission only, but when the data transmission is necessary the other connection ports for the connection with the frame relay are used. The router consequently interrogates the output registers of the coding devices of all m ports. The interrogation period is determined by the packetization time h. Thus, the time moments at which the packet can arrive for the transmission are discrete. The active packet arrival probability at the output of each codec is determined by the current state of the given user voice signal.

The statistical modelling method is used to estimate the operation quality of the designed network node and to choose the optimal parameters for adjustment. This method in essence consists of an imitation model development and obtaining with its help the node probabilistic characteristics [22]. The software should provide a collection of the statistical information on the number of arriving, serviced (transmitted), delayed and lost packets, and also on the delay of each packet. On the basis of the collected information the following variables are determined:

- the probability of the occurrence at the output of the packet with delay $d(t)$, equal to the number of packets delayed by t divided by the number of received active packets at the input;
- the probability of occurrence at the output of the packet with the delay that does not exceed $d(t)$, equal to the sum of the packet occurrence probabilities with the delay that does not exceed the given value;
- the delay average in the sample of N packets for the definite router adjustment parameters (the number of definite users at the input and the packet number at the output FR frame), equal to the sum of the products of all delays by its occurrence probabilities;
- the channel utilization as the ratio of the number of transmitted packets to the packet number, which could be transmitted potentially during the time interval, determined by the sample length.

It is assumed for the future that the router speed is so high that the query of all ports is performed simultaneously. It is obvious that if the number of input ports exceeds the packet number in frame relay frames, the router should use the buffer in which the arriving packets are saved, forming the queue. This buffer presence is the distinctive sign of the statistical multiplexing system.

The queue buffer is a one-dimensional array. The number of buffer cells in which the user packets can be recorded is strictly assigned at the beginning of simulation. During the operation the arriving packets are recorded in the free cells of the buffer, after which the FR frame is formed from the packets that are given from the beginning of the buffer queue. The following stage is the buffer contents overwriting (packets movement in the queue to the place of transmitted packets), after which the cycle is repeated. The queue service algorithm is the FIFO. The FR output frame formation time is constant. In the analysing router this time was defined by the voice packet formation time, equal to 5 ms.

As an example, Figure 5.20 illustrates the router model with the users connected to five input ports. At each time moment the active packets may arrive from all users. If at this time moment the delayed packets are already saved in the buffer, the record of the arriving active packets is executed in the queue end. Then the router forms the frame relay frame, in the informative part of which the packets that are saved in the initial buffer cells are included, and transmits it to the output port.

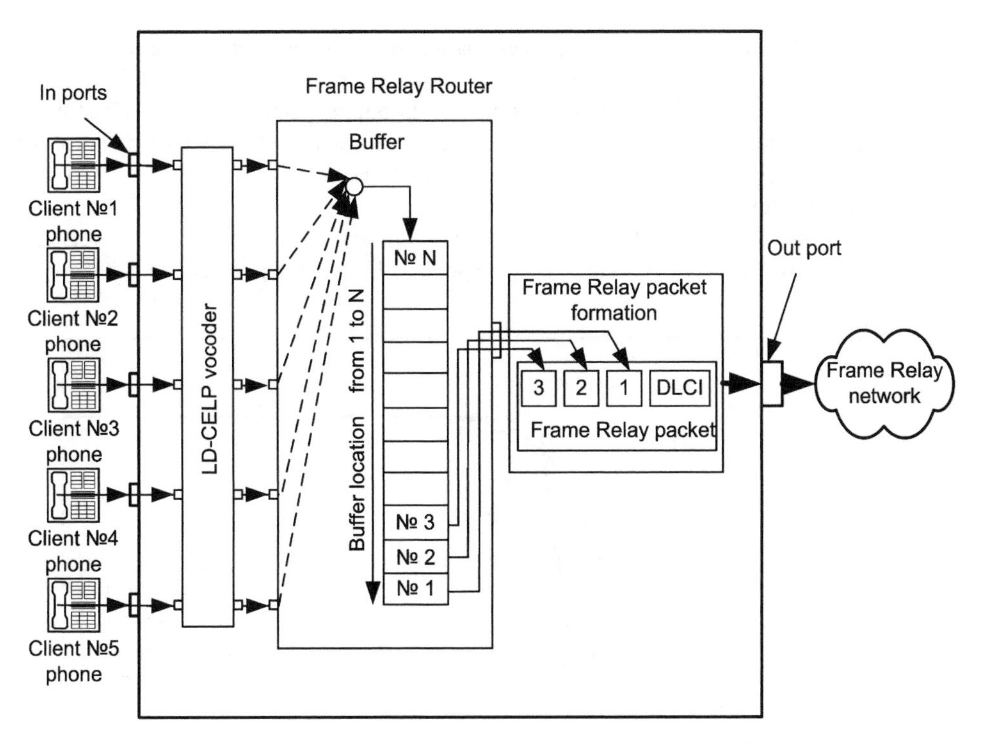

Figure 5.20 FR router model: 1, 2, 3 are packets forming the FR frame (dotted arrows designate the possible LD-CELP frame arrivals from the codec outputs)

In the analysed example the number of packets in the frame is equal to 3. The delay packet buffer is renewed, and all packets saved in the buffer move three cells forward. After that the input port interrogation is repeated and the packet that has just arrived at the input from any user is recorded at the buffer queue end. As a result, the voice packets, saved in the first three cells of the buffer, will be included in the informative part of the next frame relay frame. Afterwards the cycle is repeated.

Using the developed models of the voice packet source and of the statistical multiplexing node, relations permitting parameter choice of the real network node can be found. The generalized software algorithm that imitates the FR router operation (Figure 5.20) and the FR router imitation modelling results for codec G.728 at the input and Markovian models of the input processes are considered in some publications. At the same time, researches show [22] that the aggregated voice traffic is described more accurately by the fractal (self-similar) models.

5.4.2.1 FR multiplexer modelling for fractal traffic as input

Input Traffic Formation

The input traffic in the form of the aggregated flow is considered, where several elements are presented in the form of the LD-CELP packet number. The principle used to obtain the aggregated traffic is shown schematically in Figure 5.20. To simulate the input traffic fractal

Gaussian noise is used, where the total process parameters (the mean value, the variance, the correlation function and the Hurst exponent H characterizing the fractal (self-similar) properties) are found as a result of processing of real aggregated voice traffic [18,23]. To estimate the influence of input process fractality (self-similarity) on the characteristics of the multiplexer performance, several processes with various Hurst exponents (H=0.5, 0.6, 0.7, 0.8, 0.9) were modelled.

Since the input multiplexer flow is considered as the number of LD-CELP packets at a certain time moment, the FGN values were expressed by the integer number. In the case

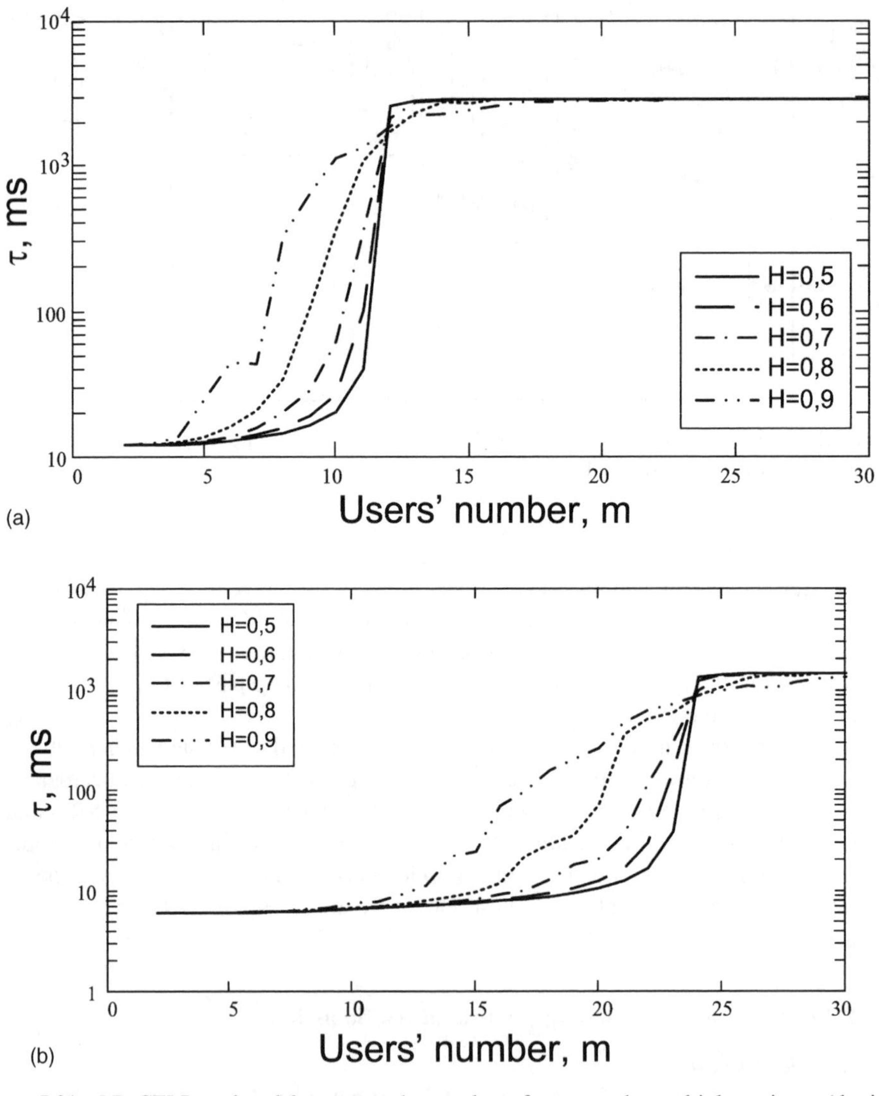

Figure 5.21 LD-CELP packet delay versus the number of users at the multiplexer input (the input traffic was obtained on the basis of FGN): (a) output channel rate 64 kb/s; (b) output channel rate 128 kb/s; (c) output channel rate 256 kb/s

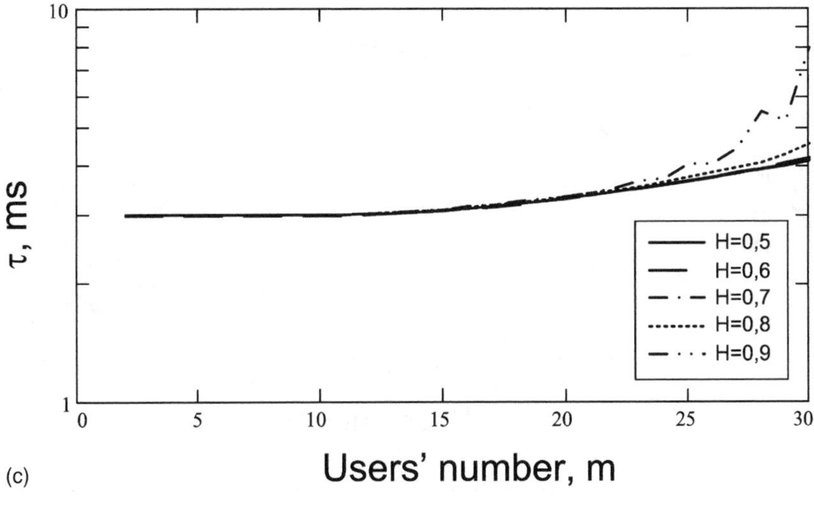

Figure 5.21 *(continued)*

when the FGN process had negative values, they are equal to zero. The FFT algorithm was used for FGN modelling.

5.4.3 Simulation Results

When the above-mentioned processes are used for modelling, the following results are obtained.

Figure 5.21 shows the plots of the LD-CELP packet delay versus the number of users at the multiplexer input. The plots are presented for various rates of the output channel. Each plot has five curves, each having its own Hurst exponent value assigned. The plot shows that as the number of users at the input grows, the LD-CELP packet average delay increases and in all cases converges to some constant value, which can be explained by the limited buffer capacity used in modelling. In the cases when the buffer was overflowed, the packets were simply dropped, which proved the existence of maximal delay. There is also the lower delay limit, which equals the transmission time of one FR frame. When the output channel rate increases, a decrease in the average delay total is observed in all cases.

The plots in Figure 5.21 show that as the Hurst exponent varies the average delays in the system also essentially change. To be more exact, the self-similarity degree increase leads to a growth in average delays. This phenomenon is especially vivid for the medium system utilization ∼0.8–0.9. However, in the overflowed state, all average delays converge to a single value due to the limitation of the buffer size.

Figure 5.22 shows the plots of the LD-CELP packet blocking probability versus the number of users at the multiplexer input. The plots are presented for various output channel rates. Each plot has five curves, each having its own Hurst exponent value assigned. The plot shows that when the number of users at the input grows, the LD-CELP packet blocking probability also increases. The blocking probability in all cases tends to some value that can be explained by the limited buffer size.

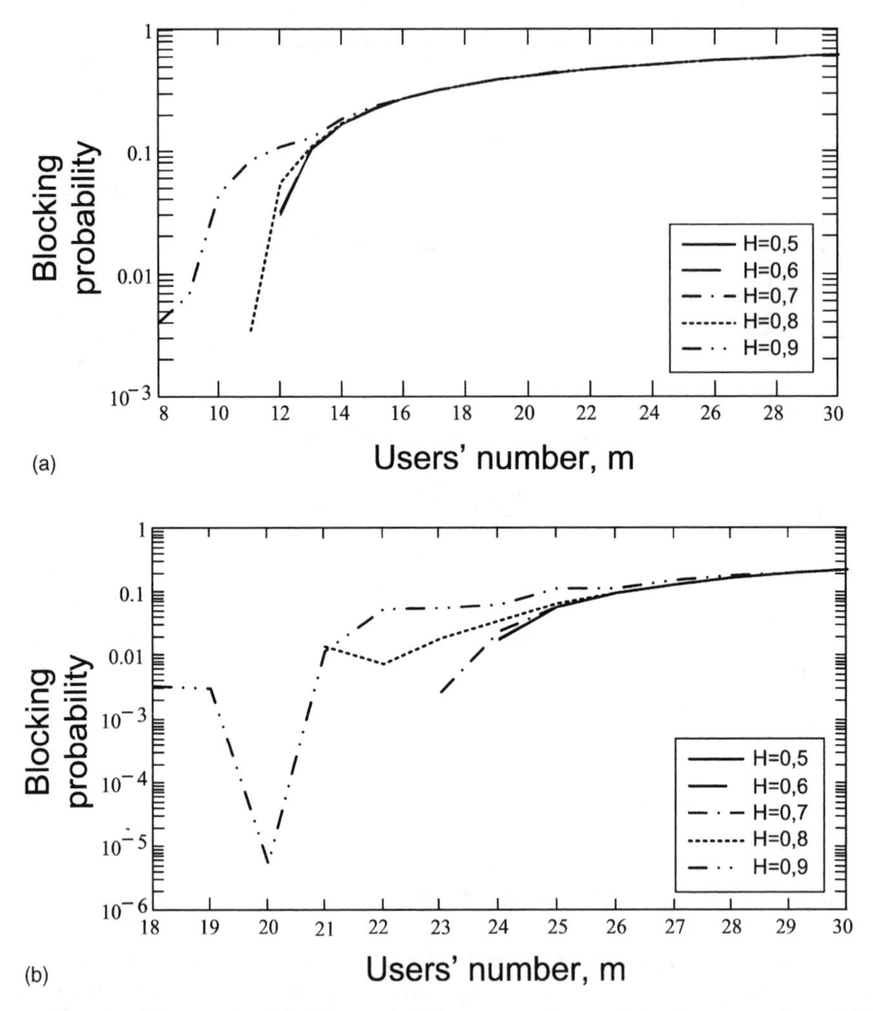

Figure 5.22 LD-CELP packet blocking probability versus the number of users at the multiplexer input (the input traffic is obtained on the basis of FGN): (a) output channel rate 64 kb/s; (b) output channel rate 128 kb/s

From the plots in Figure 5.22 it can be seen that as the Hurst exponent grows the blocking probability increases for the permanent user number at the multiplexer input. Figure 5.23 shows the channel utilization factor versus user number at the multiplexer input. It should be noted that as the Hurst exponent grows the utilization factor increases for the permanent user number at the input, although the increase is insignificant.

Figure 5.24 shows the delay distributions that are constructed for the experimental data and the exponential distributions fitted to them. The plots are presented for various rates of the output multiplexer channel and for various Hurst exponent values. It is evident that in all cases the exponential distribution factually does not correspond to the experimental data histograms. When the self-similarity exponent H increases under other equal conditions, the

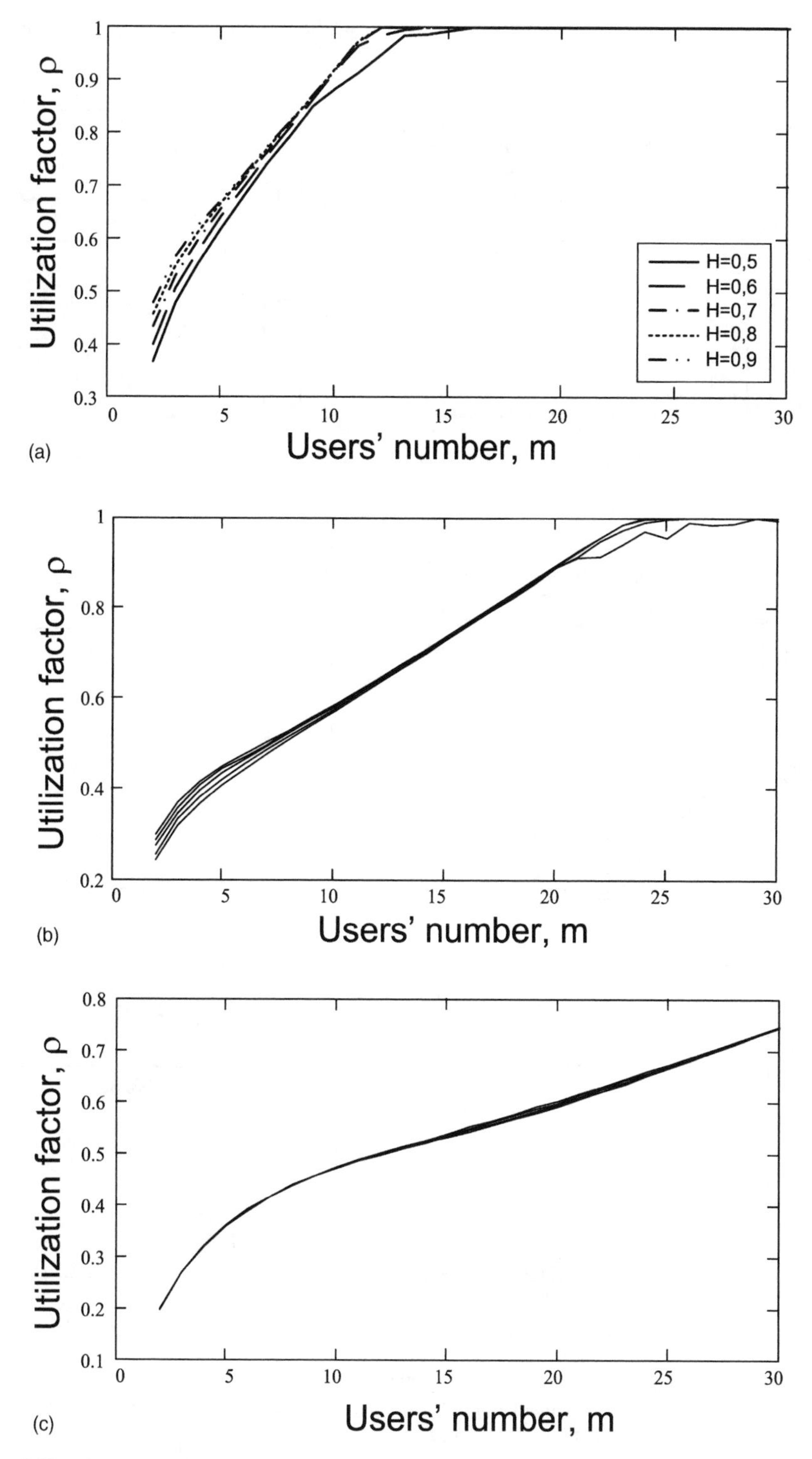

Figure 5.23 Channel utilization factor versus the number of users (the input traffic on the basis of FGN): (a) channel rate 64 kb/s; (b) channel rate 128 kb/s; (c) channel rate 256 kb/s

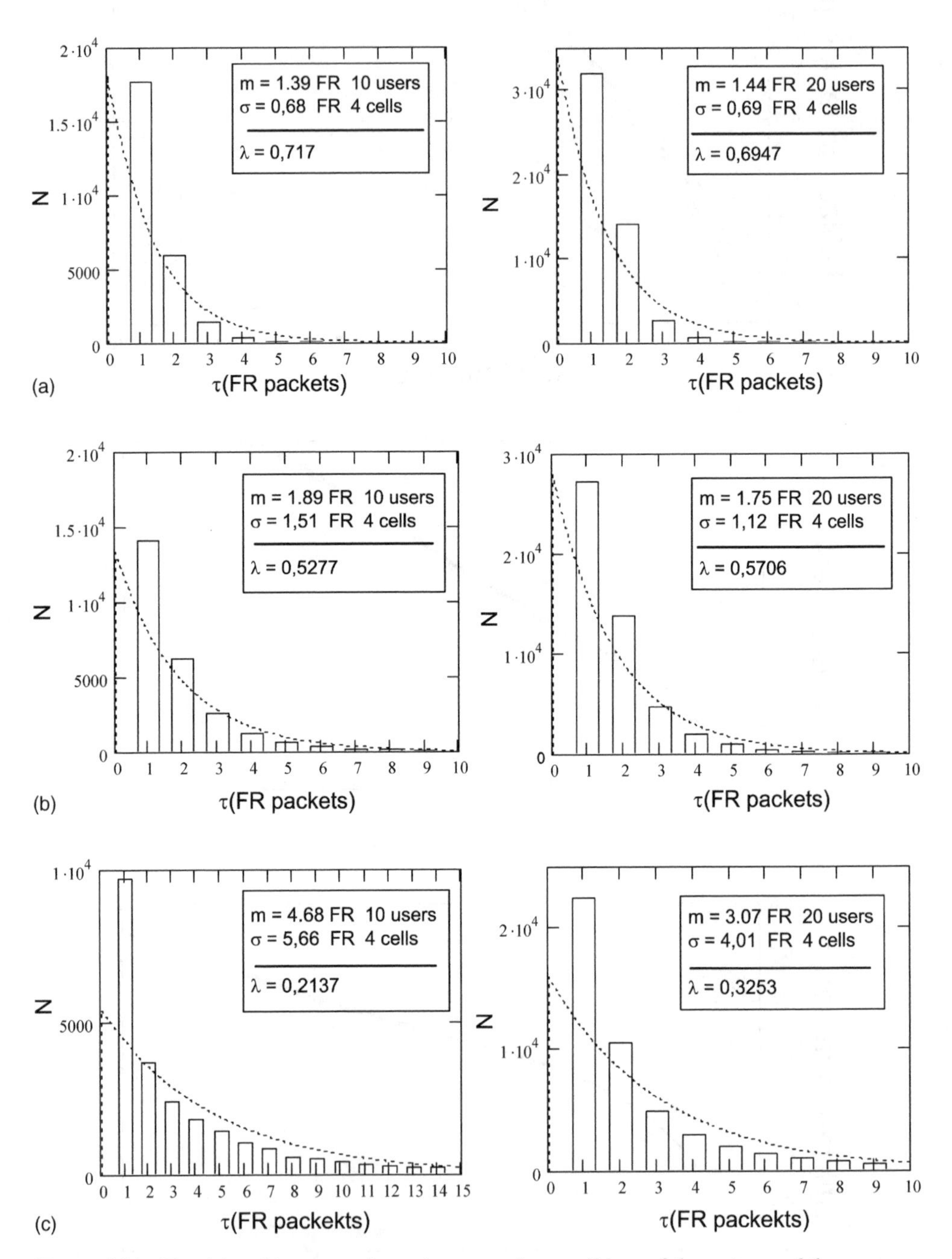

Figure 5.24 The delays histograms for various operation conditions of the system and the exponential PDF fitted to them for the channel rate 64 kb/s(left) and channel rate 128 kb/s: (a) $H = 0.5$; (b) $H = 0.6$; (c) $H = 0.7$; (d) $H = 0.8$; (e) $H = 0.9$

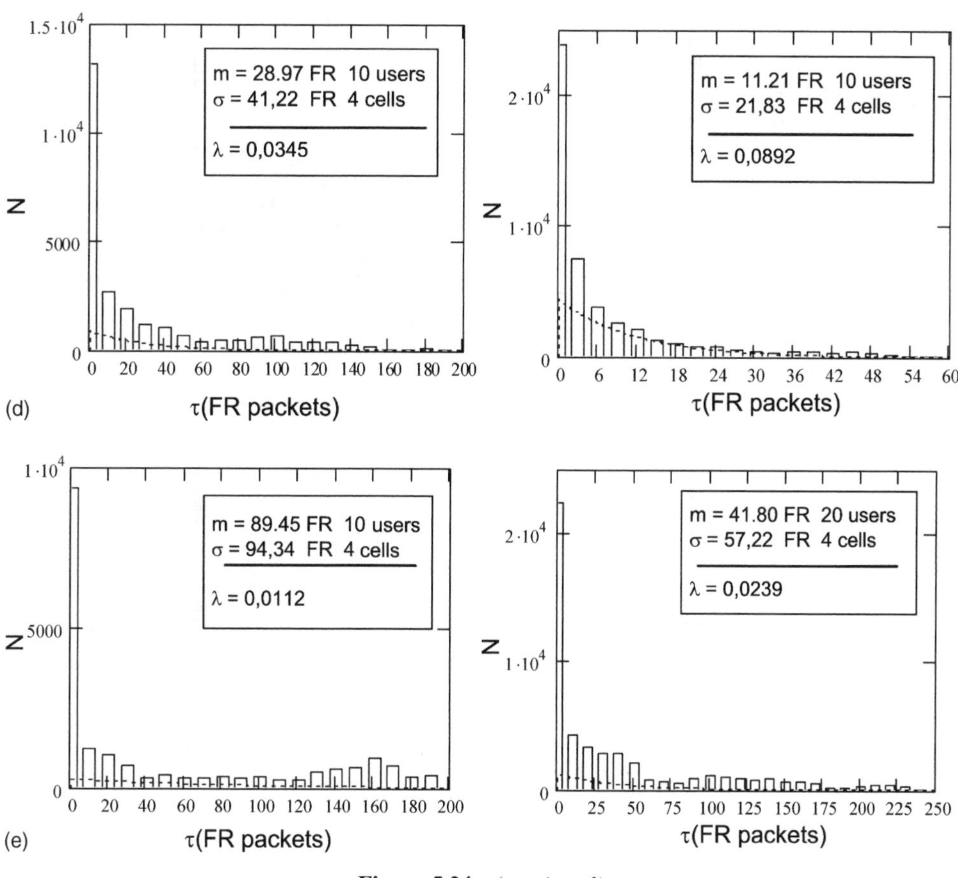

Figure 5.24 *(continued)*

tail of the delay distribution becomes 'heavier' so the exponential distribution should not be used to describe such a process.

Delay CCDFs for various Hurst exponents and for various output channel rates present the most exhaustive information on the delay statistics in the system illustrated in Figure 5.25. The modelling mode with 11 users for the output channel rate of 64 kb/s was chosen, but for a rate of 128 Kb/s there are 22 users. It is evident that as the Hurst exponent increases the system performance decreases.

5.5 Bandwidth Prediction in Telecommunication Networks

The problem of prediction and its solution is quite acute because the forecasting data concerning the carrying capacity allow additional information to be obtained to solve the control problem, namely to form the algorithm to prevent congestion. The solution of this problem, as a rule, is reduced to determination of the algorithm with the adaptive realignment mechanism of separate network components. The algorithm of the current TCP

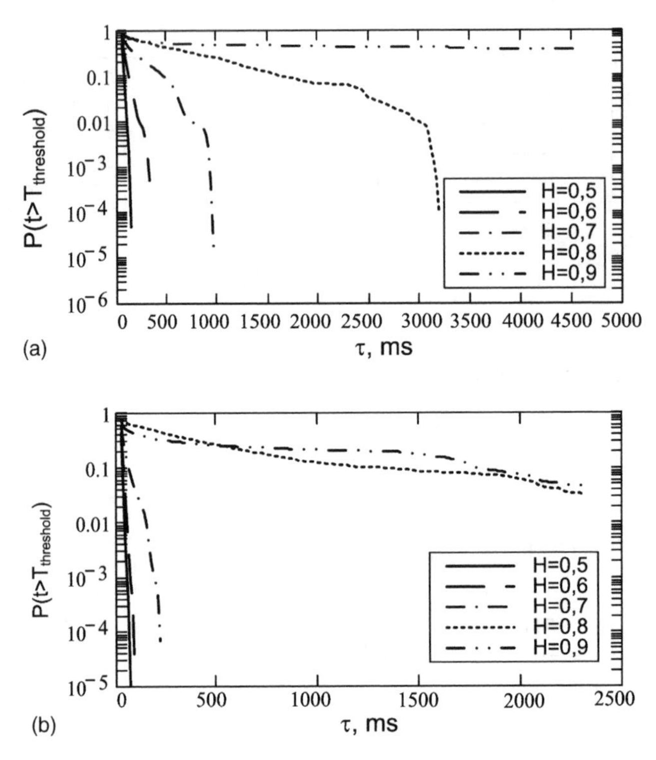

Figure 5.25 Delays CCDF for various operation conditions of the system (the number of users varies): (a) rate 64 kb/s, 8 LD-CELP packets on FR; (b) rate 128 kb/s, 8 LD-CELP packets on one FR frame

window variation (the structure to congestion and thereby to decrease the losses in the network carrying capacity) can be indicated as one of the options of using this mechanism. If the packet loss is not crucial for the information transmission process, data flow control using the UDP can be chosen through the mechanism of changing the intensity of the number of sent packets on separate network sections, while is another example of this algorithm application.

The process estimate during the prediction is formed not on the finite observation interval but outside it, on some prediction time interval. The process prediction in time moment t_2 can be designated as $\hat{x}_2^0$:

$$\hat{x}_2^0 = M(x_2^0|x_1) = \int\limits_{-\infty}^{\infty} x_2^0 \omega(x_2, t_2|x_1, t_1)\mathrm{d}x_2 \tag{5.39}$$

The expression for the prediction estimate over the forestalling interval t_2–t_1, can be written on the basis of the known value x_1^0 in time moment t_1 as

$$\hat{x}_2^0 = r(t_2, t_1)x_1^0 \tag{5.40}$$

where $r(t_2, t_1)$ is the correlation coefficient of the random point process and $x^0_{1(2)} = x_{1(2)} - \mu_{1(2)}$ are respectively the centred component and the mathematical expectation of the random process.

The prediction problem is considered for the counting characteristics of the network traffic, described by the point process, which approximates the flows of the packet series (packs). The optimal forecast implies finding the optimal (in an r.m.s. sense) estimates of the series X_{n+k} in the interval $(t_{n+k}, t_{n+k} - T)$, displaced from the last observation result (series X_n in the interval $(t_n, t_n - T)$) on kT time, where k is the displacement parameter. Assuming that the process is stationary, the intensity of the point process operation is assumed to equal the known constant value λ. For the counting characteristics the forestalling interval and the correlation coefficient become equal to kT and $r(k, T)$ respectively. Identifying t_2 and t_1 with the time moments $(k+n)T$ and nT, and also assuming that $\mu_1 = \mu_2 = \lambda T$, from expression (5.40) the optimal sample X_{n+k} prediction estimate for the known sample X_n is obtained:

$$X_{n+k} = r(k;T)(X_n - \lambda T) + \lambda T \tag{5.41}$$

The value $r(k;T)$ depends on the initial data and on the problem solution features and takes one of the forms, (see, for example, Equation (2.37))

$$r(k;T) = \frac{R(k;T)}{D(T)} = \frac{T^\alpha}{2(T^\alpha + T_0^\alpha)}\left[(k+1)^{\alpha+1} - 2k^{\alpha+1} + (k-1)^{\alpha+1}\right]$$

The prediction quality for the considered problem can be estimated on the basis of the error variance value for the given displacement parameter k:

$$\varepsilon_k^2 = M[(\hat{X}_{n+k} - X_{n+k})^2]$$

After transformation,

$$\begin{aligned}\varepsilon_k^2 &= M\{[r(k;T)(X_n - \lambda T) + \lambda T - X_{n+k}]^2\}\\ &= r^2(k;T)D - 2r(k;T)R(k;T) + D\end{aligned}$$

Taking into account the fact that $M(X_n^2) = M(X_{n+k}^2) = D + \lambda^2 T^2, M(X_n X_{n+k}) = R(k;T) + \lambda^2 T^2$, where $R(k;T) = M\{X_n X_{n+k}\} - (\lambda T)^2$ is the correlation function of sample numbers in the intervals of T duration spaced on kT time, and $R(0;T) = D; r(k;T) = R(k;T)/D$, it is finally found as a result that $\varepsilon_k^2 = D[1 - r^2(k;T)]$.

It follows from this expression that when k parameters increase, which corresponds to the prediction becoming more profound, the forecast quality worsens (the error increases) since the correlation coefficient decreases. It should be noted that due to long extensive dependence $r(k, T)$of the statistics, better prediction quality can be provided for the given k parameter compared to the short extensive statistics of the usual random process model. At the limit as $k \to \infty$, the forecast error variance tends to *a priori* variance $C(0;T) = D$. The forecast can be improved by using the series of previous predictions, besides the last one, with possibly different measurement weight coefficients having m common numbers.

An optimal prediction estimate in a simple aggregation version gives the following expression:

$$\hat{X}_{n+k} = \frac{1}{m} \sum_{j=n-m+1}^{n} [r(k+n-j;T)(X_j - \lambda T)] + \lambda T, \quad 1 \leq m \leq n \tag{5.42}$$

It can be shown that, as for the counting characteristics, the prediction error value in the analysed case depends on the character of the correlation coefficient behaviour. The stronger the statistical dependence of the process adjacent samples (to which the extensive dependence contributes to a great extent) the slower is the correlation coefficient reduction and thereby the prediction error value is less.

On the basis of the results, a formulation can be obtained of the number of suggestions concerning the informational flow control in the computer networks. The carrying capacity (the data transmission rate) of the communication channels between users is one of the most important parameters characterizing the network operation quality.

Discussion will now concentrate on the suggestions of the informational flow control on the basis of the counting characteristics prediction. This control can be executed, for example, with the help of the modified UDP version.

Since the first- and second-order counting characteristics (the intensity and the correlation function of the point process) can only be determined experimentally on the network separate parts, the information flows between the separate nodes of the network virtual connection become subject to optimization and control. Two nodes will be chosen, of which the ith node is the source and the jth is the receiver. It is assumed that the flow intensity (the network section carrying capacity between the ith and jth nodes) is determined by the queue in node j, caused, for example, by the limited buffer memory size in this node, by the low intensity of the departures from the buffer or by the packets arriving at this node from other network connections, etc.

Due to this, the information flow intensity from node i to node j is reduced, and in the case of buffer overflowing in node j the information transmission is interrupted and is accompanied by partial packet loss. To avoid the complete loss of the carrying capacity, it is necessary to control the buffer loading level in node j. To do this, use should be made of the forecasting estimates of the counting characteristics. The sample number X_n in the interval $(t_n, t_n - T)$ is measured for the analysed network section. Simultaneously, the prediction is fulfilled over some forestalling interval kT, where k is the displacement parameter. As an example, the procedure will not be taken beyond the prediction of more than one step ahead. The forecasting estimate X_{n+1} is determined either by the last measurement result (5.41) or on the basis of some number of previous measurements, according to the aggregation procedure (5.42) at $k = 1$.

If the prediction value is more than the threshold, depending on the buffer loading level and determined by some adaptive algorithm, the information generation intensity from node i decreases by a value depending on the buffer loading level and the prediction estimate values due to the feedback signal. This causes a reduction of carrying capacity in the network section, but since the information transmission process goes on and the number of lost packets decreases, the average losses in carrying capacity of this connection can be reduced. If carrying capacity needs to be retained, the available network resources are redistributed on the basis of the feedback signal in favour of this section of the network virtual connection.

5.6 Congestion Control of Self-Similar Traffic

A study of the *control* problem for self-similar traffic is at the initial stage of its development. Self-similar traffic control means those traffic regulations that would make the network efficiency (including the carrying capacity) optimal. The scale-invariant traffic structure introduces new complexities in the general background; consequently the task to ensure a high quality of service (QoS) (together with achieving the high utilization factor) becomes much more difficult. The most important issue here is that the scale-invariant bursty structure implies the existence of density periods with high activity over the large time scale, and it adversely affects the congestion control. The bursty structure over the finite time scales is similar to the bursty structure observed for the traditional short-range dependent traffic models. The same property can be observed for rougher scales, where additional congestion intervals appear as well as the lack of ample operation and a decrease in total efficiency. However, on the other hand, the long-range dependence (by definition) implies the presence of the unusual correlation structure, which can be used for congestion control and is not used in the existing algorithms.

The possibility of 'prediction' with great reliability under self-similar traffic conditions is proved in some publications so that this information can be effectively used for congestion control:

1. It is demonstrated that for a prediction of future traffic levels and for a disputable situation on the time scale in the limits and over time scales of the overloading feedback control the long-range dependence can be determined in the real time mode.
2. The traffic regulating mechanism can be offered based on the multiscale structure of congestion control [24], in order to use this information for the common good for higher network operation efficiency, particularly to increase perfromance.

The congestion control mechanism operates selectively, applying the aggressiveness, using the predicted property when it is ensured, increasing the data rate if the reduced disputable situation level is predicted and raising the predicted low disputable situation level. The mechanism of *selective aggressiveness* is quite useful even for short-range dependent traffic, although this mechanism is much more effective for long-range dependent traffic, leading to a relatively greater advantage in performance.

Consider the strategy ideas for congestion control, referred to as *selective aggressiveness control* (SAC), and show its efficiency, when using the forecasting structure that is present in long-range dependent traffic, to improve network operation efficiency [25]. The control scheme is the type of predicted congestion control based on the multiscaled structure of congestion control. An accurate prediction of the long-term network condition l is executed in the SAC time scale (1–5 s). The specific control action $\varepsilon(l)$ is fulfilled by SAC based on information about the future, and is included in the applied scheme of congestion control in order to influence decision-making related to traffic control. The generalized structure is shown in Figure 5.26. SAC is aimed at stability, efficiency and availability so could easily be introduced into the existing schemes of congestion control.

The SAC method should supplement and raise the efficiency of the existing responding congestion control schemes. To demonstrate this, a simple, often used congestion control, based on the rate, is chosen and it is assumed that the control module uses it. SAC always

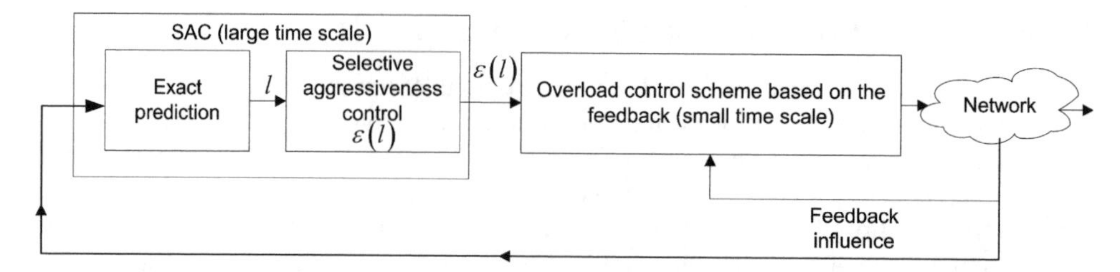

Figure 5.26 The generalized scheme structure of the predicted congestion control. The SAC module is active on the time scale (1–5 seconds), exceeding the time scale of the feedback loop for the applied scheme of congestion control

supports the decision made by the applied scheme of congestion control concerning the immediate variation of the traffic rate (increasing or decreasing); nevertheless, it can regulate the variation value. In other words, if at any time moment the applied scheme of congestion control decides to increase the traffic rate, SAC will never undertake the opposite action to decrease the transmission rate. Instead, SAC will amplify or weaken the immediate variation value on the basis of the predicted future of the network condition.

The fundamental idea of the approach consists in the following. SAC will aggressively try to occupy the carrying capacity, if it is predicted that the future network condition is underloaded, fitting the aggressiveness level as the function with respect to predicted underloading. As a result, the more long-range dependent the traffic, the higher is the performance benefit due to SAC application.

During two decades (at the end of the 1980s and at the beginning of the 1990s) congestion control was the subject of active network studies [26–35]. The problem of 'delay-carrying capacity', occurring in the networks with large carrying capacity, as well as the problem of the quality of service, resulting from the necessity to support the multimedia communication in the real time mode (on-line mode), is more up to date [36–39]. One of the lessons learnt from the congestion control research is that end-to-end rate control, based on feedback and using various types of linear increase/exponential decrease, can be effective, and to achieve the stability it is necessary to keep asymmetry in the control laws.

To demonstrate the selective aggressiveness control efficiency under the conditions of self-similar traffic, as the basic configuration a simple but typical case of congestion rate control based on feedback is considered. Applying the aggressiveness selectively (based on a network future content prediction), a reduction in the effort spent to achieve the stability is discussed.

Let λ be the packet arrival intensity and γ characterize the network performance. The applied congestion control scheme, based on feedback with linear increase/exponential decrease, uses the control law in the form

$$\frac{\mathrm{d}\lambda}{\mathrm{d}t} = \begin{cases} \delta, & \text{if } \mathrm{d}\gamma/\mathrm{d}\gamma > 0 \\ -a\lambda, & \text{if } \mathrm{d}\gamma/\mathrm{d}\lambda \end{cases} < 0 \tag{5.43}$$

where $\delta, a > 0$ are positive constants. As a result, if the data rate increase leads to performance growth (i.e. $\mathrm{d}\gamma/\mathrm{d}\lambda > 0$), the data rate increases linearly. Conversely, if the data rate increase leads to a performance decrease (i.e. $\mathrm{d}\gamma/\mathrm{d}\lambda < 0$), the data rate decreases exponentially. In fact, the condition $\mathrm{d}\gamma/\mathrm{d}\lambda > 0$ can be substituted with different congestion measures.

5.6.1 Unimodal Ratio Load/Throughput

The 'throughput' γ (meaning useful productivity) can be defined in several ways depending on the context, from *reliable throughput* (the number of bits reliably transmitted during a time unit, when the congestion mechanism reliability is taken into account) to *rough productivity* (the number of bits transmitted during a time unit), to *power* (one of the productivity measures divided by the delay). The rough productivity (denoted as v) can be easily measured (by way of simple tracking of the packet numbers, in bytes, arriving to the recipient during the time unit) and achieved (in most cases $v = v(\lambda)$ is the monotonously increasing function in respect of λ). However, it cannot adequately divide the congestion control schemes, which achieve the definite rough productivity, without experiencing high packet losses or delays.

For example, using the automatic request (ARQ) on repeated transmission with finite buffers on the sender and recipient sides, the achieved reliability requires complicated control and coordination. As a result, the high packet losses can have a great influence on the effective functioning of such control schemes (e.g. the TCP control window). In particular, if the packet loss degree is high, this can mean, for the given rough productivity, that the essential part of the rough productivity includes the duplicate packets (due to leading repeated transmissions) or, due to the packets, will be truncated on the recipient side due to 'fragmentation' and buffer overflow. Therefore, the reliable productivity will be lower, i.e. related to the combination 'rough productivity/packet loss degree'.

The used application characteristics determine how much the packet loss affects the application performance. To examine such waiste better, the productivity measure $\gamma_k = (1 - c)^k v$ will be used, which decreases (polynomially) the rough productivity v with the help of the packet loss degree $0 \leq \tilde{n} \leq 1$, where the severity can be defined by the $k \geq 0$ parameter. Therefore, the rough productivity v is the specific case of v_k, where $k = 0$.

5.6.2 Selective Aggressiveness Control (SAC) Scheme

Assuming that the network future condition is predictable with a sufficient degree of accuracy, the following question still remains: what is to be done with this information to improve network efficiency? The choice of the actions after all is restricted by the network context and by the degree of freedom that is acceptable. In traditional alignments of the congestion control end-to-end scheme the network sharing the resources can be considered as the 'black box' and only one variable, traffic intensity λ, is available to control the flow.

The offered SAC protocol is divided into two parts: the prediction and the aggression application. The type of selective aggressiveness control is able to use the correlation structure present in long-range dependent traffic with the aim of improving network operation efficiency.

The selective aggressiveness together with the predictability can lead to performance improvement compared with that achieved when using the usual congesttion control schemes based on the feedback. The relative benefit in performance resulting from SAC and the predictability use increase as the long-range dependence grows.

The self-similar bursty structure (in spite of the pernicious influence on the network performance, in particular on QoS) has the composition that can be used to reduce the negative influence. Factually, the more self-similar the traffic, the more effectively can its structure be used.

References

[1] M. E. Crovella and A. Bestavros, 'Self-similarity in world wide web traffic: evidence and possible causes', In Proceedings of the 1996 ACM SIGMET-RICS, International Conference on '*Measurement and Modeling of Computer Systems*', Philadelphia, PA, May 1996.

[2] W. Willinger, M. S. Taqqu, R. Sherman and D. V. Wilson, 'Self-similarity through high-variability: statistical analysis of Ethernet LAN traffic at the source level', *Computer Communications Review*, **25**, 1995, 100–113; also in Proceedings of the ACM/SIGCOMM'95, Boston, MA, August 1995.

[3] R.M. Metccalfe and D. R. Boggs, 'Ethernet: distributed packet switching for local computer networks', *Communications of the ACM*, **19**, 1976, 395–404.

[4] V. Paxson and S. Floyd, 'Wide-area traffic: the failure of Poisson modeling', *IEEE/ACM Transactions on Networking*, **3**, 1995, 226–244.

[5] F. Brichet, J. Roberts, A. Simonian and D. Veitch, 'Heavy traffic analysis of a storage model with long range dependent on/off sources', *Queuing Systems*, **23**, 1996, 197–215.

[6] S. Giordano, N. O'Connell, M. Pagano and G. Procissi, 'A variational approach to the queuing analysis with fractional Brownian motion input traffic', in 7th IFIP Workshop on '*Performance Modelling and Evaluation of ATM Networks*', Antwerp, Belgium, June 1999.

[7] Z. Lui, P. Nain, D. Towsley and Z.L. Zhang, 'Asymptotic behavior of a multiplexer fed by a long-range dependent process', *J. Appl. Prob.*, **36**, 1999, 105–118.

[8] I. Norros, 'A storage model with self-similar input', *Queuing Systems*, **16**, 1994, 387–396.

[9] B. Tsybakov and N.D. Georganas, 'On self-similar traffic in ATM queue: definitions, overflow probability bound, and cell delay distribution', *IEEE/ACM Transactions on Networking*, **5**(3), 1997, 397–409.

[10] B. Tsybakov and N.D. Georganas, 'Self-similar processes in communication networks', *IEEE Transactions on Information Theory*, **44**(5), 1998, 1713–1725.

[11] B.Tsybakov and N.D. Georganas, 'Buffer overflow under self-similar traffic', in Proceedings of SPIE99 Conference on *Performance and Control of Network Systems*', Boston, MA, September 1999.

[12] B.Tsybakov and N.D. Georganas, 'Overflow and losses in a network queue with a self-similar input', *Journal of Queueing Systems and Applications*, **35**(1–4), 2000, 201–235.

[13] W. Willinger, M.S. Taqqu and A. Erramilli, 'A bibliographical guide to self-similar traffic and performance modeling for modern high-speed networks', in '*Stochastic Networks: Theory and Applications*' (eds F. P. Kelly, S. Zachary and I. Ziedins), Royal Statistical Society Lecture Notes Series, Volume 4, Oxford University Press, 1996, pp. 339–366.

[14] N. Laskin, I. Lambadaris, F.C. Harmantzis and M. Devetsikiotis, 'Fractional Levy motion and its application to network traffic modelling', *Computer Networks*, **40**, 2002, 363–375.

[15] A. Karasaridis and D. Hatzinakos, 'Network heavy traffic modeling using α-stable self-similar process', *IEEE Transactions on Communications*, **49**(7), 2001, 1203–1214.

[16] Trang Dinh Dang, 'New results in multifractal traffic analysis and modeling', PhD Dissertation, Budapest, Hungary, 2002.

[17] O.I. Sheluhin, A.V. Kuyun and D.A. Luk'ancev, 'An estimation of the traffic fractality impact on the queuing in the telecommunication networks' (in Russian), *Electrotekhnicheskie i Informacionnie Kompleksi i Sistemi*, **2**(2), 2006, 47–54.

[18] O.I. Sheluhin, A.V. Osin and G.A. Urev, 'Voice traffic experimental study in VoIP networks' (in Russian), *Electrotekhnicheskie i Informacionnie Kompleksi i Sistemi*, **2**(2), 2006, 54–58.

[19] O.I. Sheluhin and A.V. Osin, 'Speech traffic self-similarity impact on QoS parameter optimization in the telecommunication network' (in Russian), *Nelineinii Mir*, **4**(3), 2006, 116–121.

[20] A.N. Tikhonov and V.Y. Arsenin, '*Techniques of The solution for Ill-Posed Problems*', Science, Moscow, 1986, p. 243.

[21] O.I. Sheluhin, A.V. Pruginin and A.V. Osin, 'Telecommunication network parameter optimization by Tihonov regularization technique' (in Russian), *Informacionno-Izmeritel'nie I Upravlyaushie Sistemi*, **4**(6), 2006, 62–72.

[22] O.I. Sheluhin and A.V. Osin, 'Evaluation of the voice traffic self-similarity influence on QoS in frame relay telecommunication networks' (in Russian), *Nelineinii mir*, **4**(10), 2006, 110–120.

[23] O.I. Sheluhin, A.V. Pruginin, A.V. Osin and G.A. Urev, 'Mathematical models and imitation modeling of the VoIP traffic aggregation' (in Russian), *Electrotekhnicheskie i Informacionnie Kompleksi i Sistemi*, **2**(1), 2006, 32–37.

[24] T.Tuan and K.Park, 'Multiple time scale congestion control for self-similar network traffic', *Performance Evaluation*, **36**, 1999, 359–386.

[25] T.Tuan and K.Park, 'Congestion control for self-similar traffic', in '*Self-Similar Network Traffic Analysis and Performance Evaluation*' (eds K. Park and W. Willinger), Wiley–Interscience, New York, 1999.

[26] J.-C. Bolot and A. U. Shankar, 'Analysis of a fluid approximation to flow control dynamics', in Proceedings of IEEE INFOCOM '92, Florence, Italy, 1992, pp. 2398–2407.

[27] L. Brakmo and L. Peterson, 'TCP Vegas: end to end congestion avoidance on a global internet', IEEE *Journal on Selected Areas in Communications*, **13**(8), 1995, 1465–1480.

[28] M. Geria and L. Kleinrock, 'Flow control: a comparative survey', *IEEE Transactions in Communications*, **20**(2), 1980, 35–49.

[29] Z. Haas and J. Winters, 'Congestion control by adaptive admission', in Proceedings of IEEE INFOCOM '91, Vol. 2, Bal Harbour, FL, 1991, pp. 560–569.

[30] V. Jacobson, 'Congestion avoidance and control', in Proceedings of ACM SIGCOMM '88, Standford, CA, 1988, pp. 314–329.

[31] S. Keshav, 'A control–theoretic approach to flow control', in Proceedings of ACM SIGCOMM '91, Zurich, Switzerland, 1991, pp. 3–15.

[32] D. Mitra and J. Seery, 'Dynamic adaptive windows for high speed data networks: theory and simulations', in Proceedings of ACM SIGCOMM '90, Philadelphia, PA, 1990, pp. 30–40.

[33] A. Mukherjee and J. Strikwerda, 'Analysis of dynamic congestion control protocols – a Fokker–*Planck approximation*', in Proceedigns of ACM SIGCOMM '91, Zurich, Switzerland, 1991, pp. 159–169.

[34] K. Park, 'Warp control: a dynamically stable congestion protocol and its analysis', in Proceedings of ACM SIGCOMM '93, San Francisco, CA, 1993, pp. 137–147.

[35] S. Shenker, 'A theoretical analysis of feedback flow control', in Proceedings of ACM SIGCOMM '90, Philadelphia, PA, 1990, 156–165.

[36] R. Dighe, C. J. May and G. Ramamurthy, 'Congestion avoidance strategies in broadband packet networks', in Proceedings of IEEE INFOCOM '91, Vol. 1, Bal Harbour, FL, 1991, pp. 295–303.

[37] Z. Haas, 'A communication architecture for high-speed networking', in Proceedings of IEEE INFOCOM '90, Vol. 2, San Francisco, CA, 1990, pp. 433–441.

[38] D. Hong and T. Suda, 'Congestion control and prevention in ATM networks', *IEEE Network Magazine*, July 1991, 10–16.

[39] Y. T. Wang and B. Sengupta, 'Performance analysis of a feedback congestion control policy under non-negligible propagation delay', in Proceedings of ACM SIGCOMM '91, Zurich, Switzerland, 1991, 149–157.

Appendix A

List of Symbols

The following symbols are used in the equations and text. The section numbers listed are the locations of their first use.

Symbol	Meaning	Section
A	Total number of inversions	1.5
	Threshold parameter	2.4.3
	Real operator	5.3.3
A_t	Variance of entrance quantity	1.2.2
$\tilde{A}$	Approximated operator	
a	Scaling coefficient	2.5.3
B	Inverse shift operator	2.3.1
	Threshold parameter	2.4.3
	Reverse shift operator	4.3
$B_N(\tau)$	Second-order moment function of random intensity	2.4.1
C_n	Worst channel carrying capacity value	5.2
D_f	Hausdorrf dimension; fractal dimension	1.1
D_q	Renji dimension; spectrum of generalized fractal Renji dimensions	1.1
D_1	Informational dimension	1.1
$d_x(j,k)$	Wavelet coefficients	1.4.1
d_1	Gaussian random number	2.1.1
$\boldsymbol{e}$	Unit vector	5.2
$F(T)$	Normalized variance of counts (Fano factor)	2.4.1
$F_d(\cdot)$	Integral distribution function	2.5.3
f	Frequency	1.2.1
$f(\alpha)$	Function of the multifractional spectrum; spectrum of multifractal singularities	1.1
$f(\lambda)$	Power spectrum for X	1.4.1
$G_H(t)$	Centralized fractal Gaussian noise with the Hurst exponent	3.6.4
H	Hurst exponent	1.2.1
$h(t)$	Impulse response	2.4.3
$I_N(\omega)$	Periodogram	1.4.2
K_I	Curvature parameter	3.8
K_j	Wavelet coefficient number for scale j	3.9

Self-Similar Processes in Telecommunications O. I. Sheluhin, S. M. Smolskiy and A.V. Osin
© 2007 John Wiley & Sons, Ltd

k	Normalizing constant	2.4.3
	Second shape parameter of Pareto distribution	3.8
k_D	Diffusion factor	2.1
L	Size in Euclidean space	1.1
	Function slowly variable at infinity	1.2.1
L^{D_f}	Volume in D-dimensional space	1.1
$L_W(\theta)$	Cost function	2.3.5
L_α	Levy motion	2.5.1
L_2	Slowly varying function at 0	1.2.1
l_{min}	Minimal length of the naturally originated fractal	1.1
l	Length scale	1.1
l_{max}	Typical geometrical size of the object	1.1
£	Boundary area	1.1
M	Number of summed FRPs	2.4.3
$M(l)$	Minimum number of nonempty cubes	1.1
m	Initial moment (mean value); averaging operation; mathematical expectation	1.2.1
	Sample size	1.3
	Average arrival intensity	2.5.3
N	Total number of test points over $1/L$ intervals	1.1
	Vanishing moment of the wavelet function	1.4.1
	Time series length	1.4.2
N_i	Number of points in i-th set	1.1
$N(\varepsilon)$	Total quantity of the occupied cells depending on the size of cell side ε	1.1
N_n	VoIP services user number	5.2
N_1	Number of outcomes (+)	1.5
N_2	Number of outcomes (−)	1.5
$n_i(\varepsilon)$	Number of points in all occupied cells	1.1
P	Conditional probability	1.2.2
$p(\varepsilon)$	Relative occupancy	1.1
p^i	Probability of the test point presence in the ith cell	1.1
$p_i(\varepsilon)$	Probability that the point chosen at random from the set is located in cell i	1.1
$\hat{p}_k$	Noise variance estimate for the AR model of the kth order	3.9.3
$Q_{i,j}(t)$	Transition function	3.6.2
q	Power of the Renji dimension	1.1
$R(k;T)$	Covariance function for counts	2.4.1
$R(t_1, t_2)$	Covariance function	1.2.1
$r(k)$	Correlation factor	1.2.1
$r(\tau)$	Normalized covariation function	2.2
r_τ	Decaying degree	2.3.4
S	Summation operator	2.3.4
	Scene number in the sequence	3.8
$S(t)$	Cumulative process	2.3.5
$S^m(q)$	Fragmentation functions	4.3
$S(\varepsilon)$	Entropy of the fractal set	1.1
$S(\omega)$	Spectral density	1.2.1
T	Window width	2.4.1
T_0	Fractal onset time	2.4.2
t	Time instant	2.1
U	i.i.d. homogeneously distributed random variable	2.3.4
u	Order statistics of the distribution	1.3.1

V	Delay in ADC	5.4
v	Freedom degree	1.5
$w(X)$	Probability density function (PDF)	1.5
$\mathbf{w}$	Vector reflecting the weights of each component	5.2
$\overline{X}$	Sampling mean value	1.4.1
$X(t), t \in Z$	Random process discrete in time	1.2.1
$X(t)$	Traffic volume	1.2.1
Y_k	Sequence of increments	1.2.1
$Y^{(m)}$	m aggregated time series	1.2.1
Y_t	Total number of packets generated by the sources active at moment t	5.1.2
Z	Data vector	3.5
Z	Connection duration or life time	1.3
$Z(q, \varepsilon)$	Generalized statistical sum	1.1
Z_n	Cell number in frame n	3.8
$1(t-\tau_i)$	Unit function	2.4
α	'tail' index or form parameter	1.3
	Levy parameter	2.5
	Shape parameter of Pareto distribution	3.7.1
α_i	Some exponent	1.1
$\hat{\alpha}$	Hill's estimation	1.3.1
α_A	Tail index	2.4.3
β	Parameter characterizing the distribution asymmetry	2.5.3
$\Gamma(.)$	Gamma function	1.2.2
γ	Circular frequency order	1.2.1
	Fractal parameter	2.4.3
∇^2	Operator of the second central difference	2.4.2
Δ_k	Difference between the maximal and minimal deviations	
$\delta(x)$	Delta function (Dirac function)	2.4.2
ε	cell size	1.1
ε^{D_f}	Volume of the cubes	1.1
ε_n	Independent and identically distributed normal random variables with zero mean value	3.8
η	Asymptotic decay degree	5.1
$\hat{\eta}$	Whittle estimation	1.4.2
$\hat{\theta}$	Maximum likelihood estimation	2.3.5
$\Lambda(\omega)$	Spectral window	3.8
λ	Point process intensity	2.4.1
	Averaged input intensity	5.1.1
μ	Location parameter	2.4.3
	Discretization measure	3.5
μ_j	Consistent estimation for $M[d_X(j,\cdot)^2]$	1.4.1
ξ	Total number from Gaussian distribution	2.1
$\xi(t)$	Continuous Markovian chain	3.6.2
ξ_t	Source number for which $\omega_s = t$	5.1.2
ρ	Queue usage coefficient	5.1.3
σ	Scaling parameter	2.5.1
σ^2	Variance (dispersion)	1.2.1
τ	Nonlinear function defining the behaviour of the statistical sum	1.1

	Mass parameter (scale function)	1.2.1
$\tau_i^{(j)}$	ith time between arrivals for the jth FRP stream	2.4.3
τ_{ON}, τ_{OFF}	Duration of ON/OFF states respectively	2.4.3
$\phi(B)$	Autoregressive operator	2.3.4
$\Psi(\lambda)$	Power spectrum for the Fourier transform for the wavelet function $\Psi_0(\cdot)$	1.4.1
ψ	Orthonormal wavelet in the $L^2(R)$ space (Haar wavelet)	3.9
ψ_0	Mother wavelet	1.4.1
$\Omega(p)$	Normalizing functional	5.3.4
ω	Circular frequency	1.2.1

Appendix B

List of Acronyms

ABR	Available bit rate
AEVT	Asymptotic extreme value theory
AFRP	Alternative FRP
AR	Autoregressive (model)
ARMA	Autoregressive moving average (model)
ATM	Asynchronous transfer mode
B-ISDN	Broadband ISDN
BM	Brownian motion
CCDF	Complementary cumulative density function
CDF	Cumulative density function
CPU	Central processing unit
CR	Coincident rate
CTS	Critical time scale
DCT	Discrete cosine transform
DSPP	Doubly stochastic Poisson point process (method)
DTFT	Discrete time Fourier transform
DWT	Discrete wavelet transform
EAFRP	Extended AFRP
ESS	Exactly second-order self-similar (process)
FARIMA	Fractional autoregressive integrated moving average (model)
FBM	Fractional brownian motion
FBN	Fractal binomial noise
FBNDP	Fractal binomial noise-driven Poisson process
FDA	Frequency domain approach
FDDI	Fibre distributed data interface
FFT	Fast fourier transform
FGN	Fractional gaussian noise
FIFO	First input first output
FLM	Fractional Levy motion
FPP	Fractal point process
FRP	Fractal renewal process
FSLN	Fractional stable Levy noise
FSN	Fractal shot noise

Self-Similar Processes in Telecommunications O. I. Sheluhin, S. M. Smolskiy and A.V. Osin
© 2007 John Wiley & Sons, Ltd

FSNDP	Fractal shot noise-driven Poisson process
FTP	File transfer protocol
GOP	Group of pictures
GRN	Gaussian random number
H-sssi	self-similar process with self-similarity parameter H with stationary increments
IDC	Index of dispersion for counts
IDI	Index of dispersion for intervals
iid	Identical and independent distributed
IP	Internet protocol
IPP	Interrupted Poisson process
ISDN	Integrated service digital network
LAN	Local area network
LLCD	Log–log complementary distribution
LRD	Long-range dependence
MA	Moving average (model)
MLE	Maximum likelihood estimate
MMPP	Markov-modulated Poisson process
MPEG	Motion Pictures Experts Group
MPLS	Multiprotocol label switching
MRA	Multiresolution analysis
OS	Operating system
PDF	Probability density function
PDH	Plesiochronous digital hierarchy
PMR	Peak-to-mean ratio
POP	Post office protocol
PSD	Power spectrum density
PT	Power tail (distribution)
QoS	Quality of service
RMD	Random midpoint displacement (method)
SDH	Synchronous digital hierarchy
SMDS	Switched multimegabit data service
SONET	Synchronous optical network
SRA	Successive random additional (method)
SRD	Short-range dependence
STM	Synchronous transport module
Sup-FRP	Superposition-FRP
TCP	Transmission control protocol
UDP	User datagram protocol
VBR	Variable bit rate (traffic)
WAN	Wide area network
WIG	Wavelet-domain independent gaussian (model)
WSS	Wide sense stationary (quotient)
WWW	World wide web

Index

© 2007 John Wiley & Sons, Ltd